FUNDAMENTALS OF TUNNEL FIELD-EFFECT TRANSISTORS

FUNDAMENTALS OF TUNNEL FIELD-EFFECT TRANSISTORS

Sneh Saurabh

Indraprastha Institute of Information Technology, Delhi, India

Mamidala Jagadesh Kumar

Indian Institute of Technology, Delhi, India

CRC Press
Taylor & Francis Group
Boca Raton London New York

CRC Press is an imprint of the
Taylor & Francis Group, an **informa** business

CRC Press
Taylor & Francis Group
6000 Broken Sound Parkway NW, Suite 300
Boca Raton, FL 33487-2742

First issued in hardback 2019

ISBN-13: 978-1-4987-6713-2 (hbk)

Library of Congress Cataloging-in-Publication Data

Names: Saurabh, Sneh, author. | Kumar, Mamidala Jagadesh, author.
Title: Fundamentals of tunnel field effect transistors / Sneh Saurabh, Mamidala Jagadesh Kumar.
Description: Boca Raton, FL : CRC Press, Taylor & Francis Group, [2016] | Includes bibliographical references and index.
Identifiers: LCCN 2016022619| ISBN 9781498767132 (hardback ; alk. paper) | ISBN 1498767133 (hardback ; alk. paper) | ISBN 9781498767163 (e-book) | ISBN 1498767168 (e-book)
Subjects: LCSH: Tunnel field-effect transistors. | Integrated circuits--Design and construction. | Nanostructured materials. | Low voltage integrated circuits.
Classification: LCC TK7871.95 .S28 2016 | DDC 621.3815/284--dc23
LC record available at https://lccn.loc.gov/2016022619

Visit the Taylor & Francis Web site at
http://www.taylorandfrancis.com

and the CRC Press Web site at
http://www.crcpress.com

Dedicated to Saraswati, the Goddess of Learning.

Contents

Preface

For more than five decades, the miniaturization of circuits, achieved by the scaling down of metal-oxide field-effect transistors (MOSFETs), has been the principal driver for the semiconductor industry. However, the stringent power constraints of integrated circuits and the non-scalability of the subthreshold slope in a conventional MOSFET, now pose a serious threat to the continued scaling of field-effect transistors. Therefore, intensive research is being carried out on novel steep subthreshold-slope devices that can replace the conventional MOSFET in future integrated circuits. Hence, Tunnel Field Effect Transistors (TFETs) are very promising devices as they exhibit a subthreshold swing lower than 60 $mV/decade$, which is the theoretical limit for a conventional MOSFET. Consequently, TFET-based circuits can be highly energy efficient and TFETs are often regarded as "Green Transistors."

An exhaustive research of TFETs was triggered around 2004 with the successful demonstration of the subthreshold swing below 60 $mV/decade$ and the necessity for a device that can replace conventional MOSFETs for low-power energy-efficient circuits. Since then, there has been a thorough investigation of various aspects of TFETs and their applications in integrated circuits. The number of research papers on TFETs has been growing exponentially. Various research groups are actively working on improving the characteristics of TFETs and have been reporting consistent improvements in the electrical parameters. Different types of materials are being explored to realize TFETs and have yielded encouraging results. Our understanding of TFETs has considerably improved over the years. The technological advances in realizing better TFETs will make them ideal candidates in commercial low-power applications.

Since TFETs have a huge potential for employment in energy-efficient circuits, engineers from diverse backgrounds and different levels of understanding of electronic devices need to comprehend the fundamental aspects of TFETs. However, since TFETs are aggressively studied, and due to the lack of a comprehensive textbook, research papers are currently the primary source of knowledge on TFETs. With a plethora of papers appearing on TFETs, gaining a basic understanding of a TFET and keeping track of the latest research is a challenge. Therefore, we believe that a book describing the fundamental aspects of a TFET will be useful, not only for engineers who need to understand the basics of TFETs, but also for researchers who are innovating and applying the TFET concepts. This book, *Fundamentals of Tunnel Field Ef-*

fect Transistors, is intended to fulfill this requirement for the device research community.

The material presented in this book is based primarily on research papers and draws benefits from our research work on TFETs carried out for more than a decade. In this book, as an illustration, we have included several device architectures that have been proposed by different research groups. This will help readers appreciate the techniques for improving the characteristics of the TFETs. Additionally, these techniques will encourage readers to develop their own device improvement techniques and carry out research on TFETs to make them suitable for future circuit applications. Throughout this book, we have emphasized the qualitative aspects of the devices rather than focusing on the mathematical equations governing the devices. This has been done so that readers can gain an intuitive understanding of the devices rather than being bogged down by complicated equations. However, we have provided appropriate references so that interested readers can delve deeper into the subject and appreciate the quantitative aspects of TFETs as well.

This book starts by discussing CMOS scaling in Chapter 1. This provides the context in which the research on TFETs is occurring. The fundamental aspects of quantum tunneling, and the physical principle that forms the basis of TFET operation are discussed in Chapter 2. The basic operating principle and the characteristics of a TFET are explained in Chapter 3. Since TFETs, in general, suffer from a low ON-state current, the techniques to boost the ON-state current in a TFET are crucial and are described in Chapter 4. TFETs based on III-V materials, carbon and nanowires are discussed in detail in Chapters 5, 6, and 7. These novel material systems are important for realizing TFETs since the characteristics of a TFET using these materials are significantly better than silicon-based TFETs. In Chapter 8, different analytical models for TFETs are described, and in Chapter 9, the applications of TFETs are presented. Lastly, in Chapter 10, the current state of research on TFETs is summarized including future prospects for TFETs.

We are hopeful that this book will be very useful for students, researchers, and practicing engineers in the industry.

1

CMOS Scaling

1.1 Introduction

Ever since the demonstration of the first Field-Effect Transistor (FET) in 1960 and Complementary Metal-Oxide-Semiconductor (CMOS) configuration in 1963, there has been continuous advancement in the CMOS technology [1–4]. These advancements are driven and sustained by uninterrupted CMOS scaling or miniaturization of transistors following an exponential trend that is commonly known as Moore's law [5]. It is remarkable that despite several difficult technical challenges, the trend of reducing dimensions of transistors and the resultant exponential growth of the semiconductor industry have continued unabated for more than five decades. The obstacles associated with forthcoming miniaturization of transistors that looked unsurmountable at first, were finally overcome using ingenious techniques before the trend could be upset.

As the transistors are miniaturized below 10 nm, the continued scaling down of transistors faces a real threat due to the inability of the transistor to switch from the OFF-to the ON-state without leading to increased power dissipation [6–10]. With CMOS scaling, the architecture of transistors and the technology to realize them have undergone numerous changes over the years, though the basic mechanism of operation of the transistors has remained the same. The physics of current transport in a conventional transistor imposes certain restrictions that prohibit further lowering of supply voltages in an integrated circuit (IC). As a result, the current trend of miniaturizing transistors, while keeping the supply voltage constant does not allow improving the energy efficiency of the ICs, which could have been obtained by decreasing the supply voltage [11]. This necessitates exploring transistors based on new operating principles, improved architectures, and perhaps novel non-silicon-based material systems that can overcome the limitations of the conventional transistors [10]. In this respect, Tunnel Field-Effect Transistors (TFETs), operating on the basis of quantum mechanical tunneling, have been demonstrated to possess excellent switching properties which are beyond the theoretical capabilities of a conventional transistor [7, 10, 12]. As a result TFETs can be employed in ICs with low supply voltages, resulting in highly energy-efficient circuits.

This makes TFETs quite fascinating for researchers and has spurred intense experimental and theoretical exploration of TFETs during the last ten years.

In this chapter, the concepts of CMOS scaling will be discussed, highlighting the need for novel devices such as TFETs in the near future. This will help readers in understanding the context in which the current research on TFETs is being carried out and in appreciating the remaining chapters of this book.

This chapter starts by describing the fundamentals of Metal-Oxide-Semiconductor Field-Effect Transistors (MOSFETs), CMOS and CMOS scaling. This is followed by a description of the historical perspective and the current trends of CMOS scaling. Finally, major challenges to the continued scaling of CMOS are discussed. A short review of the emerging devices that can potentially improve the existing CMOS technology or replace conventional transistors is also presented.

1.2 Basics of MOSFET and CMOS

In this section, we look at the basics of the MOSFET and the CMOS circuits, focusing on the topics that are relevant for understanding CMOS scaling and subsequent chapters of this book.

1.2.1 MOSFET Structure and Operation

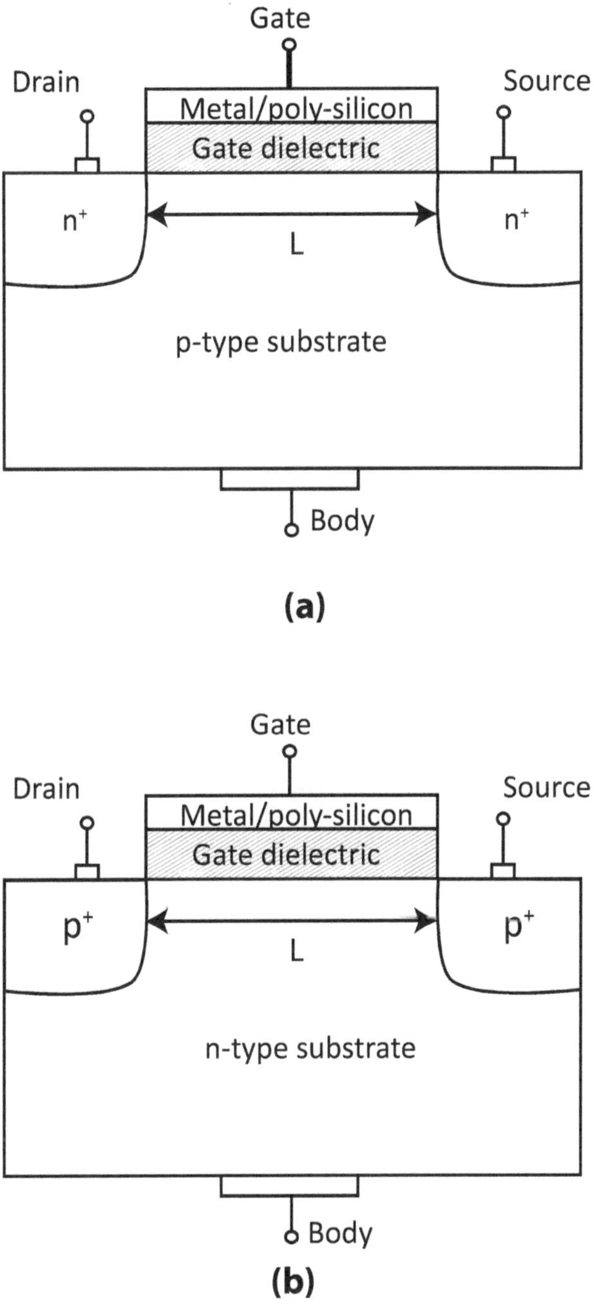

FIGURE 1.1
Schematic cross-sectional view of (a) an NMOS and (b) a PMOS.

Figure 1.1(a) and Figure 1.1(b) show cross-sectional views of an n-channel MOSFET (NMOS) and a p-channel MOSFET (PMOS), respectively. In an NMOS, the source and the drain are doped n^+ and the substrate is a p-type region. In a PMOS, the source and the drain are doped p^+ and the substrate is an n-type region. The substrate and the source terminals are often tied together. A MOS (Metal-Oxide-Semiconductor) structure is obtained by growing or depositing a layer of silicon dioxide or other dielectric material over a silicon substrate or other semiconductor material. This is followed by depositing a layer of metal or polycrystalline silicon over the gate dielectric to form the gate electrode.

A MOS structure is similar to a planar capacitor. The gate and the semiconductor form the electrodes and the gate dielectric material is sandwiched between them. When a voltage is applied across a MOS structure using the gate terminal, the distribution of charges in the semiconductor is changed due to field-effect. When gate voltage greater than the threshold voltage of the MOSFET is applied, a strong conducting channel is induced in the silicon body below the gate. The threshold voltage of a MOSFET primarily depends on the properties of the materials forming the MOS structure. The channel formed under the gate contains electrons in an NMOS and holes in a PMOS and extends between the source and the drain, for low drain voltages. Therefore, when the channel is formed under the gate, current can flow between the source and the drain through the channel if a potential difference between the source and the drain exists. This is a simplistic description of the mechanism of operation of a MOSFET. Readers are directed to [13–15] for an in-depth understanding the MOSFET. The behavior of a MOSFET can be divided into three separate regions: the cut-off region, the linear region, and the saturation region, depending on the voltage of the gate, source and drain terminals. We describe the behavior of an NMOS with the source and the substrate grounded in the following paragraphs of this section.

When the gate voltage V_{GS} is less than the threshold voltage V_T of the NMOS, the transistor is said to be in the cut-off region. Ideally, no current flows between the source and the drain in the cut-off region. However, some electrons with higher energies can flow from the source to the drain resulting in subthreshold current. The subthreshold current can be given as

$$I_d \approx I_{D0} \exp\left(\frac{V_{GS} - V_T}{n(\frac{kT}{q})}\right), \tag{1.1}$$

where I_D is the drain current, I_{D0} is the drain current when $V_{GS} = V_T$, k is the Boltzmann constant, T is the temperature in Kelvin, q is the electronic charge, and n is the slope factor of the transistor determined by the control of gate voltage on the channel potential with the constraint that $n \geq 1$.

When $V_{GS} > V_T$ and $V_{DS} < V_{GS} - V_T$, the NMOS is in the linear region and behaves as a resistor. The drain current I_D in this region of operation can be given as

$$I_D = \mu_n C_{ox} \left(\frac{W}{L}\right) \left((V_{GS} - V_T)V_{DS} - \frac{V_{DS}^2}{2}\right), \tag{1.2}$$

where μ_n is the electron mobility in the channel, C_{ox} is the oxide capacitance per unit area, W is the width of the gate, and L is the length of the gate. The oxide capacitance per unit area C_{ox} can be written as

$$C_{ox} = \frac{\epsilon_{ox}}{t_{ox}}, \tag{1.3}$$

where ϵ_{ox} is the dielectric constant of the gate dielectric and t_{ox} is the thickness of the gate dielectric.

When $V_{GS} > V_T$ and $V_{DS} \geq (V_{GS} - V_T)$, the channel gets pinched off from the drain side and the NMOS is said to operate in the saturation region. The drain current in the saturation region can be given as

$$I_D = \frac{\mu_n C_{ox}}{2} \left(\frac{W}{L}\right) (V_{GS} - V_T)^2. \tag{1.4}$$

The above equation shows that the NMOS behaves as a perfect current source in the saturation region and is independent of the drain voltage. However, the effective length of the channel is modulated by the drain voltage by a phenomenon called channel length modulation. When channel length modulation is taken into account, the drain current depends on the drain voltage and can be written as

$$I_D = \frac{\mu_n C_{ox}}{2} \left(\frac{W}{L}\right) (V_{GS} - V_T)^2 (1 + \lambda V_{DS}), \tag{1.5}$$

where λ is the channel length modulation parameter and, in general, is proportional to the inverse of the channel length.

The behavior of a PMOS in which the source and the substrate terminals are connected to the supply voltage are similar to the NMOS described above, except that the polarity of all the voltages and current reverse in the PMOS. Additionally, for a PMOS, the drain current primarily depends on the mobility of holes in the channel rather than the mobility of electrons.

1.2.2 Operation of MOSFET as a Switch

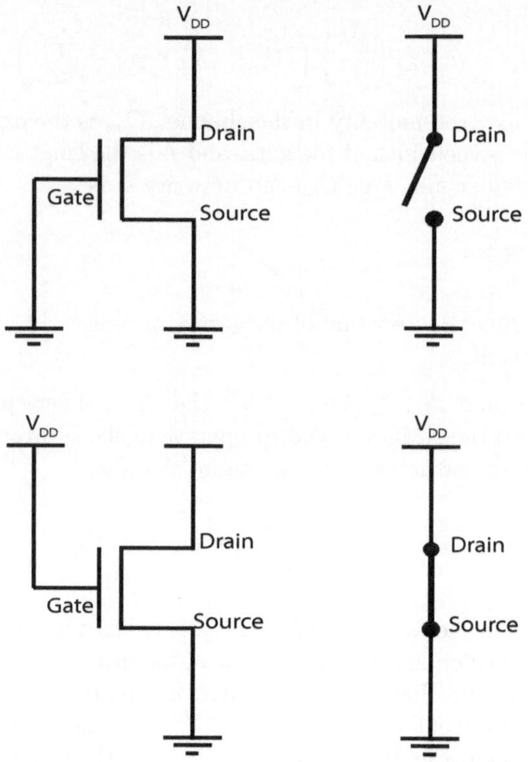

FIGURE 1.2
Operation of an NMOS as a switch.

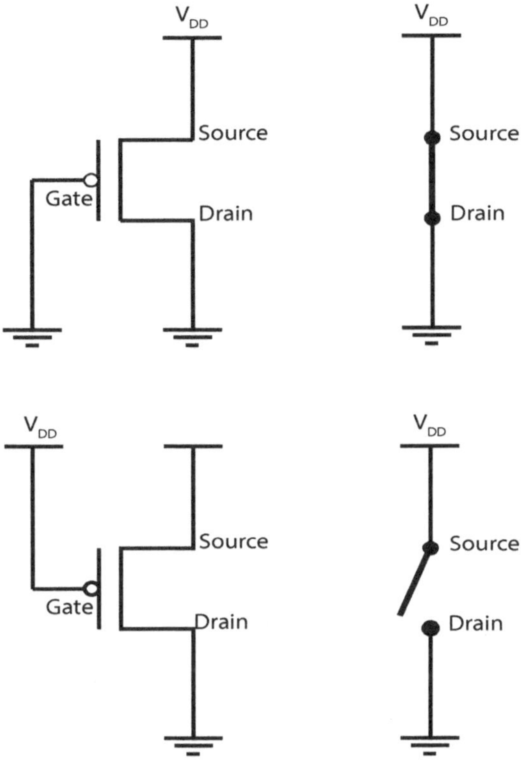

FIGURE 1.3
Operation of a PMOS as a switch.

In simple terms, a MOSFET can be considered as a switch. The operation of an NMOS and a PMOS as switches is shown in Figure 1.2 and Figure 1.3, respectively. When the gate of an NMOS is connected to the ground, then the NMOS can be considered as an open switch, disconnecting the source and the drain. When the gate of an NMOS is connected to the supply voltage, then the NMOS can be considered as a closed switch, connecting the source and the drain terminals. For an NMOS, the source is typically tied to ground and is used to pull down the drain voltage to the ground potential. For a PMOS, the switch is open when the gate is connected to the supply voltage and the switch is closed when the gate is connected to the ground. For a PMOS, the source is typically connected to the supply voltage and is used to pull up the drain voltage to the potential of the supply voltage.

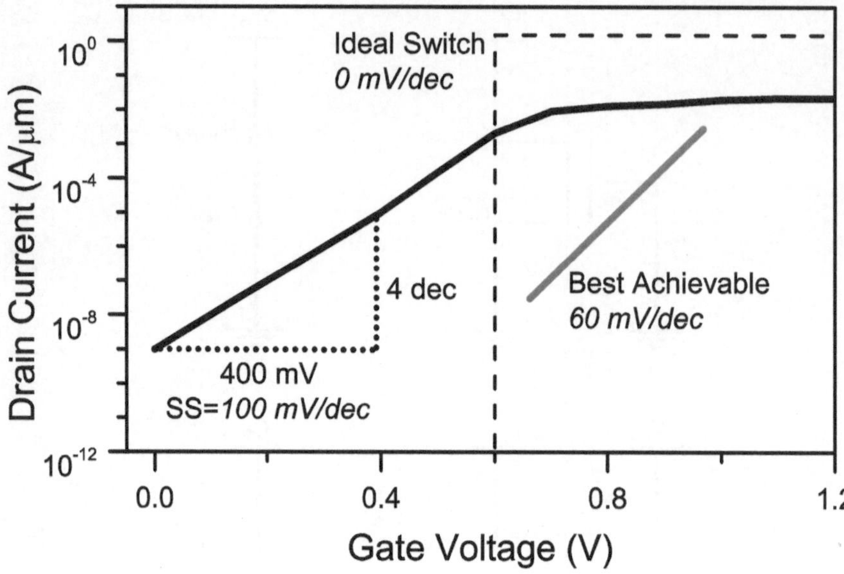

FIGURE 1.4
Transfer characteristics of a typical NMOS and comparison with an ideal switch (shown in dashed line). The minimum subthreshold swing in a MOSFET is limited to 60 $mV/decade$ at room temperature.

An ideal switch allows the current to abruptly flow/stop between its terminals as the switch is changed from the closed position to the open position or vice versa. Therefore, ideally, an NMOS should allow the current to flow between the drain and the source terminals when the gate voltage is greater than the threshold voltage and no current should be allowed to flow between the drain and the source terminals when the gate voltage is smaller than the threshold voltage. However, a MOSFET does not behave like an ideal switch. The drain current in a MOSFET rises exponentially with the application of the gate voltage as shown in Figure 1.4 and is represented by Equation (1.1). The abruptness of the transition from the OFF-state to the ON-state in a transistor is measured by a quantity called the subthreshold swing, which is defined as the gate voltage required to increase the drain current by one decade or ten times. The minimum subthreshold swing attainable in a conventional MOSFET is approximately 60 $mV/decade$ at room temperature. This implies that to increase the drain current by four orders of magnitude, at least a 240 mV increase of gate voltage is required in a conventional MOSFET. If the subthreshold swing of a MOSFET could have been made smaller, say 5 $mV/decade$, then 10^4 times increase in the drain current could have been obtained by just a $20mV$ increase in gate voltage and the circuit could have been made to work at smaller supply voltages. Therefore, for low voltage operation, it is desirable that the transistors have as small a subthreshold swing as possible. The minimum achievable subthreshold swing of 60 $mV/decade$ in a MOSFET is a bottleneck in reducing the supply voltages of the circuits. This limitation on the minimum subthreshold swing is determined by the physics of current transport in a conventional MOSFET and will be discussed in detail in Section 3.5.

1.2.3 Short-Channel Effects in a MOSFET

Ideally, the threshold voltage of a MOSFET does not depend on the channel length and the drain voltage. However, as the channel length of a MOSFET is reduced, the fraction of charge in the channel controlled by the gate potential reduces. As a result, the gate loses its full control over the channel potential in short-channel devices. The threshold voltage starts decreasing with the increase in the drain-source voltage due to a phenomenon called Drain Induced Barrier Lowering (DIBL). Additionally, the threshold voltage decreases with a decrease in the channel length for short-channel devices due to a phenomenon called V_T roll-off. These effects are called short-channel effects in a MOSFET and can cause serious problems in reducing the dimensions of the MOSFETs [16].

1.2.4 CMOS Inverter

Among all the IC technologies currently available, CMOS technology is the dominant technology, replacing NMOS and bipolar technology due to its very low static power dissipation, excellent noise immunity, very high input impedance and advanced processing technology [17]. In a CMOS circuit, the PMOS and the NMOS transistors are combined in a complementary symmetry configuration drawing close to zero power in standby mode [2].

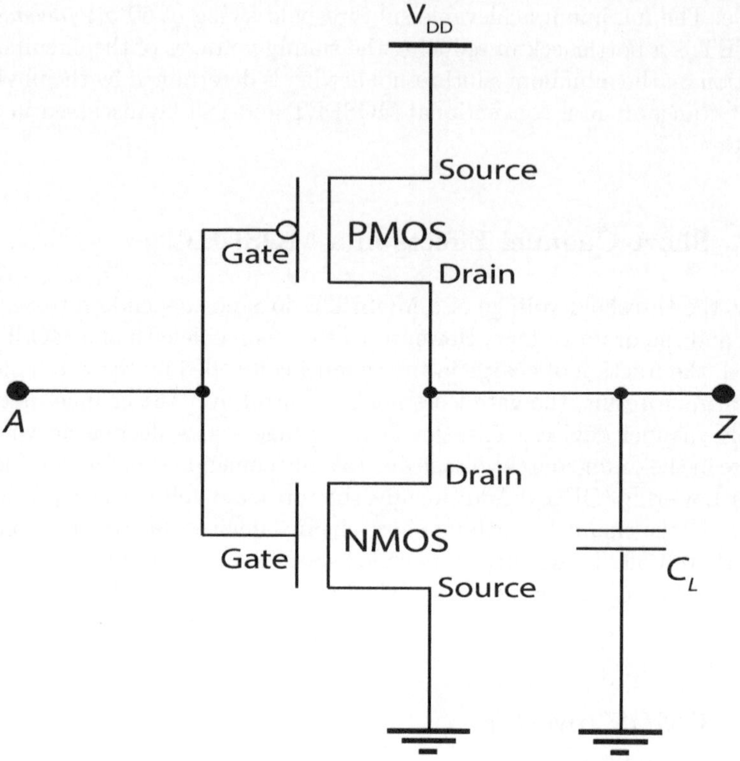

FIGURE 1.5
Circuit diagram of a CMOS inverter.

Figure 1.5 shows a schematic diagram of a CMOS inverter. When the voltage of input A is low, the NMOS is OFF and the PMOS is ON. Therefore, when the input A is low, the output Z is pulled up to supply voltage V_{DD}. On the other hand, when the voltage of input A is high, the NMOS is ON and the PMOS is OFF. Therefore, when the input A is high, the output Z is pulled down to the ground potential. This is a simplistic description of the mechanism of operation of a CMOS inverter. Readers are directed to [18–20] for further information.

1.2.5 Power Dissipation in a CMOS Circuit

There are two types of power dissipation in a CMOS circuit: static power dissipation and dynamic power dissipation. The static power dissipation is the power consumed by the circuit in the absence of switching. In a CMOS circuit, since there is no low-resistance path between the power supply V_{DD} and the ground in any of the states, the static power consumption in a CMOS circuit is ideally zero. However, due to the subthreshold current of the MOSFETs, the static power dissipation in a CMOS becomes significant, especially at advanced technology nodes. The dynamic power dissipation is the power consumed by the circuit when switching occurs. A CMOS inverter operating between the supply voltage V_{DD} and the ground, driving a load capacitance C_L and switching with a frequency f dissipates dynamic power P_D given by

$$P_D = C_L V_{DD}^2 f. \tag{1.6}$$

The above equation shows that decreasing the supply voltage of the circuit results in a quadratic decrease in the dynamic power dissipation. Therefore, decreasing the supply voltage is the most effective technique to reduce power consumption in the circuits.

1.3 CMOS Scaling

The process of reducing the dimensions of the MOSFETs and the wires connecting them in an IC such that the functionality of the IC remains unchanged is called CMOS scaling. It is based on the scaling theory proposed by Dennard et al. in 1974 showing that the basic operating characteristics of a MOSFET can be well-preserved if the parameters of the device are scaled in accordance with a given criterion [21]. The scaling down of the dimensions of transistors are done such that the geometric ratios that are important for the functioning of the IC remain unchanged. For example, if the length of the transistor is scaled by the factor S, then the width of the transistor is also scaled by the same factor S. The advantages of CMOS scaling are as follows:

1. With the reduction in transistor size, more transistors can be packed in a given area of an IC. Therefore, more functionalities can be realized in a given area of an IC or a given functionality can be realized in a smaller area of an IC.

2. With the reduction in the sizes of transistors and the area of ICs, more ICs can be produced per wafer. Since the cost of fabrication of a wafer is relatively fixed, the cost per IC is reduced with scaling.

3. With scaling, the delay and power consumed are expected to decrease, as will be shown in the following sections.

1.4 Types of CMOS Scaling

TABLE 1.1
Impact of the scaling of device dimensions and voltage on device and circuit parameters for constant-field scaling and constant-voltage scaling [15, 18–22].

Device/Circuit Parameters		**Constant-field scaling**	**Constant-voltage scaling**
	Scaled Parameters		
Length	L	$1/S$	$1/S$
Width	W	$1/S$	$1/S$
Gate Oxide Thickness	t_{ox}	$1/S$	$1/S$
Doping Density	N_A,N_D	S	S^2
Supply Voltage	V	$1/S$	1
	Effect on Device Parameters		
Electric field across gate oxide	$E = V/t_{ox}$	1	S
Depletion Layer Thickness	x_j	$1/S$	$1/S$
Capacitance	$C = WL/t_{ox}$	$1/S$	$1/S$
Current	$I = 1/t_{ox}(W/L)V^2$	$1/S$	S
Area per Device	$A = WL$	$1/S^2$	$1/S^2$
	Effect on Circuit Parameters		
Circuit Delay	$\tau = CV/I = L^2/V$	$1/S$	$1/S^2$
Power dissipation per device	$P = VI$	$1/S^2$	S
Power Density	$\rho_{power} = P/A = VI/WL$	1	S^3
Power Delay Product per device	$E = \tau_P = CV^2$	$1/S^3$	$1/S$
Current Density	$J = I/A$	$1/S$	S^3

There are two major types of scaling schemes: constant-field scaling and constant-voltage scaling [15, 18–22]. Table 1.1 shows the scaled parameters and their impact on the device and circuit parameters based on first-order analysis of these two scaling schemes. In this chapter, the scaling factor S is taken to be greater than one. Therefore, when the length of a MOSFET is scaled by S, it implies that if the initial length of the MOSFET is L, then the length of the miniaturized MOSFET is L/S. The typical value of S, i.e., the ratio of dimensions of two successive generations of MOSFETs, is $\sqrt{2}$. With $S = \sqrt{2}$, the channel length of a MOSFET reduces to half every two generations of CMOS technology.

1.4.1 Constant-Field Scaling

Constant-field scaling implies that the dimensions of the transistors and the supply voltage are both scaled by the factor S such that the electric fields in the vertical and in the horizontal directions remain essentially constant. This is the original scaling method described by Dennard et al. [21]. In order to reduce the source/drain depletion region depths by the factor S, the doping densities are increased by the factor S. With constant-field scaling, the circuit delay reduces by S and the power consumed by the device reduces by S^2 while the power density remains unchanged. The power delay product reduces by S^3, indicating that the constant-field scaling leads to a more energy-efficient device. It must be mentioned that with the reduction in the supply voltage, the threshold voltage of the MOSFET also needs to be reduced to maintain the same gate overdrive and the performance of the circuit. However, the threshold voltage of a MOSFET depends on the properties of the material that constitute the MOSFET, which do not scale. Additionally, the leakage current that strongly depends on the threshold voltage prohibits decreasing the threshold voltage beyond a certain limit. Therefore, in practice, scaling supply voltage has been difficult compared to scaling the dimensions of the transistor.

1.4.2 Constant-Voltage Scaling

Constant-voltage scaling implies that the dimensions of the transistors are scaled by the factor S keeping the supply voltage constant. As a result, the electric fields in the device increase due to constant-voltage scaling. The doping densities of the devices are increased by the factor S^2 to reduce the source/drain depletion region depths by the factor S. In this scheme, the logic levels do not change and the compatibilities with the original interfaces and the peripheral circuits are maintained. The circuit delay reduces by S^2. However, the power consumed by the device increases by a factor S, increasing the power density by S^3. The increase in power density creates problems of localized heating and heat dissipation. Additionally, an increase in current density by the factor S^3 can cause issues of electro-migration. In general, an increase in the electric fields in the device can cause issues of reliability when constant-voltage scaling is employed.

1.5 Historical Perspective of CMOS Scaling

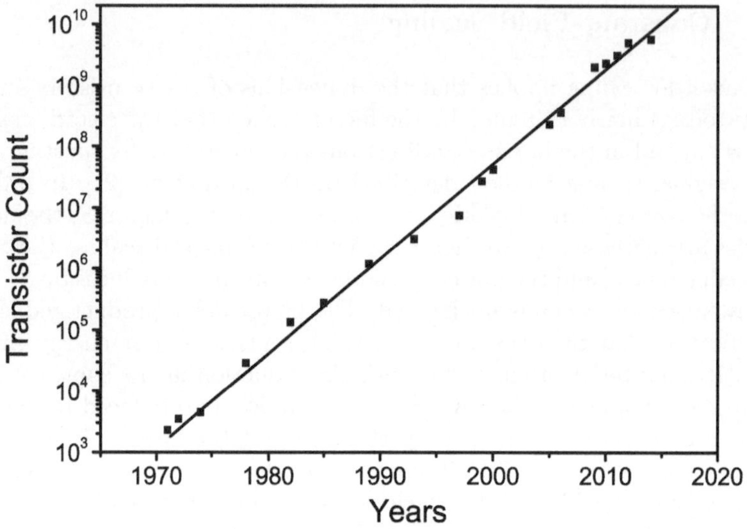

FIGURE 1.6
Trend of increase in transistor count for Intel's microprocessors [23–26].

The exponential trend in the miniaturization of transistors in an IC has continued for more than five decades [5, 27]. This trend was first observed by Gordon E. Moore and originally described in 1965 as follows: the number of components in an IC realized at the minimum cost doubles every year, this was revised in 1975 to double every two years [5, 27]. In the original paper of Moore, components represented transistors, diodes, resistors, and capacitors in an IC however, with the dominance of CMOS technology, now the meaning of components is taken as the transistors. Figure 1.6 shows the trend of exponential increase in the transistor count for Intel's microprocessors [23–26].

TABLE 1.2
Impact of scaling of device dimensions and voltage on device and circuit parameters when Dennard's constant-field scaling and Moore's law are applied **(taking** $S = \sqrt{2}$).

Parameter	**Factor by which parameter changes**	**% change in parameter**
Dimension of transistor (L, W, t_{ox})	$1/S = 1/\sqrt{2} \approx 0.7$	30% decrease
Voltage (V)	$1/S = 1/\sqrt{2} \approx 0.7$	30% decrease
Number of transistors in circuit (N)	2 *(by Moore's Law)*	100% increase
Area of transistor (A)	$1/S^2 = 0.5$	50% decrease
Area of circuit (A_{total})	$A \times N = 1$	No change
Density of transistors in circuit (D)	$N/A_{total} = 2$	100% increase
Circuit delay (τ)	$1/S = 1/\sqrt{2} \approx 0.7$	30% decrease
Frequency of circuit (f)	$1/\tau = S = \sqrt{2} \approx 1.4$	40% increase
Power dissipation in transistor (P)	$1/S^2 = 0.5$	50% decrease
Power dissipation in circuit (P_{total})	$P \times N = 0.5 \times 2 = 1$	No change
Power-delay product in transistor (E)	$P \times \tau = 1/S^3 \approx 0.35$	65% decrease
Power-delay product in circuit (E_{total})	$E \times N \approx 0.35 \times 2 = 0.7$	30% decrease

Using Dennard's constant-field scaling theory (Table 1.1) and Moore's law, the impact of scaling on various circuit parameters can be computed. The results are shown in Table 1.2. The scaling of transistors results in a 100% increase in transistor density, a 40% increase in the clock-frequency and a 30% reduction in the power delay product (a measure of energy dissipated per switching operation), with no change in the area and the power dissipation of the circuit. Additionally, since the cost of fabrication depends primarily on the area of the circuit, which has remained fairly constant with scaling, the cost of fabrication per transistor decreases with scaling. Therefore, scaling allows the addition of more transistors to the circuits and the scaled transistors are cheaper, faster, and consume less energy, with apparently no engineering trade-offs [6]. These unprecedented technological and economic incentives to shrink the transistors have been the driving force behind the Moore's law and the semiconductor industry for last five decades. However, Moore's law has another interesting aspect – it is a self-fulfilling prophecy [6, 28]. The predictions made using Moore's law become production goals. When these goals are achieved, the legitimacy of Moore's law as a benchmark of industry progress is reinforced [28]. The strong economic incentives to continue Moore's law ensure that considerable effort is devoted to move on the predicted trajectory [6].

It should be noted that Table 1.1 and Table 1.2 have been derived using first-order analysis of the devices. If phenomena such as velocity saturation, channel-length modulation, and short-channel effects of the transistors are taken into account, there will be less improvement in the device and circuit parameters with scaling. Additionally, the strong dependency of the maximum attainable clock frequency in a circuit on the interconnect delays that do not scale similar to the device scaling, prohibits exploiting full benefits of miniaturization of devices in a given circuit.

Another important point to note is that, historically, there has been a greater inertia in reducing the supply voltages of the circuit, despite transistors shrinking continuously [29]. From the early period of CMOS technology to the late 1980s, constant-voltage scaling predominated [29]. The detrimental effects of increased electric fields due to constant-voltage scaling were handled by improvements in the processing of gates that allowed larger electric fields and by employing ingenious transistor designs [29]. Later on, starting in the early 1990s, power constraints of the ICs and reliability issues forced migration to lower supply voltages [29]. Technologically, the following are some of the chronic problems of continued CMOS scaling:

1. Fabricating devices with smaller feature sizes maintaining adequate yield.
2. Controlling the short channel effects in the transistors.
3. Reducing the supply voltage and threshold voltage of the transistors retaining the overall electrical performance of the device.
4. Reducing interconnect delays.
5. Taking care of localized heat generated due to increased transistor density and associated problems in the reliability of the circuit.
6. Handling impact of increased process-induced variations at smaller device dimensions.

During the course of five decades of scaling, these technological challenges have been overcome by some ground-breaking inventions and engineering ingenuity. Roadmaps are prepared by International Technology Roadmap for Semiconductors (ITRS) outlining technology progression in terms of technology "nodes." The ITRS roadmap anticipates the evolution of the semiconductor market and helps in the planning of technological needs of IC production.

It identifies critical challenges in moving to future technology "nodes" and provides its assessment of emerging technologies and solutions. Therefore, researchers from the industry and academics have a clear vision of the impending problems of CMOS scaling and can develop a timely solution for them.

Though the success of Moore's law has become folklore in the semiconductor industry, another interesting trend of energy efficiency in computing has been identified by Koomey et al. [30, 31]. Koomey identified that, for a given computing load, the power consumption has decreased by half every 1.6 years. Alternatively, the number of computations per Joule of energy dissipated has been doubling every 1.6 years. This rate of increase in energy efficiency has been steady since the 1950s and is still continuing unabated [31]. Since smaller transistor sizes also tend to reduce electricity consumption, improvements in the energy efficiency are observed at similar rates. The current generation of mobile computing devices has become a reality due to continuous improvement in the computational energy efficiency over the decades. In the future, energy efficiency will become even more critical as the computing platforms become more mobile, end users become increasingly concerned with the battery life and billions of devices interact and analyze real-time information as embodied by the Internet of Things (IoT) [32, 33].

1.6 Current Trends in CMOS Scaling

The historical trend of shrinking transistors and exponential increase in the number of components is continuing relentlessly even today. Some recent technological advancements in transistors that have enabled continued miniaturization are [27, 33–35]:

1. **High-κ/metal gate stack:** The higher permittivity of the high-κ dielectric materials allow the same electrical oxide thickness with a dielectric layer that is physically thicker than the silicon dioxide layer. This reduces the leakage through the gate dielectric. Additionally, metal gates can dynamically screen electrons in the high-k dielectric. This leads to an improvement in the mobility of the carriers in the channel. Therefore, high-κ/metal gate stacks are currently being employed.

2. **Non-planar transistors:** Three-dimensional transistor structures such as FinFETs or tri-gate transistors are being employed to improve the gate control of the transistors, obtaining better subthreshold swing, reduced short-channel effects like DIBL, and improved scalability compared to the planar transistors.

3. **Strained silicon:** The enhancement of channel mobility is achieved by introducing tensile strain for the NMOS devices and compressive strain for the PMOS devices.

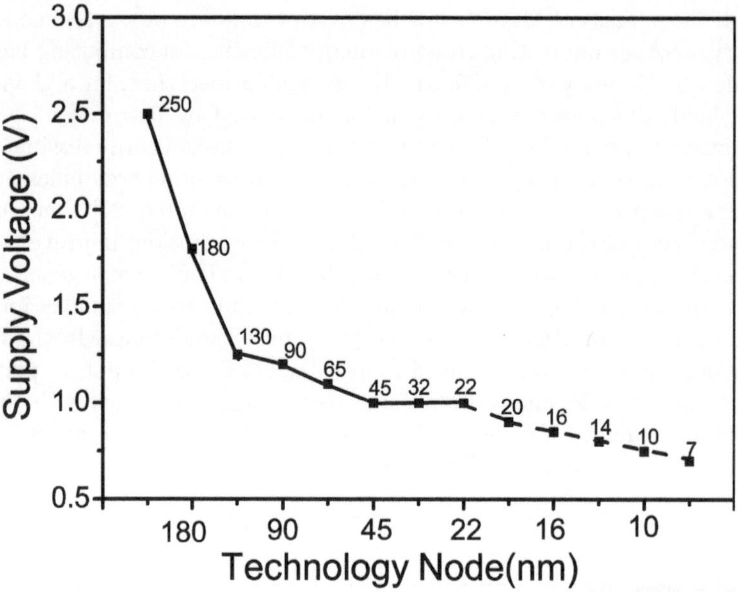

FIGURE 1.7
Supply voltages at different technology nodes [11, 36]. For the technology nodes 20 *nm* and beyond, the values are based on prediction [36].

In addition to advancement in the devices, there have been several important developments in the area of CMOS processing technology, circuit design techniques, and architectural improvements that have contributed to the continued shrinking of transistors. In recent times, although the advancements in technology have enabled "cramming" more components into the ICs, it has not translated into proportionate increase in the circuit performance or energy efficiency. The Dennard scaling trends, i.e., scaling the dimensions of the transistors as well as the supply voltage of the transistors, were upset circa 2005. Figure 1.7 shows the supply voltages employed for different technology nodes. Beginning with the 65*nm* technology node, the supply voltage could not be reduced, sacrificing the dynamic energy efficiency improvements obtained due to supply voltage reduction [11, 36]. The leakage currents in the transistors have also continued to increase [11]. The supply voltage reduction requires a decrease in the threshold voltage of the MOSFET. However, due to the non-scalability of the subthreshold swing, the threshold voltage of the

MOSFET cannot be reduced further without an unacceptable increase in the leakage current.

The breakdown of Dennardian scaling has driven the IC manufacturers to multicore scaling, i.e., increasing the number of cores per die to obtain continuing performance growth with the advancement in the technology node [9, 37]. However, power constraints prevent turning on all the cores at their maximum speed. As a result, some of the cores need to be turned off resulting in a phenomenon called "dark silicon" or some of the cores may be underclocked or operating at low voltages, sacrificing performance and resulting in a phenomenon called "dim silicon" [8, 9, 37]. Therefore, the current trend in CMOS scaling has created a peculiar situation in which more transistors can fit on a given die, but a growing fraction of all the transistors on the die cannot be fully utilized due to stringent power constraints [11]. It is expected that the low utilization due to "dark silicon" will gradually lead to poor economic feasibility of multicore scaling [37]. In this context, Moore's measure of the number of transistors in an IC, which is still growing exponentially, becomes less relevant. Koomey's measure of energy efficiency in terms of the number of computations performed per Joule of energy dissipated, which have also shown exponential improvement with time, starting from the first computer, ENIAC, is a critical parameter and will dictate whether scaling of transistors can be sustained.

While the trend of miniaturizing transistors continues unabated, another trend called "More-than-Moore" (MtM) is gradually being recognized [38]. MtM is the functional diversification of semiconductor devices and covers applications other than the microprocessors or application-specific integrated circuits (ASICs) that were the focus of conventional CMOS scaling. It involves heterogeneous integration of components, both digital and non-digital, from a wide range of applications such as automotive, healthcare, communication, energy, infotainment, security, etc. [10, 33, 39]. These functionalities play an important role in the overall miniaturization of the electronic systems, though they may not scale at the rate described by the Moore's law [33, 38]. MtM trends do not compete with the conventional Moore's scaling, but in most cases complement the conventional scaling. It is expected that MtM components will have a larger share in the future evolution of the electronic industry and will be based not just on device scaling. The growth of MtM will depend on innovations at the systems, technology, device and circuit levels [38]. With CMOS scaling now yielding diminishing returns, MtM may open up new markets and growth opportunities [10, 39].

1.7 Challenges in Continued CMOS Scaling

Intuitively, no exponential growth can continue forever and the same would apply to Moore's law. Theoretically, the energy limit is set by Landauer's principle on the maximum achievable energy efficiency in an irreversible computation. As per Landauer's principle, the minimum energy required to erase one bit of information is $kT \ln 2$, where k is the Boltzmann constant and T is the absolute temperature. The implication of this principle is that the energy efficiency of an irreversible computation cannot be improved forever. There are other theoretical limits on the minimum possible energy used per computation based on different models and constraints, each implying that there is still a long way to go with the current technology before theoretical limits become effective [30].

In the near term, some of most serious challenges to the continued scaling of transistors are as follows[6]:

1. There should be a sustained demand for more silicon devices to provide the economic pull to Moore's Law.
2. The increasing IC power consumption has become a bottleneck to further increases in the complexity and the density of the ICs.
3. Keeping the manufacturing costs low, especially that of lithography at smaller device dimensions.

In order to overcome the challenges to the continued scaling of transistors, the smaller transistors must also become "better" in terms of energy efficiency [6, 8]. Novel devices such as TFETs might provide a solution to this problem as we shall see in the remaining chapters of this book [8]. The economic pull for the continued scaling of transistors might come from newer electronic applications, IoT, or heterogeneous integration and scaling as embodied by MtM.

It is worth mentioning that suspicions regarding the end of scaling, for example due to the "lithographic barrier" or "oxide scaling barrier," have been raised many times over the last five decades [40]. However, engineering ingenuity and some ground-breaking inventions have repeatedly overcome these hurdles and it will be fascinating to see how the semiconductor industry evolves over the next few years to meet the current challenges of scaling [40, 41].

1.8 Emerging Research Devices

Over the years, the predominant technology employed in electronic circuits has undergone several changes: vacuum tubes, BJT, NMOS and the current CMOS technology were adopted successively. One important observation that can be made from these historical change-overs is that the prevalent technologies were replaced by their successors when their energy consumption became unaffordable [11]. We have seen in the previous sections that the increasing power density, heat dissipation and energy overheads of the currently employed CMOS technology pose a serious challenge to the continued scaling of transistors [39, 40]. Therefore, a number of emerging research devices are being explored to improve the CMOS technology or as a potential successor to the conventional MOSFET [8, 10, 12, 39].

Novel material systems such as III-V semiconductors, germanium and graphene are being explored as channel materials replacing silicon to improve the carrier mobility. Nanowires and carbon nanotubes (CNTs) are being investigated for MOSFETs since these materials provide improved mobility, better gate control and ameliorate short-channel effects. These novel materials can possibly extend the CMOS scaling, though incorporation of non-silicon materials into the CMOS technology is expected to be a disruptive change. These material systems are being explored for TFETs also and we will look at these materials in detail in the following chapters.

Since the conventional transistors based on modulation of the energy barrier suffer from a minimum achievable subthreshold swing of 60 $mV/decade$, inhibiting low-voltage operations, devices based on novel mechanisms, or devices based on non-charge-state variables are being actively investigated. In addition to the TFETs that are based on quantum-mechanical tunneling for current transport, a single electron transistor (SET) based on Coulomb blockade, a Mott-FET based on phase transition, and a nano-electro-mechanical switch (NEMS) based on mechanical switching are being actively explored. Additionally, ionization-MOS (IMOS) and negative-gate-capacitance transistors are also being investigated for their ability to switch more abruptly than the conventional transistors.

There are several emerging devices that operate on alternative state variables and novel mechanisms. Some of the non-charge-based devices that are being actively explored are a spin wave device (SWD) exploiting collective spin oscillation, nanomagnetic logic (NML) using fringing field interactions between magnetic islands, and all-spin logic (ASL) using magnets to represent non-volatile binary data [10]. These devices differ from the conventional MOSFETs in terms of logic representation, materials, device structures and architecture [10]. Adopting these devices into the existing technology will be highly disruptive and can be justified only if significant advantages in terms of performance and energy efficiency are attainable.

Several quantitative benchmarking results suggest that some of the emerging devices, such as TFETs, may achieve lower switching energy than the existing CMOS, demonstrating their potential to be employed in future electronic circuits. However, the overall performance of all the emerging devices is still poorer than that of existing CMOS devices [39]. Therefore, we are still a long way from a replacement for the conventional MOSFET becomes a reality.

1.9 Summary

In this chapter, we looked at the basics of MOSFET, CMOS, and their scaling. We discussed the historical and current trends of CMOS scaling. We saw that the energy efficiency of the conventional MOSFET poses a serious threat to the continued scaling of transistors. We briefly discussed the emerging research devices that could address the issues of energy efficiency of existing CMOS technology, although none of the emerging devices have shown overall results that are better compared to the existing CMOS. The future of CMOS scaling looks uncertain, including the timeframe during which the current technology can be pushed and extended. From the perspective of device researchers, this creates an extremely exciting opportunity since novel devices and technologies need to be invented that will inevitably reinvigorate or replace the conventional MOSFET. TFETs are highly energy efficient, though they are still far from being able to replace the conventional MOSFET. With appropriate innovation, possibly in the device architecture and/or material systems, TFETs could be a substitute for conventional MOSFETs. With this perspective, we will look at the strengths and weaknesses of TFETs in detail in the following chapters. Several different architectures and material systems employed for TFETs will be presented in the following chapters, with the objective of encouraging readers to pursue their own research on TFETs, which might finally end up as the next milestone in the CMOS technology.

REFERENCES

[1] D. Kahng and M. M. Atalla, "Silicon-silicon dioxide field induced surface device," *Solid State Device Research Conf.*, June 1960.

[2] F. Wanlass and C. Sah, "Nanowatt logic using field-effect metal-oxide

semiconductor triodes," *IEEE International Solid State Circuits Conference Digest of Technical Papers*, pp. 32–33, Feb. 1963.

[3] D. Kahng, "A historical perspective on the development of MOS transistors and related devices," *IEEE Trans. on Electron Devices*, vol. 23, pp. 655–657, Jul. 1976.

[4] W. F. Brinkman, D. E. Haggan, and W. W. Troutman, "A history of the invention of the transistor and where it will lead us," *IEEE Journal of Solid-State Circuits*, vol. 32, pp. 1858–1865, Dec. 1997.

[5] G. E. Moore, "Cramming more components onto integrated circuits," *Electronics*, vol. 87, pp. 114–117, Apr. 1965.

[6] C. A. Mack, "Fifty years of Moore's law," *IEEE Transactions on Semiconductor Manufacturing*, vol. 24, pp. 202–207, May 2011.

[7] A. M. Ionescu and H. Riel, "Tunnel field-effect transistors as energy-efficient electronic switches," *Nature*, vol. 479, pp. 329–337, Nov. 2011.

[8] M. Taylor, "Is dark silicon useful? Harnessing the four horsemen of the coming dark silicon apocalypse," *Design Automation Conference (DAC)*, pp. 1131–1136, June 2012.

[9] K. Swaminathan, E. Kultursay, V. Saripalli, V. Narayanan, M. Kandemir, and S. Datta, "Steep-slope devices: From dark to dim silicon," *IEEE Micro*, vol. 33, pp. 50–59, Sept. 2013.

[10] International Technology Roadmap for Semiconductors (ITRS), "Emerging research devices." `http://www.itrs.net`, 2013.

[11] R. G. Dreslinski, M. Wieckowski, D. Blaauw, D. Sylvester, and T. Mudge, "Near-threshold computing: Reclaiming Moore's law through energy efficient integrated circuits," *Proceedings of IEEE*, vol. 8, pp. 253–256, Feb. 2010.

[12] A. C. Seabaugh and Q. Zhang, "Low-voltage tunnel transistors for beyond CMOS logic," *Proceedings of the IEEE*, pp. 2095–2110, Dec. 2010.

[13] Y. Tsividis and C. McAndrew, *Operation and Modeling of the MOS Transistor.* Oxford University Press, 3rd ed., 2011.

[14] S. M. Sze, *Physics of Semiconductor Devices.* John Wiley and Sons, 2nd ed., 1981.

[15] D. A. Neaman, *Semiconductor Physics and Devices.* Tata McGraw-Hill Education, 3rd ed., 2002.

[16] A. Chaudhry and M. J. Kumar, "Controlling short-channel effects in deep-submicron SOI MOSFETs for improved reliability: A review," *IEEE Transactions on Device and Materials Reliability*, vol. 4(1), pp. 99–109, 2004.

[17] F. Semiconductor, "CMOS, the ideal logic family." `https://www.fairchildsemi.com/application-notes/AN/AN-77.pdf`, Jan. 1983. Fairchild Semiconductor Application Note 77.

[18] N. H. E. Weste and K. Eshraghian, *Principles of CMOS VLSI Design: A Systems Perspective.* Pearson, 2nd ed., 1993.

[19] J. M. Rabaey, A. Chandrakasan, and B. Nikolic, *Digital Integrated Circuits.* Prentice Hall, 2nd ed., 2003.

[20] A. S. Sedra and K. C. Smith, *Microelectronic Circuits.* Oxford University Press, 5th ed., 2008.

[21] R. Dennard, V. Rideout, E. Bassous, and A. LeBlanc, "Design of ion-implanted MOSFET's with very small physical dimensions," *IEEE Journal of Solid-State Circuits*, vol. 20, pp. 256–268, Oct. 1974.

[22] G. Baccarani, M. R. Wordeman, and R. H. Dennard, "Generalized scaling theory and its application to a $\frac{1}{4}$ micrometer MOSFET design," *IEEE Trans. on Electron Devices*, vol. 31, pp. 452–462, Apr. 1984.

[23] M. Bohr, "A 30 year retrospective on Dennard's MOSFET scaling paper," *IEEE Solid-State Circuits Society Newsletter*, vol. 12, pp. 11–13, 2006.

[24] Intel Corporation, "Intel previews Intel Xeon 'Nehalem-EX' processor." `http://www.intel.com/pressroom/archive/releases/2009/20090526comp.htm`.

[25] Intel Corporation, "The Intel Xeon Phi coprocessor: What is it and why should I care? Pt 1: Fitting it all on one slab of silicon." `https://software.intel.com/en-us/blogs/2013/03/22/the-intel-xeon-phi-coprocessor-what-is-it-and-why-should-i-care-pt-1-fitting-it-all`.

[26] Intel Corporation, "A 32-nm, 3.1-billion transistor, 12-wide issue Itanium Processor for mission-critical servers." `http://www.intel.in/content/dam/www/public/us/en/documents/white-papers/itanium-poulson-isscc-paper.pdf`.

[27] Intel Corporation, Backgrounder, "Over 6 decades of continued transistor shrinkage, innovation." `http://www.intel.com/content/www/us/en/silicon-innovations/standards-22-nanometers-technology-backgrounder.html`.

[28] E. Mollick, "Establishing Moore's law," *IEEE Annals of the History of Computing*, vol. 28, pp. 62–75, Jul. 2006.

[29] D. L. Critchlow, "MOSFET scaling: The driver of VLSI technology," *Proceedings of IEEE*, vol. 87, pp. 659–667, Apr. 1999.

[30] J. Koomey, S. Berard, M. Sanchez, and H. Wong, "Implications of historical trends in the electrical efficiency of computing," *IEEE Annals of the History of Computing*, vol. 33, pp. 46–54, Mar. 2010.

[31] J. Koomey and S. Naffziger, "Moore's law might be slowing down, but not energy efficiency," *IEEE Spectrum*, Mar. 2015. `http://spectrum.ieee.org/computing/hardware/moores-law-might-be-slowing-down-but-not-energy-efficiency`.

[32] V. De, "Energy-efficient computing in nanoscale CMOS," *IEEE Design Test*, vol. 33, pp. 68–75, Apr. 2016.

[33] S. S. Iyer, "Heterogeneous integration for performance and scaling," *IEEE Transactions on Components, Packaging and Manufacturing Technology*, vol. 6, pp. 973–982, July 2016.

[34] R. Chau, S. Datta, M. Doczy, B. Doyle, J. Kavalieros, and M. Metz, "High-κ/metal-gate stack and its MOSFET characteristics," *IEEE Electron Device Letters*, vol. 25, pp. 408–410, June 2004.

[35] S. Datta, "Recent advances in high performance CMOS transistors: From planar to non-planar," *The Electrochemical Society Interface*, vol. 22, pp. 41–46, 2013.

[36] S. Sinha, B. Cline, G. Yeric, V. Chandra, and Y. Cao, "Design benchmarking to 7 nm with FinFET predictive technology models," in *Proceedings of the 2012 ACM/IEEE International Symposium on Low Power Electronics and Design*, ISLPED '12, (New York, NY, USA), pp. 15–20, ACM, 2012.

[37] H. Esmaeilzadeh, E. Blem, R. S. Amant, K. Sankaralingam, and D. Burger, "Power challenges may end the multicore era," *Communications of the ACM*, vol. 56, pp. 3–102, Feb. 2013.

[38] International Technology Roadmap for Semiconductors (ITRS), "More-than-Moore, white paper." `http://www.itrs.net`.

[39] A. Chen, "Emerging research device roadmap and perspectives," *IEEE International Conference on IC Design and Technology (ICICDT)*, pp. 1–4, May 2014.

[40] W. Haensch, E. J. Nowak, R. H. Dennard, P. M. Solomon, A. Bryant, O. H. Dokumaci, A. Kumar, X. Wang, J. B. Johnson, and M. V. Fischetti, "Silicon CMOS devices beyond scaling," *IBM Journal of Research and Development*, vol. 50, pp. 339–361, Jul. 2006.

[41] W. M. Holt, "1.1 Moore's law: A path going forward," in *2016 IEEE International Solid-State Circuits Conference (ISSCC)*, pp. 8–13, IEEE, Jan. 2016.

2

Quantum Tunneling

2.1 Introduction

The operation of Tunnel Field-Effect Transistors (TFETs) is based on quantum mechanical tunneling. In order to develop an intuitive understanding of TFETs, it is essential that readers are familiar with the basic concepts of quantum tunneling. In this chapter, we briefly describe the key elements of quantum mechanical tunneling that are relevant to understand TFETs. Readers are, however, advised to refer to the basic textbooks on quantum mechanics [1–3] to explore more on the physics of tunneling.

2.2 Quantum Mechanics

Several experiments performed at atomic levels during the early 20th century had startling observations for physicists. For example, the experiments on black-body radiation and photo-electric effect showed wave-particle duality, the atomic spectrum showed quantization of energy and the radioactivity exhibited tunneling. These observations were impossible to explain and reconcile with the existing notion of physics. As a result, new theories and models such as Max Planck's quantum hypothesis, Bohr's atomic model, Broglie's wave-particle duality principle, the Heisenberg uncertainty principle, and the Schrödinger equation were proposed to describe these extraordinary phenomena. These theories were dramatically different from classical physics and were, at times, in conflict with the intuitive understanding of nature. However, strong experimental evidence suggesting inadequacies in classical physics led scientists to gradually accept and consolidate these theories into quantum mechanics.

The most distinguishing feature of quantum mechanics is that it describes nature probabilistically. In fact, it is the probabilistic behavior of nature that makes the tunneling phenomenon possible. In quantum mechanics, the value of a physical quantity is predicted probabilistically in contrast to classical mechanics. However, certain analogies exist between classical mechanics and quantum mechanics. In classical mechanics, given an initial state of a particle and the forces working on it, the state of the particle at any other instant can be determined using Newton's laws of motion. In quantum mechanics, the Schrödinger equation plays a role similar to Newton's laws. Using Schrödinger equation, the quantum state of a system can be determined. The Schrödinger equation is a partial differential equation that describes the wave function or the quantum state of a system and how it changes with time. Using the wave function, other quantum mechanical quantities can be evaluated. However, it should be noted that the wave function only represents the probabilities or the expectation value of the physical quantities and does not deterministically predict the value of the physical quantities. The Schrödinger equation for a particle constrained along the x-axis is given by

$$i\hbar\frac{\partial\Psi}{\partial t} = -\frac{\hbar^2}{2m}\frac{\partial^2\Psi}{\partial x^2} + V\Psi, \tag{2.1}$$

where i is the imaginary unit with the property $i^2 = -1$, $\hbar$ is the reduced Planck constant, Ψ is the wave function, t is time, m is the mass of the particle, x is the position along the x-axis, and V is the potential energy function. Using the Schrödinger equation, we can determine the wave function Ψ. The wave function is the primary variable in quantum mechanics and the expectation value of other physical quantities can be computed by using the wave function. For example, the expectation value $\langle x \rangle$ of the position x of a particle whose wave function is given as Ψ can be computed as follows:

$$\langle x \rangle = \int_{-\infty}^{+\infty} x|\Psi(x,t)|^2 dx. \tag{2.2}$$

When the potential energy function V is independent of time, the solution of the Schrödinger equation $\Psi(x,t)$ can be worked out by separation of variables. The solution $\Psi(x,t)$ of the Schrödinger equation can be represented as

$$\Psi(x,t) = \psi(x)\varphi(t), \tag{2.3}$$

where $\psi(x)$ represents the spatial component of $\Psi(x,t)$ and is a function of only x, and $\varphi(t)$ represents the temporal component of $\Psi(x,t)$ and is a function of only t. Using the Schrödinger equation, we can derive the following equation:

$$\varphi(t) = e^{-\frac{iEt}{\hbar}} \tag{2.4}$$

$$\frac{\hbar^2}{2m}\frac{\partial^2\psi}{\partial x^2} + (E - V)\psi = 0, \tag{2.5}$$

where E corresponds to the energy of the particle. Equation (2.5) is the steady-state form of the Schrödinger equation. When the potential profile is independent of time, the steady-state form of the Schrödinger equation can be used to find the solution of the Schrödinger Equation (2.1). It should be noted that the solution $\psi(x)$ for the Equation (2.5) is not unique. Each solution $\psi_n(x)$ corresponds to a specific value of the energy E_n. The values of energy E_n for which the Schrödinger Equation (2.5) can be solved are called the eigenvalues and the corresponding wave functions $\psi_n(x)$ are called the eigenfunctions.

2.3 Quantum Mechanical Tunneling

In its simplest form, quantum mechanical tunneling can be described as a phenomenon in which a particle can escape a potential barrier even when it has insufficient energy to overcome the barrier. For example, in order to throw a ball to the other side of a hill, energy sufficient to overcome the potential barrier of the hill must be imparted to the ball. However, according to quantum mechanics, even when the energy imparted to the ball is smaller than the potential barrier of the hill, there is a finite probability that the ball could cross the barrier and reach to the other side of the hill [4, 5]. This may seem counter-intuitive in the classical sense, but experiments have proven that this phenomenon of quantum mechanics does occur. In fact, the tunnel effect observed in the emission of alpha particles was one of the first experimental proofs that the principles of quantum mechanics are obeyed in nature.

Figure 2.1 shows a particle of mass m and energy E approaching a potential barrier $V(x)$. The potential energy function $V(x)$ is independent of time and is described as follows:

$$V(x) = 0, \text{ for } x < 0 \text{ and } x > a$$

$$V(x) = U_0, \text{ for } 0 \le x \le a.$$

We further assume that the particle has energy lower than the potential barrier ($E < U_0$). Classically, the particle will be reflected. However, according to quantum mechanics, there is a finite probability that the particle will go through the barrier and be transmitted to the other side of the barrier. This phenomenon is the operating principle for tunneling devices, as we shall see in the chapters to follow. In this section, using a steady-state Schrödinger equation, we shall determine the probability that the particle will be transmitted [2]. The space in Figure 2.1 can be divided into three regions as shown. A time independent Schrödinger equation for the three regions can be written as

Region 1 and Region 3: $(x < 0)$ and $(x > a)$, $V(x) = 0$

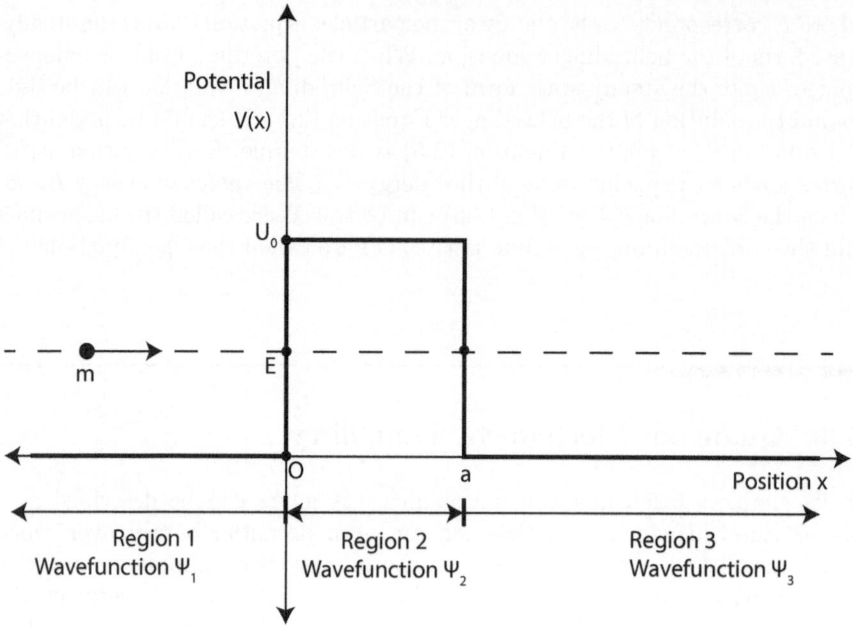

FIGURE 2.1
Rectangular tunneling barrier of height U_0 and width a. A particle of mass m having energy E approaches the barrier where $E < U_0$. Classically, the particle would be reflected, but it has a finite probability of tunneling through the barrier, as per quantum mechanics.

$$\frac{\hbar^2}{2m}\frac{\partial^2 \psi}{\partial x^2} + E\psi = 0 \tag{2.6}$$

Region 2: $0 \le x \le a$, $V(x) = U_0$

$$\frac{\hbar^2}{2m}\frac{\partial^2 \psi}{\partial x^2} + (E - U_0)\psi = 0. \tag{2.7}$$

The solution of the differential Equation (2.6) in Region 1 can be given as

$$\psi_1(x) = A_1 e^{ik_1 x} + B_1 e^{-ik_1 x}, \tag{2.8}$$

where $k_1 = \frac{\sqrt{2mE}}{\hbar}$, and A_1, B_1 are constants that need to be evaluated. Here, the first term $A_1 e^{ik_1 x}$ can be interpreted as the component of a wave function incident on the potential barrier moving in the positive x direction, while the second term $B_1 e^{-ik_1 x}$ can be interpreted as the component of a wave function reflected by the potential barrier moving in the negative x direction [2]. Similarly, the solution of the equation in Region 3 can be given as

$$\psi_3(x) = A_3 e^{ik_1 x}. \tag{2.9}$$

In Region 3, the wave function consists only of the transmitted wave and the component of the wave function moving in the negative x direction is non-existent, since there is nothing in Region 3 that can change the direction of the transmitted waves [2]. The transmission probability P_t for the particle to penetrate the barrier is given by [2]

$$P_t = \frac{A_3 A_3^* v_3}{A_1 A_1^* v_1}, \tag{2.10}$$

where v_1 is the group velocity of the incoming wave in Region 1 and v_3 is the group velocity of the transmitted wave in Region 3. The term $A_1 A_1^* v_1$ is the flux of the particles that arrive at the barrier and $A_3 A_3^* v_3$ is the flux of the particles transmitted through the barrier. We shall compute the transmission probability P_t in the rest of this section. The solution of the differential Equation (2.7) in Region 2 can be given as

$$\psi_2(x) = A_2 e^{-k_2 x} + B_2 e^{k_2 x}, \tag{2.11}$$

where $k_2 = \frac{\sqrt{2m(U_0 - E)}}{\hbar}$, and A_2, B_2 are constants that need to be evaluated. It should be noted that the exponents in $\psi_2(x)$ are real numbers and hence are non-oscillating. The wave functions ψ_1, ψ_2 and ψ_3 must satisfy the boundary condition: $\psi(x)$ and $\frac{d\psi}{dx}$ must be continuous throughout Regions 1, 2 and 3.

Applying boundary conditions at $x = 0$ yields

$$A_1 + B_1 = A_2 + B_2 \tag{2.12}$$

$$ik_1 A_1 - ik_1 B_1 = -k_2 A_2 + k_2 B_2. \tag{2.13}$$

Applying boundary conditions at $x = a$ yields

$$A_2 e^{-k_2 a} + B_2 e^{k_2 a} = A_3 e^{ik_1 a} \tag{2.14}$$

$$-k_2 A_2 e^{-k_2 a} + k_2 B_2 e^{k_2 a} = ik_1 A_3 e^{ik_1 a}. \tag{2.15}$$

Using the boundary conditions of Equations (2.12)–(2.15), the ratio $\frac{A_1}{A_3}$ is computed as [2]

$$\frac{A_1}{A_3} = \left\{\frac{1}{2} + \frac{i}{4}\left(\frac{k_2}{k_1} - \frac{k_1}{k_2}\right)\right\} e^{(ik_1+k_2)a} + \left\{\frac{1}{2} - \frac{i}{4}\left(\frac{k_2}{k_1} - \frac{k_1}{k_2}\right)\right\} e^{(ik_1-k_2)a}. \tag{2.16}$$

Let us assume that the barrier height U_0 is much higher compared to the energy of the particle E. This means that $k_2 \gg k_1$ and $(\frac{k_2}{k_1} - \frac{k_1}{k_2}) \approx \frac{k_2}{k_1}$. Therefore,

$$\frac{A_1}{A_3} = \left\{\frac{1}{2} + \frac{ik_2}{4k_1}\right\} e^{(ik_1+k_2)a} + \left\{\frac{1}{2} - \frac{ik_2}{4k_1}\right\} e^{(ik_1-k_2)a}. \tag{2.17}$$

When the energy barrier is wide enough such that $k_2 a \gg 1$, we can write [2]

$$\frac{A_1}{A_3} = \left\{\frac{1}{2} + \frac{ik_2}{4k_1}\right\} e^{(ik_1+k_2)a} \tag{2.18}$$

$$\frac{A_1 A_1^*}{A_3 A_3^*} = \left\{\frac{1}{4} + \frac{k_2^2}{16k_1^2}\right\} e^{2k_2 a}. \tag{2.19}$$

Therefore, the transmission probability in Equation (2.10) can be expressed as follows (taking $v_1 = v_3$) [2]:

$$P_t = \frac{A_3 A_3^*}{A_1 A_1^*} = \left\{\frac{16}{4 + (k_2/k_1)^2}\right\} e^{-2k_2 a}. \tag{2.20}$$

Using the definition of k_1 and k_2

$$P_t = \frac{A_3 A_3^*}{A_1 A_1^*} = \left\{\frac{16}{3 + (U_0/E)}\right\} e^{-2k_2 a}. \tag{2.21}$$

Equation (2.21) is the relationship between the transmission probability and the height and the width of the tunneling barrier and the mass and the energy of the particle. We can make the following observations from the above equation:

1. The tunneling probability is exponentially related to the tunneling width. If the tunneling width is decreased, the tunneling probability increases exponentially.
2. The tunneling probability is exponentially related to $k_2 = \frac{\sqrt{2m(U_0-E)}}{\hbar}$. If the barrier height U_0 is decreased or the energy of the particle E is increased, the tunneling probability increases.

3. If the mass of the particle is decreased, the tunneling probability increases.

These observations are of great importance in optimizing the performance of the tunneling devices, as we shall see in the subsequent chapters.

2.4 Solving the Tunneling Problem

To understand the behavior of the tunneling devices at a given bias voltage, the tunneling probability in the device needs to be calculated for the associated potential profile. If the potential profile in a device is of ideal rectangular form, the tunneling probability is given analytically by Equation (2.21). However, the potential profiles in real devices are often non-rectangular. The tunneling probability in these devices is estimated using an approximate solution or using numerical methods. In this section, we describe the approximation method and the numerical methods that are widely employed for simulation of TFETs and will be used in the subsequent chapters.

2.4.1 Analytic Approximation Methods

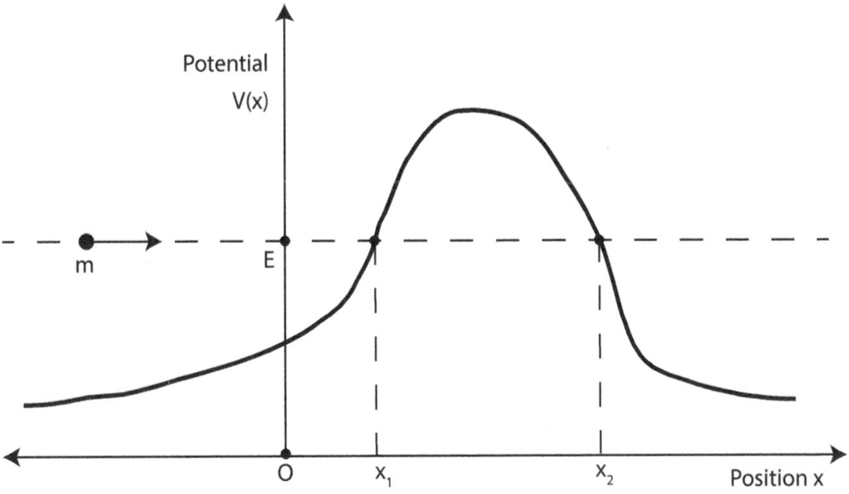

FIGURE 2.2
A particle of mass m having energy E approaches a potential barrier where $E < V(x)$ for $x_1 < x < x_2$.

Let us consider a particle of mass m and energy E approaching a potential barrier $V(x)$ as shown in Figure 2.2. The potential energy function $V(x)$ is greater than the particle energy E between x_1 and x_2. Classically, the particle approaching the barrier from the left would stop and start moving in the opposite direction at $x = x_1$ and the particle approaching the barrier from right would get reflected at $x = x_2$. These points are called the classical turning points. We also assume that the quasi-classical condition is satisfied, i.e., the de Broglie wavelength of the particle is less than the characteristic length of the problem. It should also be noted that the quasi-classical approximation is not valid at very small momentum of the particle. Under these conditions, the tunneling probability can be computed in a general form using the WKB approximation [1, 6]

$$P_t \cong \exp\left[-2\int_{x_1}^{x_2} |k(x)|dx\right], \tag{2.22}$$

where $k(x)$ is the wave vector of the particle inside the barrier, and x_1 and x_2 are the classical turning points. It should be noted that the above approximation is valid only when the exponent is a large number, which implies that the tunneling probability is small.

We illustrate the application of Equation (2.22) by computing the tunneling probability when the potential barrier is of triangular shape.

Figure 2.3 shows a particle with mass m and energy E approaching a potential barrier $V(x)$. The potential energy function $V(x)$ is independent of time and is described as follows:

$$V(x) = 0, \text{ for } x < 0$$

$$V(x) = U_0 - Fx, \text{ for } x \geq 0,$$

where F defines the negative gradient of the electric potential or the electric field in the region $x \geq 0$.

The x-coordinate $x = 0$ and $x = L$ are the classical turning points for the barrier. For $x = L$, we can write $V(L) = U_0 - FL = E$.

The wave vector inside the barrier can be written as

$$k(x) = \sqrt{\frac{2m}{\hbar^2}\{V(x) - E\}} = \sqrt{\frac{2m}{\hbar^2}(U_0 - Fx - E)}. \tag{2.23}$$

Using Equations (2.22) and (2.23):

$$P_t \cong \exp\left[-2\int_0^L \sqrt{\frac{2m}{\hbar^2}(U_0 - Fx - E)}dx\right] \tag{2.24}$$

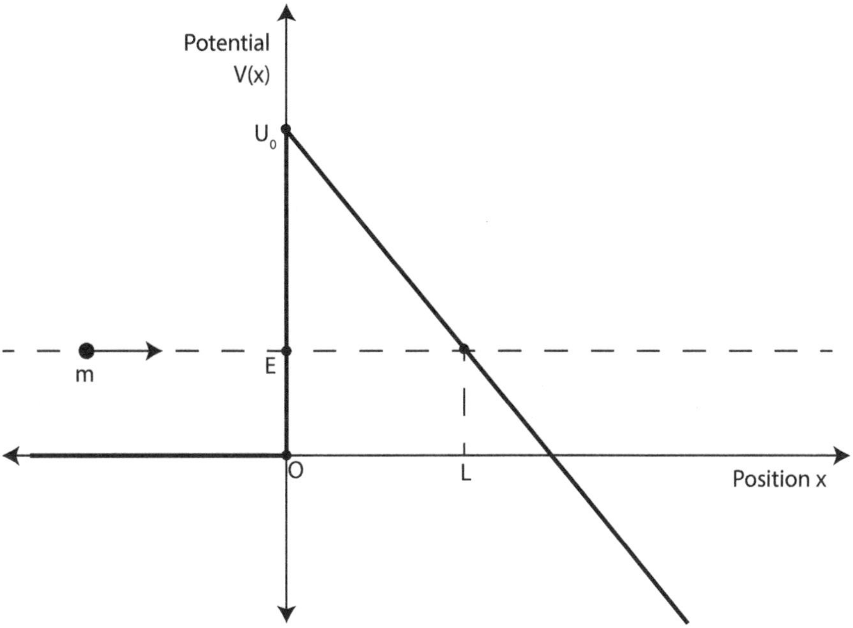

FIGURE 2.3
A particle of mass m having energy E approaches a triangular potential barrier where $E < V(x)$ for $0 < x < L$. The triangular barrier is defined as $V(x) = U_0 - Fx$, for $x \geq 0$ where F is a constant defining the negative gradient of the barrier.

$$= \exp\left[-\frac{4\sqrt{2m}}{3\hbar F}\left\{(U_0 - E)^{\frac{3}{2}} - (U_0 - FL - E)^{\frac{3}{2}}\right\}\right] \tag{2.25}$$

$$= \exp\left[-\frac{4\sqrt{2m}}{3\hbar F}(U_0 - E)^{\frac{3}{2}}\right]. \tag{2.26}$$

Equation (2.26) is the relationship between the tunneling probability and the energy and the mass of the particle, the height of the potential barrier, and the electric field. We can make the following observations from Equation (2.26). The tunneling probability increases

1. as the height of the potential barrier $(U_0 - E)$ decreases and
2. as the electric field F increases.

These observations are quite useful in designing TFETs, as we shall see in the subsequent chapters.

2.4.2 Numerical Methods

The potential profiles in TFETs are not simple geometrical forms. Therefore, computing the tunneling probability and deriving analytical expressions for tunneling currents are very difficult in TFETs, even using approximation methods. As a result, researchers regularly employ numerical simulators to investigate the behavior of the TFETs. Since tunneling is a quantum mechanical phenomenon, rigorous quantum transport computational methods give a precise result. Therefore, rigorous quantum transport formalism, like the non-equilibrium Green's function (NEGF), are important for simulating TFETs especially at small device dimensions. Various researchers have employed rigorous quantum treatment to compute the tunneling current in TFETs [7–19]. However, full quantum treatment is computationally expensive, especially for 2-D and 3-D structures. Therefore, tunneling models in TCAD simulators that work semi-classically using a drift-diffusion mechanism are widely employed and often give acceptable results. There are different tunneling models available in TCAD tools with different degrees of accuracy and computational resource requirements.

The simplest of the tunneling models is the local band-to-band tunneling model. When the electric field inside a device is high enough, then there is sufficient bending of bands such that the tunneling of electrons from the valence band to the conduction band is facilitated. The tunneling of an electron from the valence band to the conduction band can be viewed as a generation of an electron in the conduction band and a generation of a hole in the valence band. In the local tunneling models, the effect of tunneling of electrons is modeled as the generation of electron–hole pairs. The increase in the electric field at a particular location results in greater tunneling at that location, which is modeled as an increase in the generation of electrons. A general form of the expression for the generation rate, based on Kane's derivation, that is widely used in simulators is [7, 20]

$$G_{BTBT} = AF^{\alpha} \exp\left(-\frac{B}{F}\right), \tag{2.27}$$

where F is the electric field and A, B, and α are the fitting parameters. The advantage of the local tunneling model is that it is simple, computationally robust and can easily be plugged into semi-classical device simulators. However, it ignores the recombination of carriers due to tunneling and can yield physically wrong results [7].

For the tunneling models to be accurate enough for TFETs, at least the following physical aspect of band-to-band tunneling must be captured [7, 21]:

1. The rate of tunneling depends on quantities at locations that are

spatially separated. Therefore, only a non-local tunneling model can accurately represent the physics of tunneling.

2. The trajectory of the tunneling paths can change with the applied bias in a TFET. Therefore, the tunneling paths must be identified accurately at different bias points.

In non-local tunneling models, the possible tunneling paths are identified by the simulator or are given as an input to the simulator by the user. The tunneling probabilities, for the given tunneling path and the given profile of the energy bands are computed by the simulator by employing techniques like WKB approximation or more rigorous quantum treatment. For example, the tunneling probability in a strip of tunneling device in the tunneling path can be computed using WKB approximation as [7]

$$P_t = \exp\left[-2\int_C^V \frac{\sqrt{2m^*\{V(r)-E\}}}{\hbar}dr\right], \tag{2.28}$$

where m^* is the tunneling effective mass, r is the position in the device, $\{V(r) - E\}$ is the barrier height at r, and the numerical integration is done along the tunneling path. The tunneling effective mass is a fitting parameter for the tunneling model and attempts to capture the band structure inside the device. Formulations other than Equation (2.28), which consider the band structure inside the device more rigorously, are also used in numerical simulations [7]. The non-local tunneling models provide quite a handy tool for a quick and reasonably accurate analysis of TFETs and are widely used. We shall present the result of numerical simulation of TFETs using a non-local tunneling model in subsequent chapters.

2.5 Junction Breakdown Due to Tunneling

When a high electric field is applied across a reverse-biased p-n junction, the junction breaks down, and a high current flows through the junction. The junction can break down due to the following mechanisms [22]:

1. **Tunneling effect**: This mechanism of breakdown occurs when the depletion width is small and the electric field is sufficiently high to allow the electrons from the valence band on the p-side of the junction to tunnel into the empty states in the conduction band on the n-side as shown in Figure 2.4.

2. **Avalanche multiplication**: This mechanism of breakdown occurs

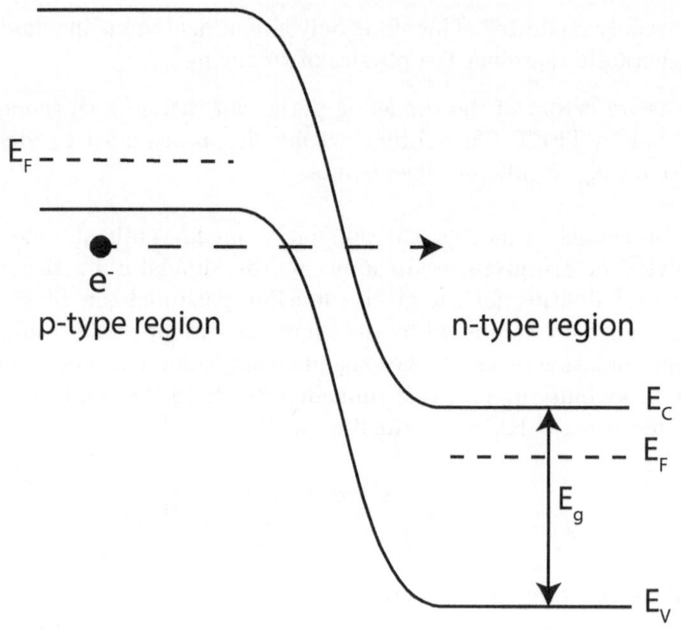

FIGURE 2.4
Tunneling mechanism in a p-n junction. Due to the high electric field, the valence band and the conduction band overlap to make tunneling possible.

when the electric field is sufficiently high such that the carriers are accelerated to energies high enough to create electron–hole pairs via collisions with the bound electrons.

3. **Thermal instability**: This mechanism of breakdown occurs because of the heat dissipated due to the reverse current at high voltages. This leads to a higher junction temperature, which further increases the reverse current and provides positive feedback leading to thermal instability.

The junction breakdown due to the tunneling effect is utilized in TFETs as we shall see in the next chapter. The tunneling current in a p-n junction is given by [22, 23]

$$I = aV_{eff}F \exp\left(-\frac{b}{F}\right), \tag{2.29}$$

where F is the electric field at the junction, V_{eff} is the tunnel-junction bias, and a and b are constants determined by the properties of the material. The constants a and b can be written as

$$a = \frac{Aq^3}{4\pi^2\hbar^2}\sqrt{\frac{2m^*}{E_g}} \tag{2.30}$$

$$b = \frac{4\sqrt{m^*}E_g^{\frac{3}{2}}}{3q\hbar}, \tag{2.31}$$

where A is the cross-sectional area of the junction, E_g is the bandgap energy, and m^* is the effective carrier mass. This equation shows that the tunneling current increases as the electric field in the p-n junction increases. For germanium or silicon p-n junctions, a significant tunneling current flows when the electric field approaches $10^6\ V/cm$.

2.6 Tunnel Diode

In order to illustrate how quantum tunneling is applied in electronic devices, we describe the working principle of a tunnel diode. The tunnel diode was invented in 1957 by Leo Esaki and is one of the earliest electronic devices where the quantum tunneling effect was studied and applied [24].

The tunnel diode consists of a degenerately doped p-n junction. As a result, the depletion layer is very thin and the tunneling of electrons across the junction is facilitated. The band diagram of a tunnel diode at $T = 0\ K$ with no applied bias is shown in Figure 2.5. The shaded region denotes the energy levels that are occupied. At $T = 0\ K$, all the energy levels below the Fermi level are occupied, while all the energy levels above the Fermi level are empty.

With the application of the forward (reverse) bias voltage, the bands on the n^+ side of the band diagram move vertically upward (downward) with respect to the bands on the p^+ side. As a result, the overlap of the conduction band on the n^+ side with the valence band on the p^+ side changes with the application of the forward or the reverse bias voltage. Therefore, the available and unoccupied states for the electrons to tunnel to the valence or conduction band changes with the application of the applied bias. This gives rise to a unique current-voltage characteristic in a tunnel diode, as shown in Figure 2.6.

We now explain the mechanism responsible for the unique current-voltage characteristics of the tunnel diode. The section of the current-voltage characteristic shown in the thicker line in Figure 2.6 is the result of the tunneling phenomenon. When a reverse biased voltage is applied, the bands on the n^+ side move downward with respect to the bands on the p^+ side. As a result,

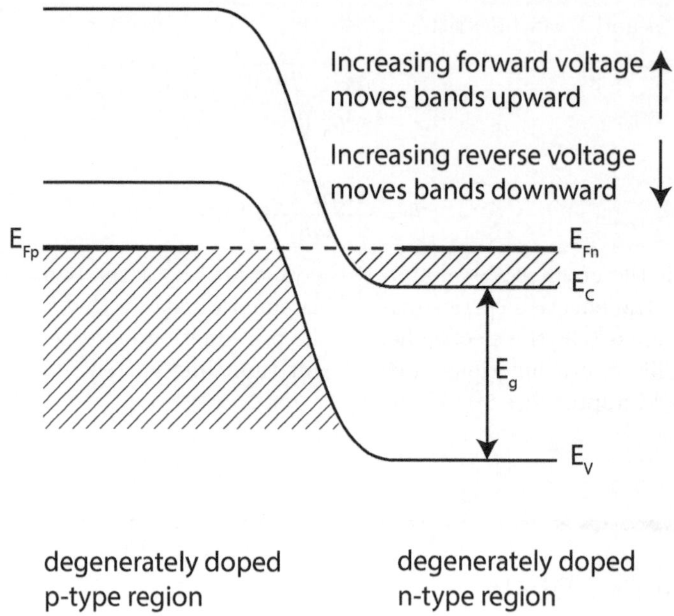

FIGURE 2.5
Band diagram of a tunnel diode at $T = 0\ K$ when no bias is applied.

the filled states in the valence band on the p$^+$ side overlap with the empty states in the conduction band on the n$^+$ side. Therefore, the electrons can tunnel from the valence band in the p$^+$ region to the conduction band in the n$^+$ region. As the reverse bias is decreased, the tunneling current decreases, since the available states to which the electrons can tunnel decreases. When there is no bias voltage ($V_R = 0$), the current reduces to zero since there are no empty states in the conduction band to which the electron in the valence band can jump. When the forward voltage is applied, the bands on the n$^+$ side move upward with respect to the bands on the p$^+$ side. As a result, the filled states in the conduction band on the n$^+$ side overlap with the empty states in the valence band on the p$^+$ side. Therefore, the tunneling current due to the tunneling of electrons from the conduction band on the n$^+$ side to the valence band on the p$^+$ side continues increasing until a peak V_P is reached. As the forward voltage is increased more than V_P, the bands on the n$^+$ side move further upward with respect to the bands on the p$^+$ side, decreasing the band overlap and the states available for tunneling, and hence the tunnel current starts decreasing. Therefore, in this region of operation, negative differential resistance is exhibited. Finally, when the forward bias voltage is increased beyond V_V, the band overlap ceases to exist, and since there are no empty states to which electrons can tunnel, the tunneling current becomes zero. However, with further increase in the applied voltage, the normal thermal current dom-

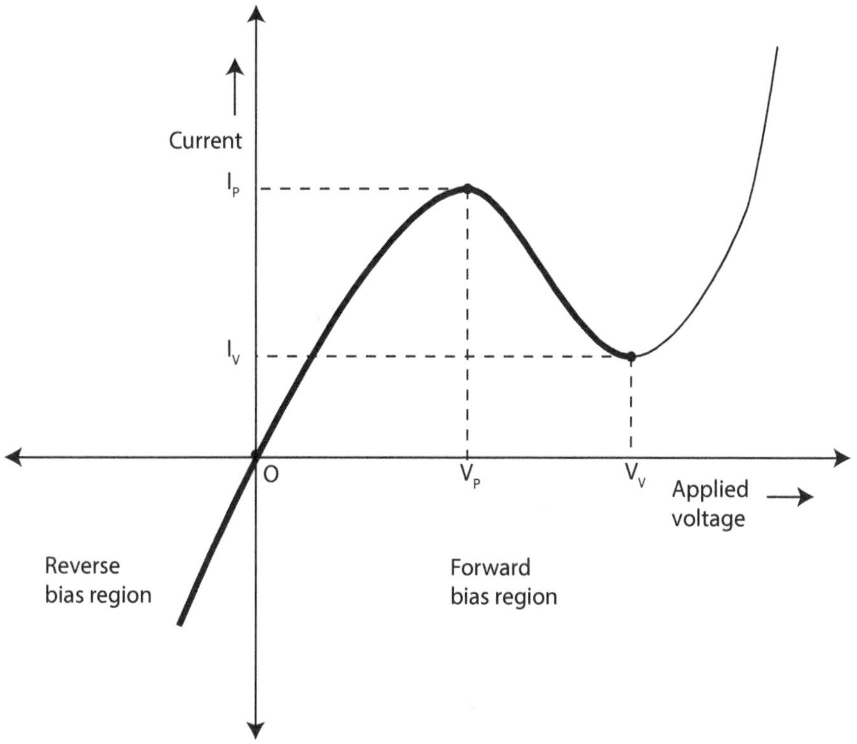

FIGURE 2.6
Current-voltage characteristics of a tunnel diode.

inates, which shows an exponential increase with the applied voltage. Thus, we see that the applied voltage changes the overlap of the conduction band and the valence band in a tunnel diode, which in turn results in its unique transfer characteristics. A similar working principle is applied in TFETs as we will see in the next chapter.

2.7 Summary

In this chapter, we reviewed the basic concepts of quantum mechanical tunneling. We have shown that even when there is a potential barrier greater than the energy of the particle, there is a finite probability that the particle can tunnel through the barrier. The tunneling probability is greater when the tunnel barrier is thin and the difference in the potential barrier height and the particle energy is smaller. We have shown that the computation of tunneling probability in electron devices can be done using approximation methods or numerical techniques. We have shown that the principle of quantum mechanical tunneling can be applied in electron devices like tunnel diodes. In the next chapter, we shall look at the working principle and the basic concepts of tunnel field-effect transistors.

REFERENCES

[1] L. D. Landau and E. M. Lifshitz, *Quantum Mechanics Non-relativistic Theory.* Pergamon Press, 2nd ed., 1958.

[2] A. Beiser, *Concepts of Modern Physics.* Tata McGraw-Hill, 5th ed., 1963.

[3] L. I. Schiff, *Quantum Mechanics.* McGraw-Hill, 3rd ed., 1955.

[4] R. W. Gurney and E. U. Condon, "Wave mechanics and radioactive disintegration," *Nature*, vol. 122(3073), p. 439, Sept. 1928.

[5] P. C. W. Davies, "Quantum tunneling time," *American Journal of Physics*, vol. 73(1), pp. 23–27, Sept. 2005.

[6] M. Tyagi, *Introduction to Semiconductor Materials and Devices.* John Wiley and Sons, 2nd ed., 1991.

[7] C. Shen, L. T. Yang, G. Samudra, and Y. C. Yeo, "A new robust nonlocal algorithm for band-to-band tunneling simulation and its application to tunnel-FET," *Solid-State Electronics*, pp. 23–30, Mar. 2011.

[8] S. Datta, *Quantum Transport: Atom to Transistor.* Cambridge University Press, 2nd ed., 2005.

[9] D. Basu, L. F. Register, M. J. Gilbert, and S. K. Banerjee, "Atomistic simulation of band-to-band tunneling in III-V nanowire field-effect transistors," *Simulation of Semiconductor Processes and Devices (SISPAD)*, pp. 1–4, 2009.

[10] M. Luisier and G. Klimeck, "Performance comparisons of tunneling field-effect transistors made of InSb, Carbon, and GaSb-InAs broken gap heterostructures," *IEDM*, pp. 1–4, 2009.

[11] Y. Gao, T. Low, and M. Lundstrom, "Possibilities for $VDD = 0.1\ V$ logic using carbon-based tunneling field effect transistors," *Symp. on VLSI Technology*, pp. 180–181, 2009.

[12] Y. Miyoshi, M. Ogawa, S. Souma, and H. Nakamura, "Analysis of geometrical structure and transport property in InAs/Si heterojunction nanowire tunneling field effect transistors," *Simulation of Semiconductor Processes and Devices (SISPAD)*, pp. 227–230, 2011.

[13] H. Da, K.-T. Lam, G. Samudra, S.-K. Chin, and G. Liang, "Graphene nanoribbon tunneling field-effect transistors with a semiconducting and a semimetallic heterojunction channel," *IEEE Transactions on Electron Devices*, pp. 1454–1461, May 2012.

[14] L. Liu, D. Mohata, and S. Datta, "Scaling length theory of double-gate interband tunnel field-effect transistors," *IEEE Transactions on Electron Devices*, pp. 2–908, Apr. 2012.

[15] L. Leem, L. A. Srivastava, S. Li, B. Magyari-Kope, G. Iannaccone, J. S. Harris, and G. Fiori, "Multi-scale simulations of partially unzipped CNT hetero-junction tunneling field effect transistor," *IEDM*, pp. 32.5.1–32.5.4, 2010.

[16] M. S. Fahad, A. Srivastava, A. K. Sharma, and C. Mayberry, "Analytical current transport modeling of graphene nanoribbon tunnel field-effect transistors for digital circuit design," *IEEE Transactions on Nanotechnology*, vol. 15, pp. 39–50, Jan. 2016.

[17] D. Verreck, A. S. Verhulst, M. L. V. de Put, B. Sore, N. Collaert, A. Mocuta, A. Thean, and G. Groeseneken, "Uniform strain in heterostructure tunnel field-effect transistors," *IEEE Electron Device Letters*, vol. 37, pp. 337–340, Mar. 2016.

[18] F. Liu, J. Wang, and H. Guo, "Atomistic simulations of device physics in monolayer transition metal dichalcogenide tunneling transistors," *IEEE Transactions on Electron Devices*, vol. 63, pp. 311–317, Jan. 2016.

[19] S.-C. Lu, M. Mohamed, and W. Zhu, "Novel vertical hetero- and homo-junction tunnel field-effect transistors based on multi-layer 2D crystals," *2D Materials*, vol. 3, no. 1, p. 011010, 2016.

[20] E. O. Kane, "Zener tunneling in semiconductors," *Journal of Physics and Chemistry of Solids*, vol. 12, pp. 181–188, 1959.

[21] A. Verhulst, B. Soree, D. Leonelli, W. Vandenberghe, and G. Groeseneken, "Modeling the single-gate, double-gate, and gate-all-around tunnel field-effect transistor," *Journal of Applied Physics*, vol. 107, 024518.

[22] S. M. Sze, *Physics of Semiconductor Devices.* John Wiley and Sons, 2nd ed., 1981.

[23] Q. Zhang, W. Zhao, and A. Seabaugh, "Low-subthreshold-swing tunnel transistors," *IEEE Electron Device Letters*, vol. 27, pp. 297–300, Apr. 2006.

[24] L. Esaki, "New phenomenon in narrow germanium p-n junctions," *Phys. Rev.*, vol. 109, pp. 603–604, 1958.

3

Basics of Tunnel Field-Effect Transistors

3.1 Introduction

The current in a conventional MOSFET is based on thermionic emission of carriers over a potential barrier. As a result, keeping the power consumption within an acceptable limit is one of the biggest challenges for the conventional MOSFETs at advanced technology nodes. Tunnel Field-Effect Transistors (TFETs), due to their different mechanism of current transport exhibit several interesting electrical characteristics and could be a feasible alternative to the conventional MOSFETs.

In this chapter, we start by describing the structure of TFETs and their operating principle. We explain the origin and the nature of some of the most distinguishing electrical properties of TFETs. We develop an approximate relationship between the tunneling current and the physical and electrical parameters of TFETs. This will help readers understand how to optimize the electrical characteristics of TFETs by adjusting the device parameters. Finally, this chapter highlights the promises and the limitations of TFETs and briefly explains why TFETs could be an alternative to the conventional MOSFETs for future applications.

This chapter forms the foundation for the following chapters in this book. At the end of this chapter, readers will become familiar with the basic operating principle of TFETs and will be able to qualitatively appreciate their distinctive electrical characteristics.

3.2 Device Structure

In simple terms, the TFET is a gated reverse-biased p-i-n structure. A simple implementation of an n-type and a p-type TFET is shown in Figure 3.1. Structurally, the most distinguishing feature of a TFET is the type of doping of the drain and the source. The doping of the drain and the source are of

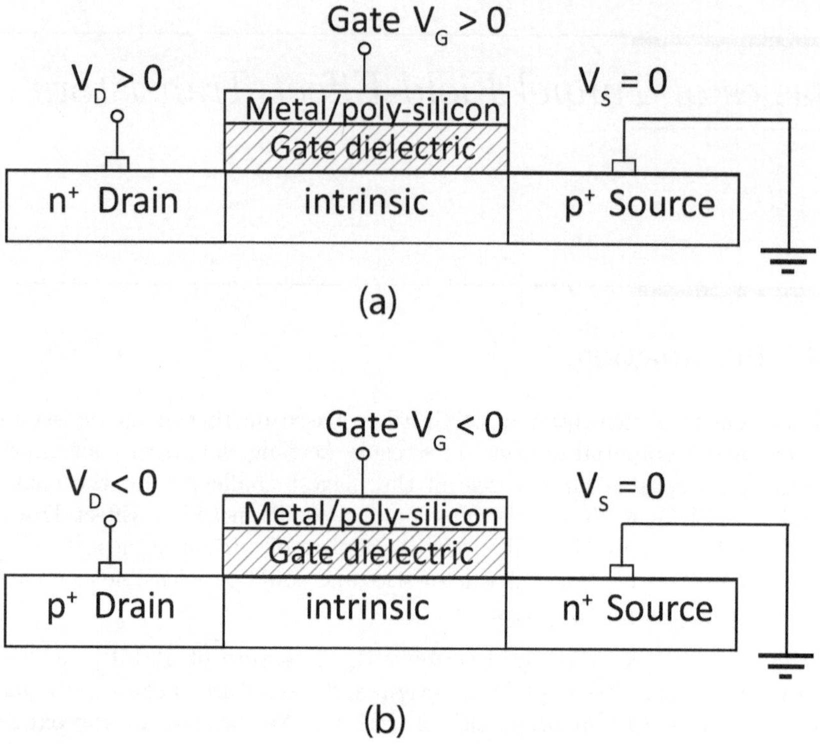

FIGURE 3.1
A simple implementation of a TFET and the biasing scheme: (a) n-type TFET (b) p-type TFET.

opposite types in a TFET. In contrast, the drain and the source doping are of the same type in a conventional MOSFET. For an n-type TFET, the drain is doped n^+ while the source is doped p^+. For a p-type TFET, the drain is doped p^+ while the source is doped n^+. The channel is an intrinsic or lowly doped p-type or n-type semiconductor. The channel is separated from the gate electrode by a dielectric, similar to a conventional MOSFET.

The biasing schemes of an n-type TFET and a p-type TFET are shown in Figure 3.1. For an n-type TFET, the source is grounded and a positive voltage is applied to the drain and the gate electrodes. For a p-type TFET, the source is grounded and a negative voltage is applied to the drain and the gate electrodes. A TFET is called an n-type TFET or a p-type TFET depending on the dominant carrier in the channel formed under the gate when the TFET is turned on. When the dominant carriers in the channel are electrons, the TFET is called an n-type TFET and when the dominant carriers in the channel are holes, the TFET is called a p-type TFET. The terminals are called source or drain depending on whether the dominant carriers enter or leave the channel through that terminal. In an n-type TFET, the electrons enter the channel through the source and leave the channel through the drain. In a p-type TFET, the holes enter the channel through the source and leave the channel through the drain. The mechanism of the dominant carriers entering the channel is band-to-band tunneling (BTBT) and is discussed in detail in the following section.

The TFET structures that are shown in Figure 3.1 suffer from several problems, extremely small ON-current being the most severe drawback. Therefore, researchers have proposed and fabricated TFETs in various other structural forms, ameliorating the problems observed in the simple implementation of the TFET. We will describe different types of TFET structures throughout this book. In this chapter, we describe a TFET called the Double Gate Tunnel Field-Effect Transistor (DGTFET) in greater detail [1]. We also provide example device simulation files of the DGTFET in Appendix A, which the readers may use to simulate and explore TFETs on their own using commercial device simulators. We have chosen DGTFET for this purpose, since a DGTFET is a simple structure and a fundamental understanding of a TFET can easily be developed using a DGTFET. It should be noted that, though different kinds of TFETs have their own peculiarities, the basic operating principle of all the TFETs remains the same. Additionally, for different types of TFETs, the dependence of the tunneling current on the device parameters like gate dielectric-thickness, body thickness, source–drain doping, etc., follow a similar trend. Therefore, we can use a DGTFET as a representative TFET to gain insight into the working of TFETs.

Figure 3.2 shows a schematic cross-sectional view of a DGTFET [1]. The device works as an n-type TFET. It should be noted that structures similar to the DGTFET have been used by several researchers in their theoretical, as well as, experimental works [1–18].

TABLE 3.1
Device parameters of DGTFET [1].

Device Parameter	**Value**
Source Doping (N_A)	1×10^{20} $atoms/cm^3$
Drain Doping (N_D)	5×10^{18} $atoms/cm^3$
Channel Doping (N_A or N_D)	1×10^{17} $atoms/cm^3$
Channel Length (L)	50 nm
Gate Oxide Thickness (t_{ox})	3 nm
Silicon Body Thickness (t_{Si})	10 nm
Gate Work Function (Φ_m)	4.5 eV

The device parameters for a typical DGTFET are shown in Table 3.1. The

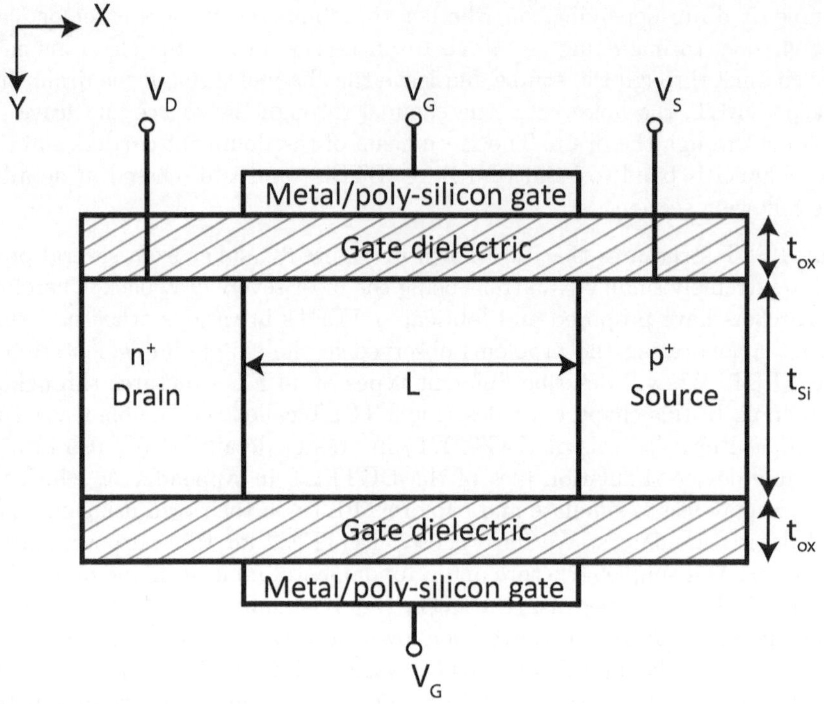

FIGURE 3.2
Schematic cross-sectional view of a DGTFET [1].

body of the transistor is made of silicon and the gate is composed of silicon dioxide. The drain is doped n^+ while the source is doped with p^+ dopants. The doping of the source is kept higher compared to the doping of the drain. This is done to suppress the ambipolar behavior of the TFET, as we shall explain in Section 3.11. In order to achieve a high ON-current, a sharp doping profile at the source is needed. The thickness of the silicon body is kept small, so that the ON-current is improved. The double-gate structure in a DGTFET improves the gate control on the channel and it is expected that it would double the tunneling current compared to the single-gate TFET structure.

3.3 Operation

The operation of a TFET is based on band-to-band tunneling (BTBT). BTBT involves tunneling of carriers from the valence band into the conduction band through the forbidden bandgap or vice versa.

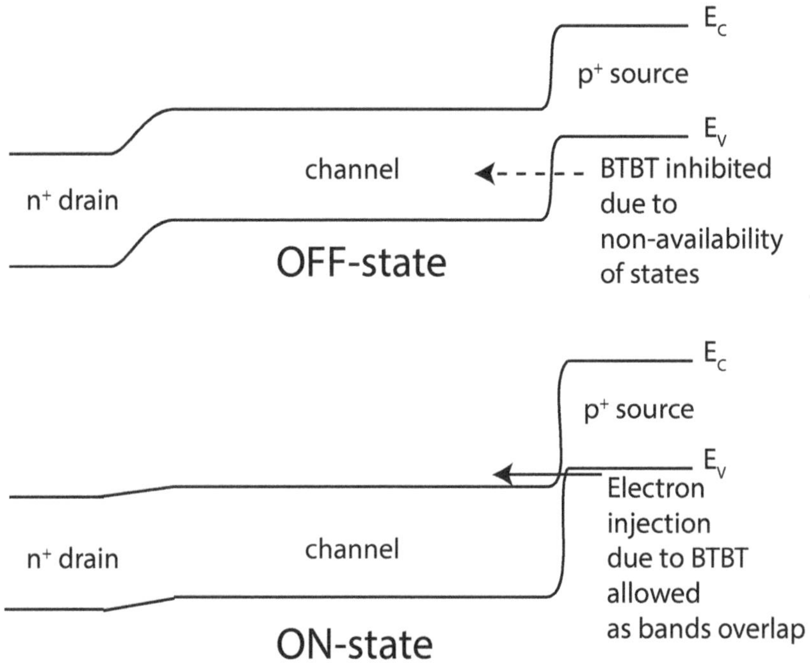

FIGURE 3.3
Band diagram of an n-type TFET in the OFF-state and in the ON-state.

Band diagrams of an n-type TFET are shown in Figure 3.3 in the OFF-state and in the ON-state. When the gate voltage is close to zero, the TFET is in the OFF-state. The conduction band in the channel lies above the valence band in the source. As a result, BTBT is inhibited and the TFET is in the OFF-state with extremely low drain current. When the gate voltage is increased, the gate voltage modulates the carrier density below the gate and the conduction band in the channel is pushed down. When sufficiently high voltage is applied to the gate, there is band bending at the source such that the valence band in the source and the conduction band in the channel get aligned, as shown in Figure 3.3. As a result, electrons in the valence band in the source can tunnel to the conduction band in the channel. The electrons that tunnel into the channel are swept to the drain terminal by the positive bias of the drain. This forms the basis of operation for an n-type TFET.

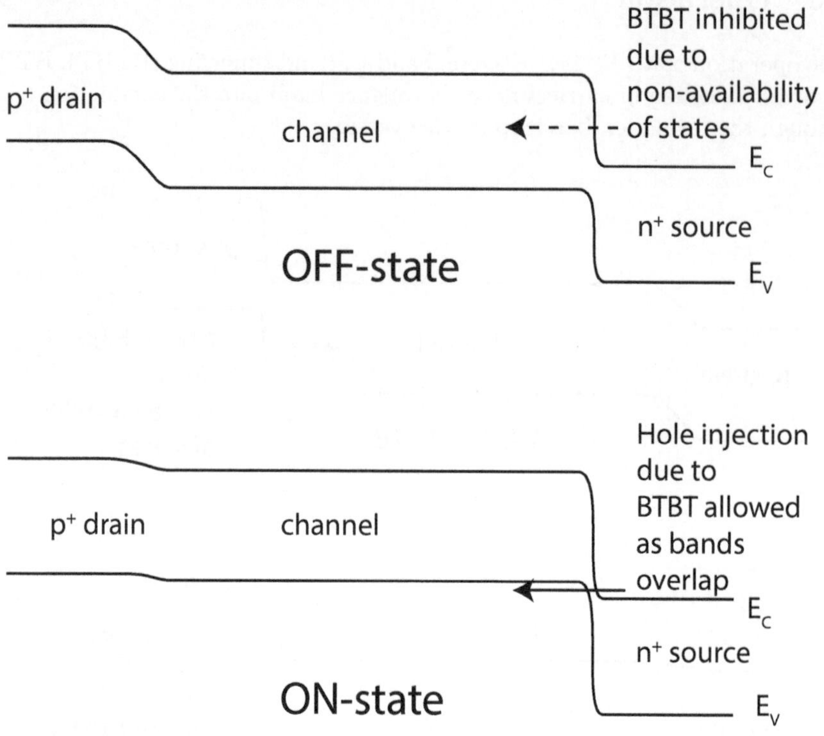

FIGURE 3.4
Band diagram of a p-type TFET in the OFF-state and in the ON-state.

The band diagrams of a p-type TFET are shown in Figure 3.4 in the OFF-state and in the ON-state. The principle of operation of a p-type TFET is similar to the n-type TFET. When the gate voltage is close to zero, the TFET is in the OFF-state. The conduction band in the source lies above the valence band in the channel. As a result, BTBT is blocked and the TFET is in the OFF-state. When sufficiently large negative voltage is applied to the gate, the valence band in the channel is pushed above the conduction band in the source as shown in Figure 3.4. As a result, holes are injected into the channel, which are subsequently swept to the drain terminal by the negative bias of the drain. This forms the basis of operation for a p-type TFET.

It should be noted that, unlike a MOSFET, a TFET is an ambipolar device. For example, an n-type TFET with the major contribution of electrons in the current transport can exhibit a p-type behavior with the major contribution of holes in the current transport, if a negative bias is applied to the gate. This ambipolar conduction of a TFET will be discussed in detail in Section 3.11.

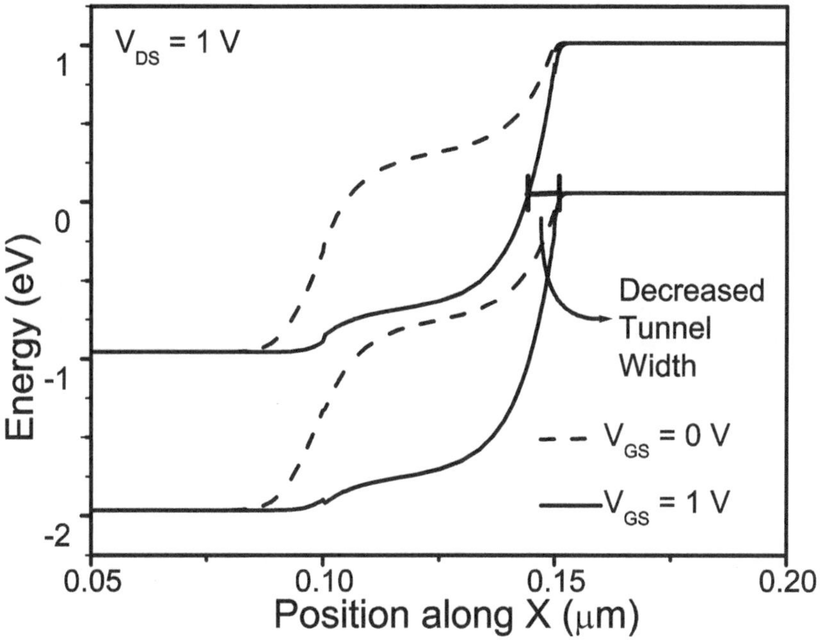

FIGURE 3.5
Band diagram of a DGTFET along a horizontal cutline close to the gate in the OFF-state and in the ON-state at $V_{DS} = 1.0\ V$. (Device parameters as in Table 3.1.)

Figure 3.5 shows a simulated band diagram of the DGTFET along a horizontal cutline close to the gate at $V_{DS} = 1.0\ V$. When $V_{GS} = 0\ V$, the conduction band in the channel is above the valence band in the source, inhibiting BTBT, and the DGTFET is in the OFF-state. When $V_{GS} = 1.0\ V$, the energy bands in the channel bend substantially aligning the valence band in the source with the conduction band in the channel, enabling BTBT. As a result, the electrons tunnel from the valence band in the source to the conduction band in the channel, and the DGTFET is turned on. In the next section, we shall discuss, in detail, how the drain current changes in a TFET with the application of the gate voltage.

3.4 Transfer Characteristics

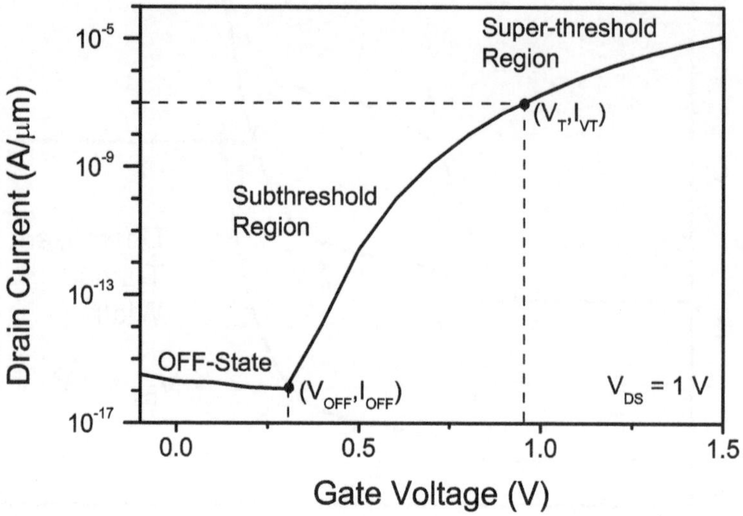

FIGURE 3.6
Simulated transfer characteristics of the DGTFET at $V_{DS} = 1.0\ V$ (device parameters as in Table 3.1).

Figure 3.6 shows the simulated transfer characteristics of the DGTFET at $V_{DS} = 1.0\ V$, obtained by sweeping the gate voltage from $V_{GS} = -0.1\ V$ to $V_{GS} = 1.5\ V$. In general, we can divide the transfer characteristics of a TFET into three regions:

1. OFF-state ($0 < V_{GS} < V_{OFF}$): In this region the TFET has an extremely low current and BTBT is inhibited. V_{OFF} is the gate voltage at which the drain current starts to take off.
2. Subthreshold region ($V_{OFF} < V_{GS} < V_T$): In this region the drain current rises rapidly. V_T is the threshold voltage of the DGTFET.
3. Super-threshold region ($V_{GS} > V_T$): In this region the drain current rises with the gate voltage at a reduced rate.

3.4.1 OFF-State

When the gate voltage is below V_{OFF}, there is a misalignment of the bands of the source and the channel (corresponding to $V_{GS} = 0\ V$ in Figure 3.5).

As a result, BTBT between the source and the channel is blocked and the tunneling current in the TFET is negligible. The drain current does not start increasing until the gate voltage reaches $V_{OFF} \approx 0.3\ V$. The drain current at $V_{GS} = V_{OFF}$ is called the OFF-current of the TFET and is denoted by I_{OFF} in this book. The I_{OFF} in the DGTFET is in the range of femtoamperes. It should be noted that, at lower drain voltage, a higher gate voltage may be required to align the bands to enable BTBT. Therefore, a higher V_{OFF} could be observed at lower drain voltages [19].

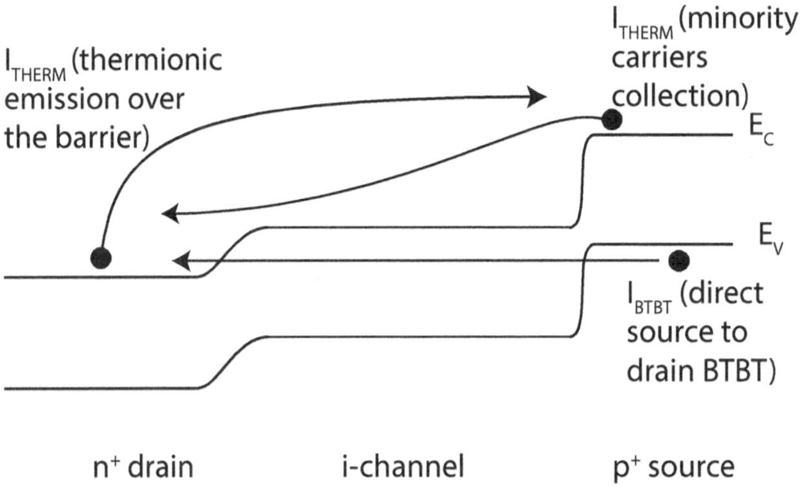

FIGURE 3.7
Components of the OFF-state current in a TFET: I_{BTBT} due to BTBT and I_{THERM} due to thermionic emission [20].

In the OFF-state, there are two major components of the drain current in a TFET, as shown in Figure 3.7 [20, 21]. The first component, I_{BTBT} in Figure 3.7, is due to the direct tunneling of the carriers from the source to the drain. This component of the I_{OFF} is normally expected to be quite low, since the probability of tunneling through the entire length of the channel would be negligible. However, when the channel length is small, I_{BTBT} can become appreciable. The second component, I_{THERM} in Figure 3.7, is due to the thermally induced flow of the carriers over the drain to the source barrier and the minority carrier collection from the source to the drain. The I_{THERM} is proportional to $\exp(-E_g/nkT)$, where E_g is the bandgap of the material, n is a junction constant, k is the Boltzmann constant, and T is the absolute temperature. Normally, I_{THERM} is found to be the dominating component of the I_{OFF} in the TFET [21]. I_{THERM} can be decreased by increasing the bandgap of the TFET material. However, increasing the bandgap of the TFET material can lead to an undesirable decrease in the ON-current of the TFET. In addition to the two components of the I_{OFF} mentioned above, direct leakage through the gate dielectric can also be significant when thin gate dielectrics are employed in the TFETs [22–24].

3.4.2 Subthreshold Region

As the gate voltage increases beyond V_{OFF}, the drain current starts to increase rapidly. In this region, the bands in the source and the channel get aligned and the BTBT is enabled. As a result, the tunneling current starts flowing in this region of operation. The drain current shows a rapid increase at around $V_{GS} = V_{OFF}$ and gradually the rate of increase of the drain current with the gate voltage decreases.

The rate of increase of the drain current with the increase in the gate voltage is quantified by the subthreshold slope or its reciprocal subthreshold swing. The subthreshold swing is defined as the change in the gate voltage required for increasing the drain current by a factor of 10 and is measured in $mV/decade$. It is desirable that a transistor exhibits a small subthreshold swing. When a transistor has a small subthreshold swing, the drain current can increase from the levels of the OFF-state to the levels of the ON-state by an application of a small gate voltage, making the transistor suitable for operating at low power-supply voltage.

The subthreshold swing of a TFET is dependent on the gate voltage, unlike a conventional MOSFET, as we can see in the transfer characteristics of the DGTFET in Figure 3.6. Therefore, two types of subthreshold swing can be identified for a TFET: the point subthreshold swing and the average subthreshold swing. The point subthreshold swing (SS_{POINT}) represents the reciprocal of the instantaneous slope of the transfer characteristics at a particular gate voltage V_{GS} and can be defined as

$$(SS_{POINT})_{V_{GS}} = \left(\frac{dV_{GS}}{d\log(I_{DS})}\right)_{V_{GS}}, \tag{3.1}$$

where I_{DS} is the drain current. The SS_{POINT} in a TFET can be very small at gate voltages where the drain current begins to rise rapidly. For the DGTFET in Figure 3.6, the $SS_{POINT} \approx 33\ mV/decade$ at $V_{GS} \approx 0.3\ V$.

It should be noted that for future low-power CMOS applications, it is not sufficient that the TFETs exhibit a small SS_{POINT} at some gate voltage V_{GS}. It is required that the TFETs show a better subthreshold swing compared to a conventional MOSFET, over a few decades of drain current. Therefore, from a practical point of view, it is essential that the average subthreshold swing is extracted for the TFETs rather than the point subthreshold swing. The average subthreshold swing (SS_{AVG}) can be defined as [1]

$$SS_{AVG} = \frac{(V_T - V_{OFF})}{\log(I_{VT}) - \log(I_{OFF})}, \tag{3.2}$$

where V_T is the threshold voltage of the TFET, V_{OFF} is the gate voltage at which the drain current begins to increase rapidly, I_{VT} is the drain current when $V_{GS} = V_T$ and I_{OFF} is the drain current at $V_{GS} = V_{OFF}$. The threshold voltage (V_T) can be defined as the gate voltage at which the drain current reaches 1×10^{-7} $A/\mu m$ at $V_{DS} = V_{DD}$, where V_{DD} is the supply voltage. From the transfer characteristics in Figure 3.6, the threshold voltage in the DGTFET is extracted as 0.9 V and $SS_{AVG} \approx 110$ $mV/dacade$. The average subthreshold swing of the DGTFET is far worse than 60 $mV/decade$, the subthreshold swing of an ideal MOSFET. It should be pointed out that the DGTFET shown in Figure 3.2 is a simple implementation of a TFET and is not suitable for low-power CMOS applications due to problems like low ON-current, high V_T, and non-competitive SS_{AVG}. In fact, the performance of the DGTFET can be improved using several novel techniques [1–4, 8, 10–15].

3.4.3 Super-Threshold Region

As the gate voltage increases beyond V_T, the drain current keeps on increasing. The drain current at $V_{GS} = V_{DS} = V_{DD}$, is known as the ON-current of the TFET. We denote the ON-current in the TFET as I_{ON} in this book. For the DGTFET in Figure 3.2, assuming that $V_{DD} = 1.0$ V, $I_{ON} \approx 0.2\mu A/\mu m$. The I_{ON} of the DGTFET is extremely low. In general, due to the mechanism of the current transport, the I_{ON} obtained in a TFET is significantly below the requirement for future low-power applications. This is the most serious drawback of the TFETs and researchers are actively exploring techniques that can improve the I_{ON} in the TFETs.

3.5 Subthreshold Swing

One of the most important electrical characteristics of a TFET, which differentiates it from a conventional MOSFET, is its excellent subthreshold swing, particularly around the gate voltage where the drain current starts to take off. In this section, we shall discuss why TFETs have a better subthreshold swing compared to a conventional MOSFET.

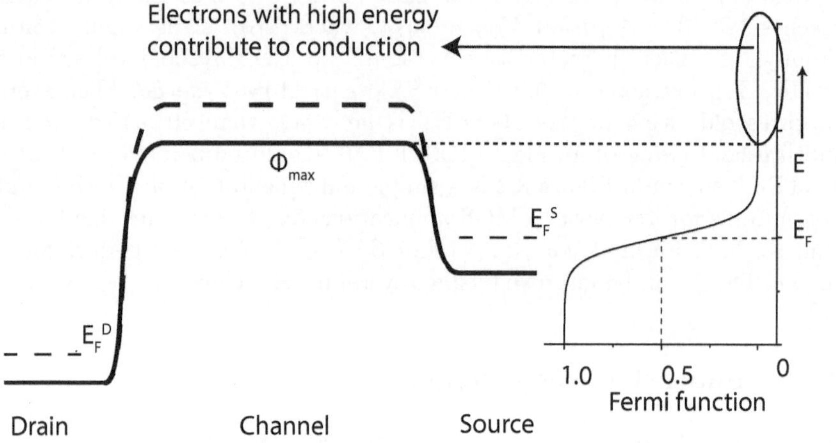

FIGURE 3.8
Conduction band of a conventional MOSFET at two different gate voltages [25]. The mechanism of current transport is the thermionic emission of the carrier over the potential barrier Φ_{max}.

Figure 3.8 shows the energy band diagram of a conventional MOSFET for two different gate voltages in the subthreshold region [25, 26]. In a conventional MOSFET, the mechanism of the current transport is the injection of the carriers from the source into the channel over a gate-modulated energy barrier, as shown in Figure 3.8. It should be noted that only the carriers with energies higher than Φ_{max} in the exponential tail of the source Fermi–Dirac distribution contribute to the current in the subthreshold region [26]. For a large bias, the drain current in a conventional MOSFET is proportional to the density of state $D(E)$, source Fermi–Dirac distribution function $f_s(E)$, and the velocity of the carriers $v(E)$, E being the energy of the electrons [25]:

$$I_d \propto \int dE.D(E).v(E).f_s(E). \tag{3.3}$$

In the exponential tail region, the source Fermi–Dirac distribution function

$f_s(E)$ can be given by a Boltzmann approximation:

$$f_s(E) \approx \exp\left(-\frac{E - E_F^S}{kT}\right), \tag{3.4}$$

where E_F^S is the Fermi energy level in the source, k is the Boltzmann constant and T is the temperature in Kelvin. For simplicity, assuming transport in one spatial direction, the product $D(E) \times v(E)$ can be taken as constant [25]. Additionally, for an ideal MOSFET, we can assume that the change in the conduction band corresponds directly to the change in the gate voltage. Therefore, we can write

$$\frac{\partial I_d}{\partial V_{GS}} = |e|\frac{\partial I_d}{\partial E}, \tag{3.5}$$

where e is the charge of an electron. Using Equation (3.4) and Equation (3.5), the subthreshold swing in an ideal MOSFET can be computed as

$$SS = \left(\frac{\partial \log(I_d)}{\partial V_{GS}}\right)^{-1} = \ln(10)\left(\frac{1}{I_d}\frac{\partial I_d}{\partial V_{GS}}\right)^{-1} \approx \ln(10)\frac{kT}{|e|}. \tag{3.6}$$

At room temperature $T = 300\ K$, $SS \approx 60\ mV/decade$. It should be noted that Equation (3.6) refers to the ideal subthreshold swing achievable in a conventional MOSFET. The subthreshold swing of a typical MOSFET would be worse than that given by Equation (3.6), since the control of the gate voltage over the channel potential and the conduction band profile would be imperfect. Equation (3.6) defines the lower limit on the subthreshold swing enforced by the physics of current transport in a conventional MOSFET.

Equation (3.6) also shows that the ideal subthreshold swing increases as the temperature increases. The Fermi–Dirac distribution function $f_F(E)$ is given by

$$f_F(E) = \frac{1}{1 + \exp\left(\frac{E - E_F}{kT}\right)}, \tag{3.7}$$

where E_F is the Fermi energy. The impact of the temperature on the Fermi–Dirac distribution function is shown in Figure 3.9. It is evident that, as the temperature increases, the probability of the carriers occupying higher energy states increases. These carriers with higher energies are responsible for the subthreshold current over the energy barrier in a conventional MOSFET. Therefore, the limit defined in Equation (3.6) is the effect of the temperature-dependent broadening of the Fermi–Dirac distribution function of the carriers [25]. Next, we see how this limitation is overcome in a TFET and why the TFET can exhibit a subthreshold swing smaller than 60 $mV/decade$.

Figure 3.10 shows the energy band diagram of a p-type TFET for two different gate voltages in the subthreshold region [26]. When a negative gate bias is applied, the valence band in the channel is pulled above the conduction band

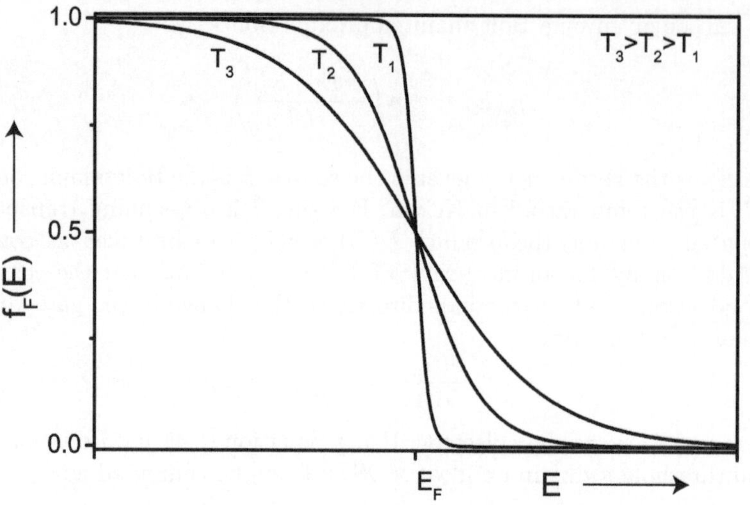

FIGURE 3.9
Fermi–Dirac distribution function at three temperatures T_1, T_2 and T_3,. The exponential tail becomes prominent as the temperature increases. The probability of occupying higher energy states increases as the temperature increases.

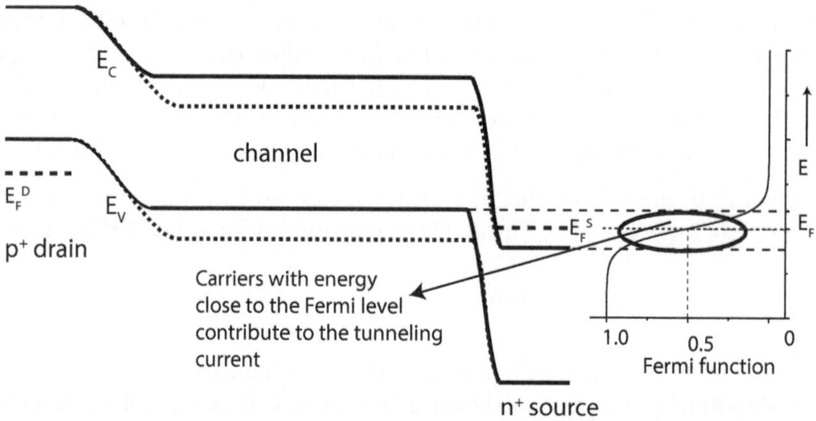

FIGURE 3.10
Band diagram of a p-type TFET at two different gate voltages in the subthreshold region [26]. The Fermi–Dirac distribution function in the source is also shown in the figure. The "band-pass filter" action on the Fermi–Dirac distribution function due to the distinct band arrangement in a TFET should be noted.

at the source, and holes tunnel into the channel. When the valence band is below the conduction band, no tunneling current flows. The channel inhibits the tunneling of the carriers with energies in the high energy tail of the Fermi–Dirac distribution. The semiconductor bandgap cuts off the low-energy tail of the Fermi–Dirac distribution $f_s(E)$ in the source contact. However, when the valence band of the channel gets aligned with the conduction band of the source, BTBT is turned on, and this makes switching from the OFF-state to the ON-state abrupt. It should be noted that the distinct band arrangement of the TFET allows only the carriers with energies around the Fermi energy level to contribute to the conduction of the current [25, 26]. The junction operates as a "band-pass filter" allowing only the carriers with energies around the Fermi energy level to be injected from the source into the channel when the valence band and the conduction band get aligned [25, 26]. This band-pass filter mechanism of the TFET can be viewed as an effective "cooling" of the Fermi–Dirac distribution. The exponential tails of the Fermi–Dirac distribution, which become dominant at elevated temperature, are prevented from contributing to the current transport in the TFET [27–31]. This explains why TFETs can exhibit a subthreshold swing lower than a conventional MOSFET having an ideal subthreshold swing of 60 $mV/decade$ at room temperature.

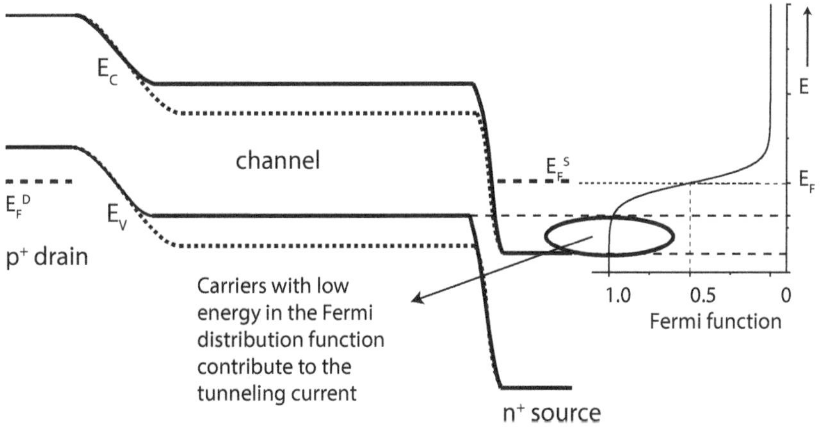

FIGURE 3.11
Band diagram in a TFET when the Fermi energy level in the source is very high [26]. The energy of the carriers that contribute to the BTBT should be noted.

It must be mentioned that in a TFET, the relative position of the Fermi energy level and the edge of the conduction band are crucial in obtaining a subthreshold swing smaller than 60 $mV/decade$, at room temperature [26]. When the Fermi energy in the source is too high, as shown in Figure 3.11, the bandgap is not able to cut off the low-energy exponential tail of the source Fermi–Dirac distribution [26]. As a result, once the valence band in the channel is pushed above the conduction band in the source by increasing the gate bias, the current is expected to increase with a subthreshold swing of not better than 60 $mV/decade$, as in a conventional MOSFET [26]. From Figure 3.9, it can be noted that the Fermi–Dirac distribution function changes abruptly around the Fermi energy level. The change in the probability in the exponential tail regions of the Fermi–Dirac distribution function is rather smooth. Therefore, in order to obtain a good subthreshold swing and an abrupt change in current, it is crucial that the carriers contributing to the conduction in the TFET have energies around the Fermi energy level and not around the exponential tail regions [26]. Using simulation, it has been shown that, when the Fermi energy level in the source is pushed farther up into the conduction band by increasing the source doping, the subthreshold swing of the TFET degrades from 18 $mV/decade$ to the thermal limit of 60 $mV/decade$ [32]. The subthreshold swing reaches the thermal limit when the contribution of the carriers with energies corresponding to the exponential "tail" of the Fermi–Dirac distribution dominates the current transport [32].

The above description illustrates that it is theoretically possible that a TFET can exhibit subthreshold swing smaller than 60 $mV/decade$, in contrast to a conventional MOSFET, which has the lowest subthreshold swing of 60 $mV/decade$ at room temperature. However, in addition to the arrangement of the bands and the Fermi levels, the subthreshold swing in a TFET depends on several other factors, for example, the control of the gate voltage over the channel potential, the BTBT probability function, the abruptness of the source–channel junction, the interface states, defect-assisted tunneling and interface roughness [33]. As a result, experimental TFET results often show a subthreshold swing greater than 60 $mV/decade$ and obtaining a subthreshold swing better than 60 $mV/decade$ at high drain current levels is a non-trivial challenge [33].

Additionally, it should be noted that, in contrast to a conventional MOSFET, the subthreshold swing in a TFET depends on the gate bias. This can be explained by the complex dependency of the drain current on the BTBT probability and the number of available states for tunneling, which is determined by the source and channel Fermi–Dirac distribution functions [30, 34]. We shall look at these dependencies in the following section.

3.6 Tunneling Current

In this section, we develop an approximate relationship between the tunneling current and various material and device parameters of the TFET.

Figure 3.12 shows the typical band profile of a p-type TFET [29]. The source is heavily doped n^+ and the drain is heavily doped p^+. A negative potential

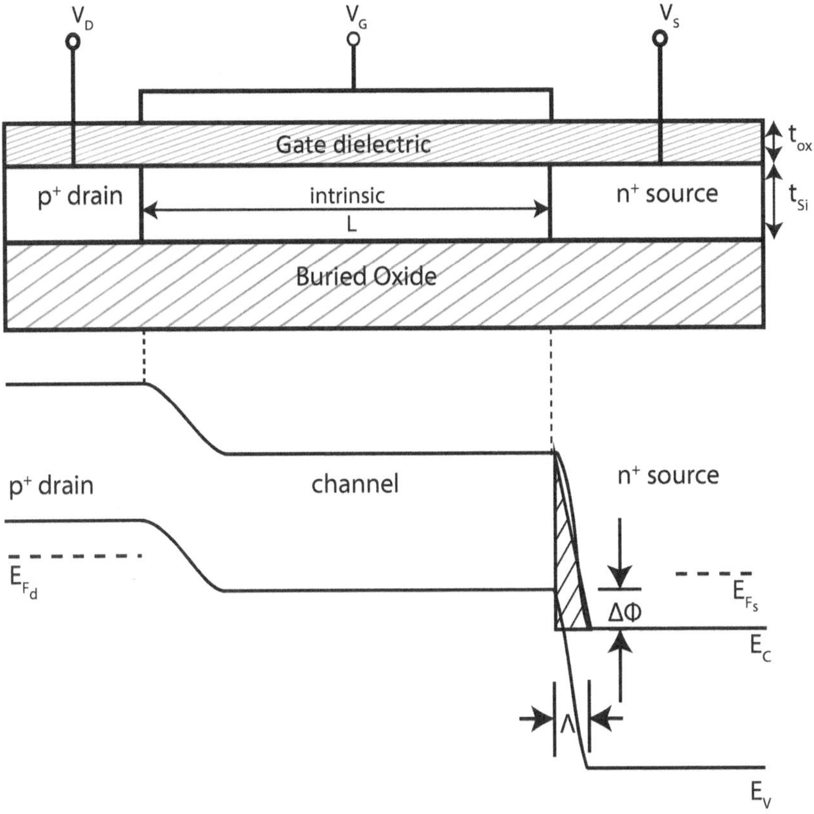

FIGURE 3.12
Band diagram in a p-type SOI TFET. The energy barrier in the source can be approximated as a triangular barrier [29].

($V_{GS} < 0$) is applied to the gate. As a result, the valence band of the channel is pushed above the conduction band of the source and band-to-band tunneling is enabled. The overlap of the valence band in the channel with the conduction band in the source is shown as $\Delta\Phi$. The transition region at the source–channel interface is shown as Λ. The Λ describes the width of the tunneling barrier and depends on the geometry and other parameters of the device, which we shall discuss later in this section.

An analytical expression for the drain current in a TFET can be obtained by using the Landauer equation [20, 28, 29, 35–37]. The Landauer equation describes one-dimensional energy-conserving transport and is widely used for computing currents in nanoscale devices. The drain current I_D in the TFET can be computed by integrating the Landauer equation over the contributing energy range. Only the electrons within the energy overlap window of $\Delta\Phi$

contribute significantly to the drain current. The summation over the transverse wave-vector k_{trans} is carried out due to the two-dimensional behavior of the SOI and results in the following expression for the drain current I_D in the TFET [28, 29]:

$$I_D = \frac{4|e|}{h} \sum_{k_{trans}} T_{tot}\left(k_{trans}\right) \int_0^{\Delta\Phi} dE \left(f_s\left(E\right) - f_d\left(E\right)\right), \tag{3.8}$$

where T_{tot} is the total transmission function, f_s and f_d are the source and the drain Fermi functions respectively, h is the Planck constant, and e is the charge of the electron.

The drain current is strongly dependent on the total transmission function T_{tot}. In a TFET, two scattering events occur in series: scattering due to tunneling and scattering within the channel. However, the dominant scattering event in a TFET is the BTBT and we can write

$$T_{tot} \approx T_{BTBT}, \tag{3.9}$$

where T_{BTBT} is the tunneling probability due to BTBT. The tunneling probability T_{BTBT} can be computed by approximating the barrier as a triangular barrier as shown in Figure 3.12 and applying the WKB approximation [26–29]. We derived an expression for the tunneling probability for a triangular barrier using the WKB approximation in Section 2.4. The expression for the tunneling probability for a triangular barrier in Equation (2.26) is shown below for convenience:

$$P_t = \exp\left[-\frac{4\sqrt{2m}}{3\hbar F}\left(U_0 - E\right)^{\frac{3}{2}}\right], \tag{3.10}$$

where P_t is the tunneling probability, F is the electric field in the region, $(U_0 - E)$ is the difference between the height of the barrier and the energy of the particle, m is the mass of the particle, and $\hbar$ is the reduced Planck constant. Using the above equation, we can compute the tunneling probability for the tunneling junction shown in Figure 3.12 as

$$T_{BTBT} = \exp\left[-\frac{4\sqrt{2m^*}}{3|e|\hbar F}\left(E_g^*\right)^{\frac{3}{2}}\right], \tag{3.11}$$

where F is the electric field in the transition region and E_g^* is the effective bandgap. The dependence of T_{BTBT} on the transverse wave-vector k_{trans} is absorbed in the effective bandgap E_g^* in Equation (3.11). The effective bandgap, assuming a direct bandgap semiconductor, can be expressed as [29]

$$E_g^* = E_g + 2\frac{\hbar^2 k_{trans}^2}{2m^*}. \tag{3.12}$$

The electric field F in the region can be given as [29]

$$F = \frac{E_g + \Delta\Phi}{\Lambda}, \tag{3.13}$$

where Λ is the width of the transition region at the source–channel interface. The spatial extent of the transition region can be described as

$$\Lambda = \lambda_{dop} + \lambda_{ch}, \tag{3.14}$$

where λ_{dop} is the screening length in the source and λ_{ch} is the effective screening length in the channel. The screening length in the source λ_{dop} can be computed as

$$\lambda_{dop} = \sqrt{\frac{\varepsilon_0 \varepsilon_{Si} kT}{e^2 N_D}}, \tag{3.15}$$

where ε_0 is the absolute permittivity of the free space, ε_{Si} is the relative permittivity of silicon, k is the Boltzmann constant, T is the temperature in Kelvin, e is the charge of an electron, and N_D is the doping in the source. The effective screening length in the channel λ_{ch} can be given as [26–29]

$$\lambda_{ch} = \sqrt{\frac{\varepsilon_{Si}}{\varepsilon_{ox}} t_{ox} t_{Si}}, \tag{3.16}$$

where ε_{Si} is the relative permittivity of silicon, ε_{ox} is the relative permittivity of the gate dielectric, t_{ox} is the thickness of the gate dielectric, and t_{Si} is the thickness of the silicon film. The expression for λ_{ch} results from the parabolic expansion of the vertical potential distribution and represents the control of the gate over the surface potential [38–40]. In a typical thin-film SOI TFET, λ_{ch} dominates over λ_{dop} in determining Λ. For example, at $T = 300\ K$ and $N_D = 1 \times 10^{20}\ atoms/cm^3$, $\lambda_{dop} \approx 0.4\ nm$. For $t_{Si} = 10\ nm$ and $t_{ox} = 3\ nm$ SiO_2 gate, $\lambda_{ch} \approx 9.5\ nm$. Therefore, we can write

$$\Lambda \approx \lambda_{ch} = \sqrt{\frac{\varepsilon_{Si}}{\varepsilon_{ox}} t_{ox} t_{Si}}. \tag{3.17}$$

Assuming a sharp doping profile at the source, such that the lateral extent of the doping is smaller than the transition region Λ, using Equations (3.11), (3.13) and (3.17), we can write

$$T_{BTBT} \approx \exp\left[-\frac{4\sqrt{2m^*} E_g^{*\frac{3}{2}}}{3|e|\hbar\left(E_g^* + \Delta\Phi\right)} \sqrt{\frac{\varepsilon_{Si}}{\varepsilon_{ox}} t_{ox} t_{Si}}\right]. \tag{3.18}$$

Equation (3.8) shows that the drain current is roughly proportional to the T_{BTBT} and the energy overlap window $\Delta\Phi$. Therefore, using Equation (3.18), we can write [25, 28]:

$$I_D \propto \exp\left[-\frac{4\sqrt{2m^*} E_g^{*\frac{3}{2}}}{3|e|\hbar\left(E_g^* + \Delta\Phi\right)} \sqrt{\frac{\varepsilon_{Si}}{\varepsilon_{ox}} t_{ox} t_{Si}}\right] \Delta\Phi \tag{3.19}$$

It should be noted that the WKB approximation used in the above derivation may not be accurate in all situations. In the direct bandgap semiconductors, such as InAs, WKB approximation yields accurate results only if one single imaginary path that connects the valence band and the conduction band plays the dominating role in the BTBT. For Si/Ge or heterostructures, or when quantum effects and phonon-assisted tunneling become dominant, the WKB approximation has limited accuracy [41]. Nevertheless, the relationship described by Equation (3.19) can be used to gain useful insights into the parameters that affect the tunneling current and could be optimized to enhance the tunneling current.

3.7 Output Characteristics

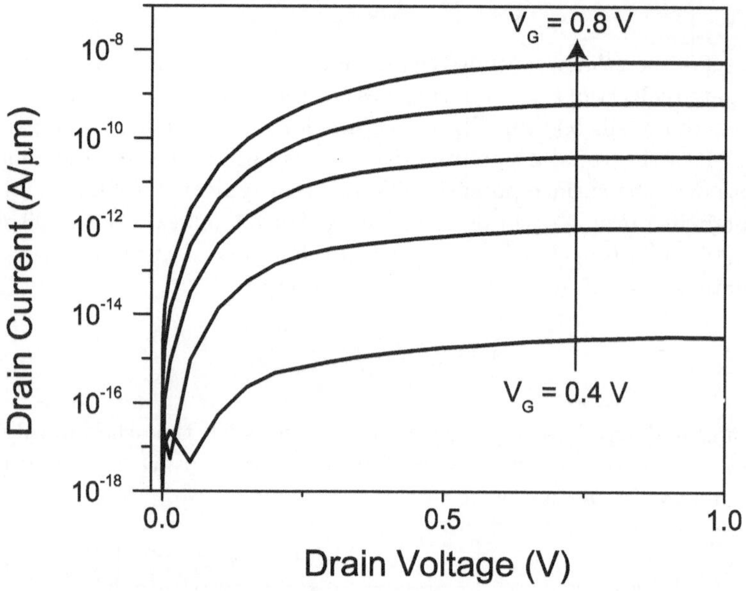

FIGURE 3.13
Simulated output characteristics in the DGTFET (device parameters as in Table 3.1).

Figure 3.13 shows the simulated output characteristics of the DGTFET at different gate voltages. The output characteristics of the TFET show an exponential onset followed by saturation. This behavior of the output characteristics has been observed, in simulation and in the experiments on TFETs [7, 16, 19, 21, 28, 29, 42–50]. Based on the rate of increase of the drain current with V_{DS}, we can divide the output characteristics into three regions:

1. Tunnel resistance–dominated region: In this region the drain current rises very rapidly with V_{DS} and the current is limited by the tunnel resistance
2. Channel resistance–dominated region: In this region the drain current rises at a reduced rate with V_{DS} and the current is limited by the channel resistance
3. Saturation region: In this region the drain current saturates.

3.7.1 Tunnel Resistance–Dominated Region

When the drain voltage is low, the application of the gate voltage, results in the formation of a channel with a high concentration of electrons. The high concentration of electrons results in a low resistance of the channel and the drain voltage is able to pin the conduction band in the channel [16]. As a result, the full drain voltage appears across the tunneling junction and the tunneling width at the source–channel junction is modulated by the drain voltage [16, 31]. In this region of operation, the onset of current is super-linear. This is a distinguishing characteristic of a TFET and has been observed both experimentally and in simulations. The super-linear onset of current can be explained by the change in the availability of carriers at the location where the tunneling path starts and the availability of empty states at the location where the tunneling path ends [48, 49]. It should be noted that, in addition to the transparency of the tunneling barrier, the occupancy of the filled states in the source and the unoccupied states in the channel play an important role in deciding the characteristics of this region of operation [48, 49]. In this region of the output characteristics, potential drop due to the resistance in the channel is negligible and the output characteristic is dominated by the source–channel tunnel resistance. Therefore, this region of operation can be referred to as the tunnel resistance–dominated region [16].

3.7.2 Channel Resistance–Dominated Region

In a TFET, the majority of the electrons in the channel are injected from the drain while a small fraction of the electrons in the channel is supplied from the source by the BTBT [19]. In fact, the contribution of the carriers injected by the source into the channel has been found to be $\approx 10^4\times$ lower than the contribution of the carriers injected by the drain into the channel [19]. Therefore, the formation of the channel in the TFET is dependent on the V_{GD} voltage rather than the V_{GS} voltage [19]. With the increasing drain voltage, V_{GD} decreases, resulting in a decrease in the electron concentration in the channel. When the V_{DS} is increased beyond 0.3 V, the electron concentration in the channel sharply decreases, since a growing fraction of electrons in the channel are pulled back to the drain, increasing the channel resistance. The increased channel resistance results in a slower increase of the drain current with the applied drain voltage [16]. In this region of the output characteristics, the conduction band of the channel no longer remains pinned to the drain potential [16]. Since the channel resistance plays a significant role in this region of operation of the output characteristics, this region can be referred to as the channel resistance–dominated region [16].

3.7.3 Saturation Region

When the drain voltage is increased beyond 0.6 V, the output characteristics of the TFET exhibit a saturating behavior. The concentration of the electrons in the channel decreases appreciably. In fact, the channel starts disappearing from the source side. This is in contrast to a conventional MOSFET, where the channel starts to get pinched off from the drain side. Since the electron concentration is very low when the drain voltage is greater than 0.6 V, the channel resistance becomes sufficiently high such that the lateral electric field from the drain cannot penetrate through the channel region [16]. The additional drain voltage appears across the drain–channel junction and the potential at the source–channel junction is controlled primarily by the gate [7, 16]. As a result, saturation in the output characteristics is expected in a TFET at high drain voltages. Since the lateral electric field from the drain cannot pass through the channel when the TFET is in saturation, a perfect saturation is expected in a TFET, in contrast to a conventional MOSFET [16]. The saturation voltage in the TFET corresponds to the drain voltage when the channel becomes devoid of the carriers [7, 51]. Similar to a conventional MOSFET, the saturation voltage V_{DSAT} in a TFET can be linearly related to the gate voltage as shown below [16, 19]:

$$V_{DSAT} = V_{GS} - V_T, \tag{3.20}$$

where V_{GS} is the gate voltage and the V_T is the threshold voltage in the TFET.

From the output characteristics shown in Figure 3.13, it can be inferred that the drain current is low compared to the standard silicon MOSFETs. This is expected, since the presence of a tunneling barrier in a TFET limits the drain current compared to a conventional MOSFET.

3.8 Threshold Voltage

There are different ways in which the threshold voltage in a TFET can be defined. The threshold voltage definition based on the constant current method is the most widely used definition of the threshold voltage in a TFET. For example, the threshold voltage can be defined as the gate voltage at which the drain current reaches $1 \times 10^{-7}\ A/\mu m$ at $V_{DS} = V_{DD}$, where V_{DD} is the supply voltage [1]. This definition of the threshold voltage is quite simple to extract from the transfer characteristics, though it does not have much physical significance.

Another method that has been proposed to define the threshold voltage in a TFET is based on the transconductance change [9]. The transconductance g_m at a given drain voltage V_{DS} is defined as follows:

$$g_m = \left(\frac{\partial I_D}{\partial V_{GS}}\right)_{V_{DS}}. \tag{3.21}$$

The g_m in a TFET changes with the change in the gate voltage V_{GS}. The threshold voltage can be defined as the gate voltage at which the maximum $\frac{dg_m}{dV_{GS}}$ is reached. The threshold voltage, computed using this method, marks the transition between the region where the drain current has a quasi-exponential dependence on the V_{GS} and the region where the drain current has a linear dependence on V_{GS} [9]. Since the drain current shows a complicated dependence on the drain voltage, as well, a threshold voltage with respect to the drain voltage can also be defined for a TFET. The drain threshold voltage is defined as the drain voltage, at which the maximum $\frac{dg_{DS}}{dV_{DS}}$ is obtained, where g_{DS} at a given gate voltage V_{GS} is defined as [9]:

$$g_{DS} = \left(\frac{\partial I_D}{\partial V_{DS}}\right)_{V_{GS}} \tag{3.22}$$

It should be noted that the extraction of the threshold voltages based on the transconductance-change method is prone to numerical errors due to the computation of higher-order derivatives.

In the modeling of TFETs, another definition of the threshold voltage based on the alignment of the bands has been proposed [35]. The threshold voltage can be defined as the gate voltage at which the Fermi energy level in the source and the edge of the conduction band in the channel get aligned [35].

The threshold voltage in a TFET can be adjusted by adjusting the work function of the gate material. Figure 3.14 shows the threshold voltage of the DGTFET at different work functions of the gate material. The threshold voltage shows a linear relationship with the work function of the gate, as in a conventional MOSFET. The transfer characteristics of the DGTFET shift along

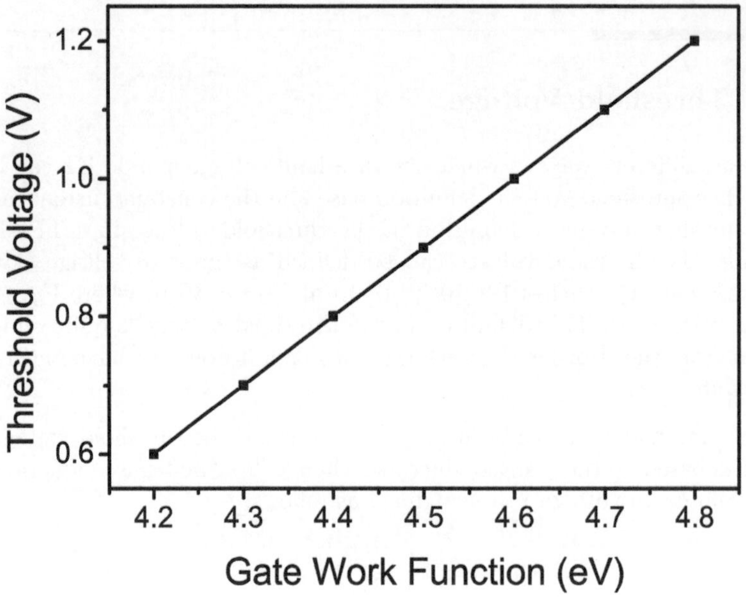

FIGURE 3.14
Effect of the gate work function on the threshold voltage in the DGTFET (other device parameters as in Table 3.1). The threshold voltage is defined as the as the gate voltage at which the drain current reaches 1×10^{-7} $A/\mu m$ at $V_{DS} = 1$ V.

the X-axis proportional to the change in the work function of the gate. The output characteristics also exhibit a similar shift along the X-axis with the change in the work function of the gate [16].

3.9 Different Types of TFETs

TFETs have been proposed and realized in various structural forms. These different TFET structures will be discussed in detail in the subsequent chapters. On the basis of the gate structure in the TFET, the TFETs can be classified as single-gate TFET, double-gate TFET, or gate-all-around TFET. The double-gate TFET and the gate-all-around TFET have better control over the channel potential compared to a single-gate TFET. TFETs have been realized with a homojunction at the source and a heterojunction at the source. The heterojunction TFET achieves a much lower effective bandgap than a homojunction TFET. Additionally, different semiconducting materials like silicon, germanium, carbon, and III-V semiconductor materials have been incorporated into the TFETs. Materials like nanowires, CNTs and graphene that exhibit quantum confinement effects are also being investigated for TFETs. These materials have special properties that can be exploited to improve the electrical characteristics of the TFETs. We will discuss these different types of TFETs, in detail, in the subsequent chapters.

3.10 Impact of Device Parameters

In this section, we describe the impact of various device parameters on the electrical properties of a TFET.

3.10.1 Gate Dielectric

We can infer from Equation (3.19) that the drain current is expected to increase with an increase in dielectric constant of the gate material or a decrease in the gate dielectric thickness. It has been widely analyzed and experimentally verified that the tighter gate control obtained by employing a high-κ dielectric material or decreasing the gate dielectric thickness results in increasing the I_{ON} of the TFET [1, 2, 29, 52–56]. For example, in an SOI-based TFET, it has been observed that the reduction of the gate oxide thickness from 4.5 nm to 3.5 nm, results in a 6$\times$ increase in the drain current [29].

3.10.2 Body Thickness

The impact of the silicon body thickness on the drain current depends on two competing effects [2]. As the silicon body thickness increases, the volume of silicon available for BTBT increases, thereby increasing the drain current. However, as the thickness of the silicon body increases, the coupling of the gate with the channel degrades, leading to a decrease in the drain current. As a result, given a TFET, an optimum body thickness can be found that can maximize the drain current. For the DGTFET, the drain current reaches maximum when the silicon body thickness is between 10 nm and 20 nm [2]. It should be noted that the subthreshold swing is also strongly dependent on the thickness of the TFET body: as the thickness of the TFET body is reduced, the subthreshold swing of the TFET improves [57]. Using atomistic simulation, it has been shown that a subthreshold swing lower than 60 $mV/decade$ can be achieved in a single-gate InAs TFET when the body thickness is smaller than 4 nm and in a double-gate InAs TFET when the body thickness $\leq 7\ nm$ [57].

3.10.3 Source Doping Concentration and Profile

When the source doping concentration is increased, the bandgap narrowing increases, which results in a decrease in the tunneling distance [29, 58]. Experimental results show that doubling the source doping concentration leads to a 2× increase in the I_{ON} of a TFET [29]. For high source doping concentration, due to the modified built-in potential at the source–channel junction, the threshold voltage is expected to reduce [29]. It should be noted that the doping concentration of the source cannot be increased arbitrarily, since increasing the doping concentration results in the raising of the Fermi level into the conduction band, which reduces the filtering effect on the Fermi tail in the TFET and degrades the subthreshold swing as described in Section 3.5 [26].

The tunneling current also depends on the abruptness of the source doping profile. With the increase in the abruptness of the source doping, the electric field at the source–channel junction increases, resulting in an increase in the tunneling current. However, even for a non-abrupt source doping profile, I_{ON} can be maximized and the subthreshold swing can be minimized by choosing an optimum gate-source overlap [12].

3.10.4 Channel Length

In contrast to a conventional MOSFET, the drain current in a TFET shows little dependence on the channel length [5, 10, 29]. Since the tunneling current depends primarily on the electric field and the band alignment near the source–channel junction, the tunneling current does not change much with the decrease in the channel length. However, when the channel length is smaller than some critical channel length L_{crit}, direct source-to-drain leakage becomes dominant. As a result, there is an appreciable increase in the I_{OFF} when the channel length is smaller than L_{crit}. Additionally, when the channel length is less than L_{crit}, the V_T of the TFET decreases and there is a degradation in the subthreshold swing of the TFET as well [5, 10, 20]. The L_{crit} for Si-based TFET is around 20 nm [5, 10]. It should be noted that for TFETs realized using materials with a smaller bandgap, the impact of the variation in channel length on the drain current could be more pronounced [59].

For the DGTFET, the dependence of the drain current on the channel length, obtained using simulation, is shown in Figure 3.15. Experimental studies have confirmed this kind of dependency of the drain current on the channel length [29]. However, it should be noted that the observed independence of the drain current on the channel length cannot be generalized for TFETs. The drain

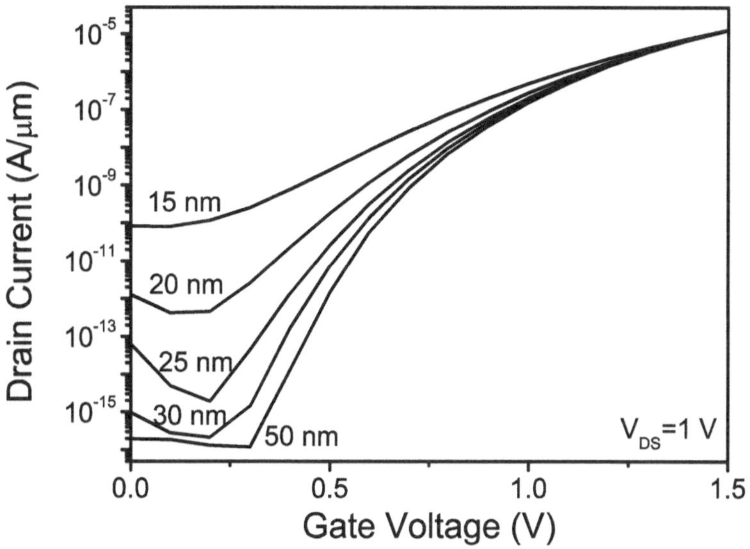

FIGURE 3.15
Simulated transfer characteristics of the DGTFET at different channel lengths (other device parameters as in Table 3.1).

current is independent of the channel length when the scattering due to BTBT is dominant and the scattering in the channel can be neglected. For future TFETs, which are expected to have a greater BTBT transmission probability, the scattering in the channel could limit the drain current.

3.11 Ambipolar Current

There are two tunneling junctions in a TFET: the source–channel tunneling junction and the drain–channel tunneling junction. Normally, the BTBT in the drain–channel junction is suppressed in a TFET by reducing the drain doping or using some other technique. However, when the TFET is symmetric, i.e., the same doping concentration but of opposite type is used on the source side and on the drain side, then the TFET exhibits ambipolar behavior. Ambipolar behavior means that the same TFET shows n-type behavior with the electrons making the major contribution in the current transportand a p-type behavior with the holes making the major contribution in the current transport, at the same drain voltage V_{DS}. An n-type TFET can conduct for the positive V_{GS} with BTBT on the source side and for the negative V_{GS} with BTBT on the drain side, keeping the V_{DS} positive. Similarly, a p-type TFET can conduct for the negative V_{GS} with BTBT on the source side and for the positive V_{GS} with BTBT on the drain side, keeping the V_{DS} negative.

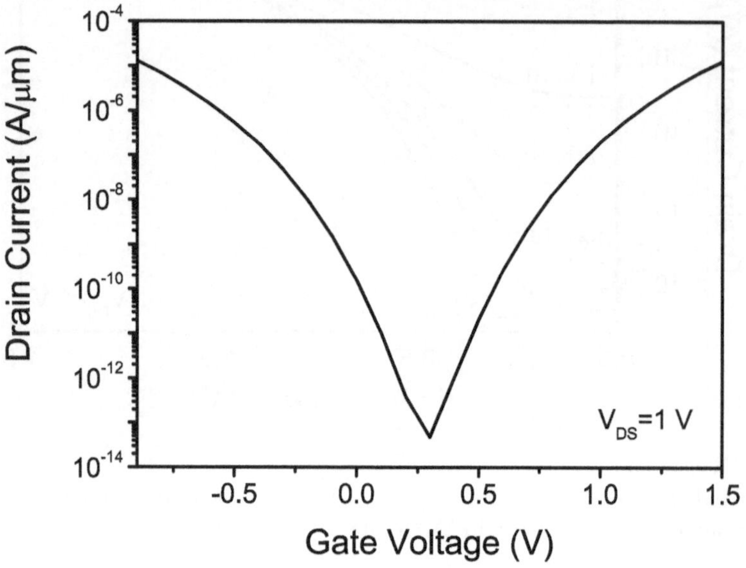

FIGURE 3.16
Transfer characteristics of the DGTFET with the same doping on the source and the drain (other device parameters as in Table 3.1).

Figure 3.16 shows the transfer characteristics of a DGTFET that has the same doping concentration $N_D = N_A = 1 \times 10^{20}$ $atoms/cm^3$ for the source and the drain. When V_{GS} is positive, the conduction band and the valence band get aligned on the source side, as in the normal n-type mode of operation of the DGTFET. However, when V_{GS} is negative, the bands get aligned on the drain side, resulting in a BTBT on the drain side. It should be noted that when V_{GS} is positive, the conduction is predominantly due to the electrons and when V_{GS} is negative, the conduction is predominantly due to the holes.

Ambipolar conduction is undesirable in a TFET for the following reasons:

1. The ambipolar current can lead to a higher OFF-state current in a TFET. The increase in the I_{OFF} due to the ambipolar current is especially exhibited in TFETs employing materials with a smaller bandgap. The ambipolar current in the OFF-state, arising due to the BTBT on the drain side, can dominate the thermal leakage current in these devices. This results in an OFF-state current that is governed by the ambipolar current, and not by the thermal leakage

current. Therefore, the TFET needs to be designed to suppress the ambipolar current.

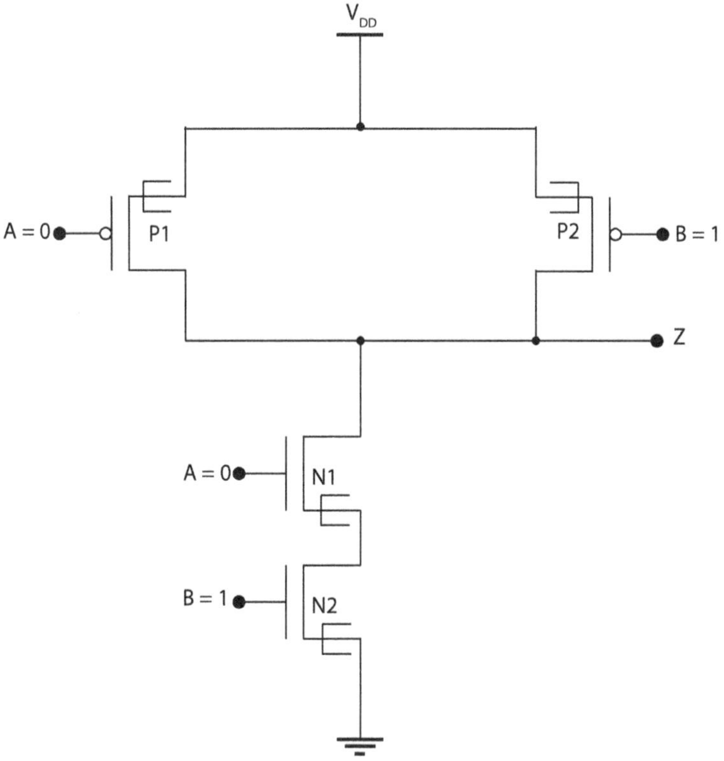

FIGURE 3.17
CMOS NAND gate designed using TFETs. Transistors P1 and P2 are p-type TFETs and transistors N1 and N2 are n-type TFETs [60].

2. The existence of ambipolar current can lead to circuit failure or high stand-by leakage for the CMOS-type circuits [60]. For example, consider the NAND gate shown in Figure 3.17, designed using p-type and n-type TFETs. Let us assume that these TFETs have appreciable ambipolar current, i.e., these TFETs can conduct for both the positive gate voltages and the negative gate voltages. When the input to the n-type TFET N1 is low ($0\ V$) and the input to the n-type TFET N2 is high (V_{DD}), the V_{GS} of the N1 could be negative and conducting, making the output Z reach low ($0\ V$) [60].

Therefore, considering the CMOS applications of the TFET, ambipolar current in TFETs needs to be eliminated, as far as practicable. This can be achieved by reducing the drain doping, creating a drain-gate underlap, employing a heterojunction, or introducing some kind of asymmetry into the TFET structure. We will look at these techniques to reduce the ambipolar current in a TFET in detail in the following chapters.

3.12 Impact of Temperature

The effect of temperature on the characteristics of a TFET has been studied in detail, both theoretically and experimentally, by several researchers [1, 11, 25, 54, 55, 61–63]. The comparison of the effect of temperature on the subthreshold swing in a conventional MOSFET and in a TFET is quite interesting. For an ideal MOSFET, we can infer from Equation (3.6) that the subthreshold swing is expected to degrade linearly with increasing temperature. Therefore, in a conventional MOSFET, with increased temperature the leakage current would increase, which in turn will increase the standby power dissipation [61]. As a result, thermal runaway can occur, which can finally destroy the circuit. However, in a TFET, due to the elimination of the exponential tail, the impact of temperature on the subthreshold slope is rather limited. The subthreshold swing in a TFET can change slightly with temperature due to the temperature-induced broadening of the Fermi–Dirac distribution close to the Fermi energy level. Additionally, the band-pass operation of the TFET on the Fermi–Dirac distribution is responsible for the non-linear relationship between the subthreshold swing and the temperature in a TFET [62]. However, compared to a conventional MOSFET, the subthreshold swing in a TFET shows much weaker temperature dependence. Therefore, TFETs can exhibit greater immunity against thermal runaway, since the leakage currents are expected to show little degradation with increasing temperature [61]. It should be noted that in the OFF-state, the BTBT in the TFET is blocked and the major component of the drain current could be due to the thermal generation of carriers in the depletion region. Therefore, the OFF-state leakage current due to thermal generation, though negligibly small, can show a strong temperature dependence [11].

The weak dependency of the subthreshold swing on temperature, expected in a TFET, is a useful measure to validate the current transport mechanism in TFETs. This technique is widely used by engineers to ascertain that the current transport mechanism in the fabricated TFET is BTBT and not the thermionic emission of the carriers over an energy barrier. It should be noted that in a TFET where BTBT is the dominant mechanism of current transport, a stronger dependency of the subthreshold swing on temperature can be occasionally observed. The reason for this could be the presence of the interface traps [54, 55, 63].

In the ON-state, the major component of the drain current in a TFET is the BTBT of the carriers over the tunneling barrier. The bandgap of the material E_g at a temperature T is given by [11]

$$E_g(T) = E_g(0) - \frac{\alpha T^2}{\beta + T}, \tag{3.23}$$

where $E_g(0)$ is the bandgap at $T = 0$ and α and β are the material constants. The bandgap of the material decreases with increasing temperature. Therefore, in the ON-state, at elevated temperatures, the drain current increases due to a greater BTBT. However, at very high drain voltages and high drain current levels, the increased channel resistance at elevated temperature can lead to a decrease in the drain current with increasing temperature [11].

3.13 Promises and Limitations

As described earlier, for a conventional MOSFET, due to the mechanism of the current transport, there is a theoretical limit on the lowest achievable subthreshold swing of 60 $mV/decade$, at room temperature. Therefore, given a maximum I_{OFF} and a minimum I_{ON}/I_{OFF} ratio, a conventional MOSFET must operate above a certain supply voltage. This is a serious handicap for the conventional MOSFET since this prevents MOSFETs from being applied in ultra low voltage circuits. However, TFETs can overcome the theoretical limitation of subthreshold swing of the conventional MOSFETs and can be employed in ultra low voltage circuits that are extremely energy efficient [30]. This is the primary reason why TFETs are being considered as a substitute for the conventional MOSFET for future applications. However, it must be mentioned that TFETs must exhibit a subthreshold swing better than 60 $mV/decade$ for at least four decades of current to become a valid candidate for replacement of MOSFETs and this is a non-trivial challenge for device engineers [30, 33, 64].

The most crucial limitation of a TFET is its low I_{ON}, largely because BTBT supports low current levels in silicon and other materials having a bandgap similar to silicon. Unless the I_{ON} in a TFET is increased to levels comparable to that of a MOSFET, TFETs cannot be deployed in mainstream CMOS circuits. Therefore, researchers are actively working on developing techniques to boost I_{ON} in TFETs, and have achieved a fair amount of success [30, 33, 64].

3.14 Summary

In this chapter, we developed a basic understanding of the TFETs. We looked at the device structure, operating principle and the important electrical characteristics of TFETs. We also looked at the differentiating factors between a conventional MOSFET and a TFET both in terms of the operating principle and the electrical characteristics. We highlighted the problems in the simple implementation of the TFETs, especially the problem of low I_{ON}. In the next chapter, we shall look at the techniques that are used to boost I_{ON} in TFETs.

REFERENCES

[1] K. Boucart and A. M. Ionescu, "Double-gate tunnel FET with high-κ gate dielectric," *IEEE Trans. on Electron Devices*, vol. 54, pp. 1725–1733, July 2007.

[2] E.-H. Toh, G. H. Wang, G. Samudra, and Y.-C. Yeo, "Device physics and design of double-gate tunneling field-effect transistor by silicon film thickness optimization," *Applied Physics Letters*, vol. 90, no. 263507, 2007.

[3] E.-H. Toh, G. H. Wang, L. Chan, G. Samudra, and Y.-C. Yeo, "Device physics and guiding principles for the design of double-gate tunneling field effect transistor with silicon-germanium source heterojunction," *Applied Physics Letters*, vol. 91, no. 243505, 2007.

[4] A. S. Verhulst, W. G. Vandenberghe, K. Maex, and G. Groeseneken, "Tunnel field-effect transistor without gate-drain overlap," *Applied Physics Letters*, vol. 91, no. 053102, 2007.

[5] K. Boucart and A. M. Ionescu, "Length scaling of the double gate tunnel FET with a high-k gate dielectric," *Solid-State Electronics*, vol. 51, pp. 1500–1507, Nov. 2007.

[6] W. G. Vandenberghe, A. S. Verhulst, G. Groeseneken, B. Sore, and W. Magnus, "Analytical model for a tunnel field-effect transistor," in *Proc. 14th IEEE MELECON*, pp. 923–928, IEEE, 2008.

[7] C. Shen, S.-L. Ong, C.-H. Heng, G. Samudra, and Y.-C. Yeo, "A variational approach to the two-dimensional nonlinear Poisson's equation for the modeling of tunneling transistors," *IEEE Electron Device Letters*, vol. 29, pp. 1252–1255, Nov. 2008.

[8] T. Krishnamohan, D. Kim, S. Raghunathan, and K. Saraswat, "Double-gate strained-Ge heterostructure tunneling FET (TFET) with record high drive currents and $< 60\ mV/dec$ subthreshold slope," in *IEEE International Electron Devices Meeting (IEDM)*, pp. 1–3, IEEE, Dec. 2008.

[9] K. Boucart and A. M. Ionescu, "A new definition of threshold voltage in tunnel FETs," *Solid-State Electronics*, vol. 52, pp. 1318–1323, 2008.

[10] S. Saurabh and M. J. Kumar, "Impact of strain on drain current and threshold voltage of nanoscale double gate tunnel field effect transistor: Theoretical investigation and analysis," *Japanese Journal of Applied Physics*, vol. 48, June 2009.

[11] P. F. Guo, L. T. Yang, Y. Yang, L. Fan, G. Q. Han, G. S. Samudra, and Y. C. Yeo, "Tunneling field-effect transistor: Effect of strain and temperature on tunneling current," *IEEE Electron Device Letters*, vol. 30, pp. 981–983, Sept. 2009.

[12] H. S. Yu and C. O. Chui, "Insights and optimizations of tunnel field-effect transistor operation," in *Device Research Conference*, pp. 87–88, IEEE, 2009.

[13] S. Saurabh and M. J. Kumar, "Estimation and compensation of process induced variations in nanoscale tunnel field effect transistors (TFETs) for improved reliability," *IEEE Trans. Devices and Materials Reliability*, vol. 10, pp. 390–395, Sept. 2010.

[14] S. Saurabh and M. J. Kumar, "Novel attributes of a nanoscale dual material gate tunnel field effect transistor,," *IEEE Trans. on Electron Devices*, vol. 58, pp. 404–410, Feb. 2011.

[15] A. Chattopadhyay and A. Mallik, "Impact of a spacer dielectric and a gate overlap/underlap on the device performance of a tunnel field-effect transistor," *IEEE Trans. on Electron Devices*, vol. 58, pp. 677–683, Mar. 2011.

[16] A. Mallik and A. Chattopadhyay, "Drain-dependence of tunnel field-effect transistor characteristics: The role of the channel," *IEEE Trans. on Electron Devices*, vol. 58, pp. 4250–4257, Dec. 2011.

[17] Hraziia, A. Vladimirescu, A. Amara, and C. Anghel, "An analysis on the ambipolar current in Si double-gate tunnel FETs," *Solid-State Electronics*, vol. 70, pp. 67–72, Apr. 2012.

[18] Y. Yang, P. Guo, G. Han, K. L. Low, and C. Zhan, "Simulation of tunneling field-effect transistors with extended source structures," *Journal of Applied Physics*, vol. 111, no. 114514, 2012.

[19] W. Lee and W. Y. Choi, "Influence of inversion layer on tunneling field-effect transistors," *IEEE Electron Device Letters*, vol. 32, pp. 1191–1193, Sept. 2011.

[20] Y. Gao, T. Low, and M. Lundstrom, "Possibilities for VDD = 0.1 V logic using carbon-based tunneling field effect transistors," in *Symposium on VLSI Technology (VLSIT)*, pp. 180–181, IEEE, 2009.

[21] L. Wang, E. Yu, Y. Taur, and P. Asbeck, "Design of tunneling field-effect transistors based on staggered heterojunctions for ultralow-power applications," *IEEE Electron Device Letters*, vol. 31, pp. 431–433, May 2010.

[22] P. Chaturvedi and N. Goyal, "Effect of gate dielectric thickness on gate leakage in tunnel field effect transistor," in *International Caribbean Conference on Devices, Circuits and Systems*, pp. 1–4, IEEE, 2012.

[23] Y. Zhul, L. Zhangl, A. Zhangl, and M. Chan, "A gate leakage model for double gate tunneling field-effect transistors," *IEEE International Conference on Solid-State and Integrated Circuit Technology (ICSICT)*, pp. 1–3, Oct. 2014.

[24] P. Chaturvedi and M. J. Kumar, "Impact of gate leakage considerations in tunnel field effect transistor design," *Japanese Journal of Applied Physics*, vol. 53, no. 7, p. 074201, 2014.

[25] J. Knoch and J. Appenzeller, "A novel concept for field-effect transistors: The tunneling carbon nanotube FET," in *Proc. 63rd DRC*, pp. 153–156, IEEE, June 2005.

[26] J. Knoch, S. Mantl, and J. Appenzeller, "Impact of the dimensionality on the performance of tunneling FETs: Bulk versus one-dimensional devices," *Solid-State Electronics*, vol. 51, pp. 572–578, Apr. 2007.

[27] J. Appenzeller, Y. Lin, J. Konch, Z. Chen, and P. Avouris, "Comparing carbon nanotube transistors—the ideal choice: A novel tunneling device design," *IEEE Trans. on Electron Devices*, vol. 52, pp. 2568–2576, Dec. 2005.

[28] J. Knoch and J. Appenzeller, "Tunneling phenomena in carbon nanotube field-effect transistors," *Physica Status Solidi (a)*, vol. 205, no. 4, pp. 679–694, 2008.

[29] C. Sandow, J. Knoch, C. Urban, Q.-T. Zhao, and S. Mantl, "Impact of electrostatics and doping concentration on the performance of silicon tunnel field-effect transistors," *Solid-State Electronics*, vol. 53, pp. 1126–1129, Oct. 2009.

[30] A. M. Ionescu and H. Riel, "Tunnel field-effect transistors as energy-efficient electronic switches," *Nature*, vol. 479, pp. 329–337, Nov. 2011.

[31] J. Nah, E.-S. Liu, K. M. Varahramyan, and E. Tutuc, "Ge–Si_xGe_{1-x} core-shell nanowire tunneling field-effect transistors," *IEEE Trans. on Electron Devices*, vol. 57, pp. 1883–1888, Oct. 2012.

[32] C. Kshirsagar and S. J. Koester, "InAs/SiGe on Si nanowire tunneling field effect transistors," in *69th Annual Device Research Conference*, pp. 151–152, IEEE, 2011.

[33] H. Lu and A. Seabaugh, "Tunnel field-effect transistors: State-of-the-art," *Journal of the Electron Devices Society*, pp. 44–49, May 2014.

[34] A. Verhulst, W. Vandenberghe, D. Leonelli, R. Rooyackers, A. Vandooren, J. Zhuge, K.-H. Kao, B. Soree, W. Magnusa, M. Fischetti, G. Pourtois, C. Huyghebaert, R. Huang, Y. Wang, K. D. Meyer, W. Dehaene, M. Heynsa, and G. Groeseneken, "Si-based tunnel field-effect transistors for low-power nano-electronics," in *69th Annual Device Research Conference (DRC)*, pp. 193–196, IEEE, 2011.

[35] B. Bhushan, K. Nayak, and V. R. Rao, "DC compact model for SOI tunnel field-effect transistors," *IEEE Trans. on Electron Devices*, vol. 59, pp. 2635–2642, Oct. 2012.

[36] M. Najari, S. Fregonese, C. Maneux, T. Zimmer, H. Mnif, and N. Masmoudi, "Analytical modeling of the tunneling current in Schottky barrier carbon nanotube field effect transistor using the Verilog-A language," in *International Multi-Conference on Systems, Signals and Devices*, pp. 1–6, IEEE, 2009.

[37] J. T. Smith, C. Sandow, S. Das, R. A. Minamisawa, S. Mantl, and J. Appenzeller, "Silicon nanowire tunneling field-effect transistor arrays: Improving subthreshold performance using excimer laser annealing," *IEEE Trans. on Electron Devices*, vol. 58, pp. 1822–1829, July 2011.

[38] K. K. Young, "Short-channel effect in fully depleted SOI MOSFETs," *IEEE Trans. on Electron Devices*, vol. 36, pp. 399–402, Feb. 1989.

[39] R.-H. Yan, A. Ourmazd, and K. F. Lee, "Scaling the Si MOSFET: From bulk to SOI to bulk," *IEEE Trans. on Electron Devices*, vol. 39, pp. 1704–1710, July 1992.

[40] W. Lu, P. Xie, and C. M. Lieber, "Nanowire transistor performance limits and applications," *IEEE Trans. on Electron Devices*, vol. 55, pp. 2859–2876, Nov. 2008.

[41] M. Luisier and G. Klimeck, "Simulation of nanowire tunneling transistors: From the Wentzel-Kramers-Brillouin approximation to full-band phonon-assisted tunneling," *Journal of Applied Physics*, vol. 107, no. 084507, 2010.

[42] K. K. Bhuwalka, J. Schulze, and I. Eisele, "Vertical tunnel field-effect transistor with bandgap modulation and workfunction engineering," in *Proc. of 34th European Solid-State Device Research Conference*, vol. 41, pp. 241–244, IEEE, 2004.

[43] O. M. Nayfeh, C. N. Chleirigh, J. Hennessy, L. Gomez, J. L. Hoyt, and D. A. Antoniadis, "Design of tunneling field-effect transistors using strained-Silicon/strained-Germanium type-II staggered heterojunctions," *IEEE Electron Device Letters*, vol. 29, pp. 1074–1077, Sept. 2008.

[44] Y. Khatami and K. Banerjee, "Steep subthreshold slope n- and p-type tunnel-FET devices for low-power and energy-efficient digital circuits," *IEEE Trans. on Electron Devices*, vol. 56, pp. 2752–2761, Nov. 2009.

[45] Y. Yoon and S. Salahuddin, "Barrier-free tunneling in a carbon heterojunction transistor," *Applied Physics Letters*, vol. 97, no. 033102, 2010.

[46] A. S. Verhulst, D. Leonelli, R. Rooyackers, and G. Groeseneken, "Drain voltage dependent analytical model of tunnel field-effect transistors," *Journal of Applied Physics*, vol. 110, no. 024510, 2011.

[47] Q. T. Zhao, W. J. Yu, B. Zhang, M. Schmidt, S. Richter, D. Buca, J.-M. Hartmann, R. Luptak, A. Fox, K. K. Bourdelle, and S. Mantl, "Tunneling field-effect transistor with a strained Si channel and a $Si_{0.5}Ge_{0.5}$ source," *Solid-State Electronics*, pp. 97–101, Aug. 2012.

[48] L. D. Michielis, L. Lattanzio, and A. M. Ionescu, "Understanding the superlinear onset of tunnel-FET output characteristic," *Electron Device Letters*, vol. 33, pp. 1523–1525, Nov. 2012.

[49] H. Lu, J. W. Kim, D. Esseni, and A. Seabaugh, "Continuous semiempirical model for the current-voltage characteristics of tunnel FETs," in *International Conference on Ultimate Integration on Silicon (ULIS)*, pp. 25–28, IEEE, Apr. 2014.

[50] C. Wu, R. Huang, Q. Huang, C. Wang, J. Wang, and Y. Wang, "An analytical surface potential model accounting for the dual-modulation effects in tunnel FETs," *IEEE Trans. on Electron Devices*, vol. 61, pp. 2690–2696, Aug. 2014.

[51] M. Lee, Y. Jeon, K. S. Son, J. H. Shim, and S. Kim, "Comparative performance analysis of silicon nanowire tunnel FETs and MOSFETs on plastic substrates in flexible logic circuit applications," *Physica Status Solidi (a)*, vol. 209, no. 7, pp. 1350–1358, 2012.

[52] A. S. Verhulst, W. G. Vandenberghe, D. Leonelli, R. Rooyackers, A. Vandooren, G. Pourtois, S. D. Gendt, M. M. Heyns, and G. Groeseneken, "Boosting the on-current of Si-based tunnel field-effect transistors," *ECS Transactions*, vol. 33, no. 6, pp. 363–372, 2010.

[53] S. Mookerjea, D. Mohata, R. Krishnan, J. Singh, A. Vallett, A. Ali, T. Mayer, V. Narayanan, D. Schlom, A. Liu, and S. Datta, "Experimental demonstration of 100 nm channel length $In_{0.53}Ga_{0.47}As$-based vertical inter-band tunnel field effect transistors (TFETs) for ultra low-power logic and SRAM applications,," in *IEEE International Electron Devices Meeting (IEDM)*, pp. 1–3, IEEE, Dec. 2009.

[54] H. Zhao, Y. Chen, Y. Wang, F. Zhou, F. Xue, and J. Lee, "$In_{0.7}Ga_{0.3}As$ tunneling field-effect transistors with an Ion of 50 $\mu A/\mu m$ and a subthreshold swing of 86 mV/dec using HfO_2 gate oxide," *IEEE Electron Device Letters*, vol. 31, pp. 1392–1394, Dec. 2010.

[55] R. Iida, S.-H. Kim, M. Yokoyama, N. Taoka, S.-H. Lee, M. Takenaka, and S. Takagi, "Planar-type $In_{0.53}Ga_{0.47}As$ channel band-to-band tunneling metal-oxide-semiconductor field-effect transistors," *Journal of Applied Physics*, vol. 110, no. 124505, 2011.

[56] S. K. Kim and W. Y. Choi, "Impact of gate dielectric constant variation on tunnel field-effect transistors (TFETs)," *Solid-State Electronics*, vol. 116, pp. 88–94, 2016.

[57] M. Luisier and G. Klimeck, "Atomistic full-band design study of InAs band-to-band tunneling field-effect transistors," *IEEE Electron Device Letters*, vol. 30, pp. 602–604, June 2009.

[58] W. G. Vandenberghe, A. S. Verhulst, K.-H. Kao, K. D. Meyer, B. Sore, W. Magnus, and G. Groeseneken, "A model determining optimal doping concentration and material's band gap of tunnel field-effect transistors," *Applied Physics Letters*, vol. 100, no. 193509, 2012.

[59] D. Haehnel, I. A. Fischer, A. Hornung, A.-C. Koellner, and J. Schulze, "Tuning the Ge(Sn) tunneling FET: Influence of drain doping, short channel, and Sn content," *IEEE Trans. on Electron Devices*, vol. 62, pp. 36–43, Jan. 2015.

[60] J.-S. Jang and W. Y. Choi, "Ambipolarity factor of tunneling field-effect transistors (TFETs)," *Journal of Semiconductor Technology and Science*, vol. 11, pp. 272–277, 2011.

[61] S. O. Koswatta and M. S. Lundstrom, "Performance comparison between p-i-n tunneling transistors and conventional MOSFETs," *IEEE Trans. on Electron Devices*, vol. 56, pp. 456–465, Mar. 2009.

[62] Y. Yoon and S. Salahuddin, "Non-linear temperature dependence in graphene nanoribbon tunneling transistors," *eprint arXiv:0909.5445*, Sept. 2009.

[63] H. Zhao, Y. Chen, Y. Wang, F. Zhou, F. Xue, and J. Lee, "InGaAs tunneling field-effect-transistors with atomic-layer deposited gate oxides," *IEEE Trans. on Electron Devices*, vol. 58, pp. 2990–2995, Sept. 2011.

[64] International Technology Roadmap for Semiconductors (ITRS), "Emerging research devices, 2013 edition." http://www.itrs.net, 2013.

4

Boosting ON-Current in Tunnel Field-Effect Transistor

4.1 Introduction

TFETs are being considered as a substitute for the conventional MOSFET for advanced technology nodes. To achieve this, TFETs must exhibit an ON-state current (I_{ON}) of around 100 $\mu A/\mu m$, along with an ON-state current to OFF-state current ratio (I_{ON}/I_{OFF}) of around 10^6, for supply voltages lower than 0.5 V [1]. However, band-to-band tunneling (BTBT), the physical principle responsible for the current transport in a TFET, inherently supports a low I_{ON}. Therefore, the I_{ON} exhibited by a conventional TFET is extremely low and typically fails to meet the above-mentioned requirements by a few orders of magnitude. This is the most severe limitation of TFETs and unless this obstacle is overcome, TFETs cannot be deployed in the mainstream CMOS technology. As a result, a great deal of research on TFETs has been directed toward boosting their I_{ON}.

In this chapter, we present an overview of the techniques that are employed to enhance the I_{ON} in a TFET. For each of these techniques, we illustrate a few novel TFET device structures where these techniques have been applied successfully to boost the I_{ON}. At the end of this chapter, readers will be familiar with a broad range of techniques to increase the I_{ON} of the TFET. Understanding of these techniques is expected to provoke readers to further investigate achieving improved characteristics of TFETs.

4.2 Types of Techniques to Boost ON-Current

The tunneling current in a TFET is determined by the gate-source and drain-source voltages and the material and device parameters. In Section 3.6 we derived an approximate relationship between the tunneling current and the device parameters for a DGTFET, shown below for convenience (Equation 3.19):

$$I_D \propto \exp\left[-\frac{4\sqrt{2m^*}E_g^{*\frac{3}{2}}}{3|e|\hbar\left(E_g^* + \Delta\Phi\right)}\sqrt{\frac{\varepsilon_{Si}}{\varepsilon_{ox}}t_{ox}t_{Si}}\right]\Delta\Phi, \tag{4.1}$$

where m^* is the effective carrier mass, E_g is the bandgap of the material, $\Delta\Phi$ is the range of energy over which tunneling can take place, t_{ox}, t_{Si}, ε_{ox} and ε_{Si} are the oxide and silicon film thicknesses and dielectric constants, respectively, e is the charge of an electron, and $\hbar$ is the reduced Planck constant. From Equation (4.1) we can infer that the tunneling current can be increased by the following techniques:

1. Decrease the gate oxide thickness (t_{ox}) or increasing the dielectric constant of the gate oxide (ε_{ox}).
2. Decrease the silicon body thickness (t_{Si}).
3. Use material with a lower effective carrier mass (m^*) and smaller bandgap (E_g).

These observations are important for exploring the techniques that can be used to enhance the I_{ON} in a TFET. In broad terms, the approaches to boost the I_{ON} in a TFET can be classified in the following categories:

1. Gate engineering
2. Tunneling junction engineering
3. Materials engineering

We shall describe the above techniques in detail in the following sections. However, it is worth mentioning that these techniques are often combined to achieve a higher I_{ON}, as we shall illustrate, using some examples from the literature. It should also be noted that, although the low I_{ON} is the biggest hurdle preventing the wide-scale application of TFETs, the impact of the techniques to improve the I_{ON} on the other electrical characteristics of the device need to be carefully considered. For example, the increase in the I_{ON} should not be accompanied by an unacceptable increase in the I_{OFF} or appreciable degradation in the subthreshold swing.

4.3 Gate Engineering

The gate modulates the carrier density in the channel in a TFET. The modulation of the carrier density in the channel results in the alignment of the valence band and the conduction band at the source–channel junction, triggering the tunneling of the carriers from the source into the channel. Therefore, the coupling of the gate voltage with the channel potential has a profound impact on the I_{ON} of the TFET. This has motivated researchers to engineer the gate design to improve the I_{ON} in a TFET using the techniques described below.

4.3.1 Thickness and Dielectric Constant of Gate Oxide

If the thickness of the gate oxide is decreased or the dielectric constant of the gate material is increased, assuming all other parameters in Equation (4.1) are the same, the tunneling current is expected to increase and has been reported by several research groups [1–8]. Since the thickness and the dielectric constant of the gate oxide are exponentially related to the tunneling current, the I_{ON} improves super-linearly in a TFET due to gate dielectric engineering. It should be noted that I_{ON} increases linearly in a conventional MOSFET by decreasing the gate oxide thickness or increasing the dielectric constant of the gate material. Therefore, TFETs are expected to have a greater enhancement in the I_{ON} due to gate dielectric engineering compared to the improvement obtained in the conventional MOSFETs using the same technique [1, 2].

It should be noted that the thickness of the gate dielectric cannot be decreased beyond a certain limit, since the direct tunneling of carriers through the gate dielectric will impose a lower limit on the gate dielectric thickness. Therefore, the technique of employing high-κ gate material is crucial for obtaining a high I_{ON} in a TFET. The advantage of using a high-κ material in a TFET is that a low equivalent oxide thickness can be obtained without decreasing the physical thickness of the gate, thus avoiding the problem of direct tunneling of the carriers through the gate. During recent times, the high-κ materials based on Hafnium oxide have been investigated comprehensively and have been successfully deployed in mass production for the conventional MOSFETs. The on-going advancements in the high-κ gate materials, targeted for the existing CMOS technology, are also expected to benefit the performance of TFETs.

4.3.2 Multiple Gates

In a MOSFET, the primary motivation of employing multiple gate structures like double gate, FinFETs, or gate-all-around is to obtain tighter gate control over the channel potential and to ameliorate the short-channel effects. In a TFET, employing multiple gate structures leads to better control of the gate over the channel potential and an increase in the effective tunneling area, and an improvement in the I_{ON}. In general, the tunneling area in a TFET is just around the edge of the channel, near the source region. In multiple gate configurations, the channels are formed below all the gates. Therefore, the effective tunneling area increases in multiple gate TFETs. For example, in the DGTFET described in Chapter 3, there are two regions where tunneling can occur: one beneath each of the two gates at the source–channel junction. It should be noted that for the DGTFET with a thin body, the increase in the I_{ON} due to employing two gates instead of a single gate is more than double, due to the combined effect of the two gates over the potential throughout the DGTFET body [8]. It should also be noted that the gate control on the channel is already tighter in a double gate and gate-all-around TFET structures, and therefore, the impact of engineering gate oxide thickness is less in double gate or gate-all-around TFET compared to a single gate TFET [9]. Another current enhancing technique that has been experimentally demonstrated is to use wrap-around gate structure in a TFET [10]. The wrap-around gate structure results in a stronger electrical field at the edge of the channel and significantly boosts the drain current [10].

4.3.3 Spacer Engineering

The fringe fields, emanating from the gate electrode to the source and the drain regions, have a significant impact on the electrostatics of a TFET. Since the fringe fields pass through the spacer, the properties of the spacer modulate the energy band profiles and the tunneling current in the TFET [11–16].

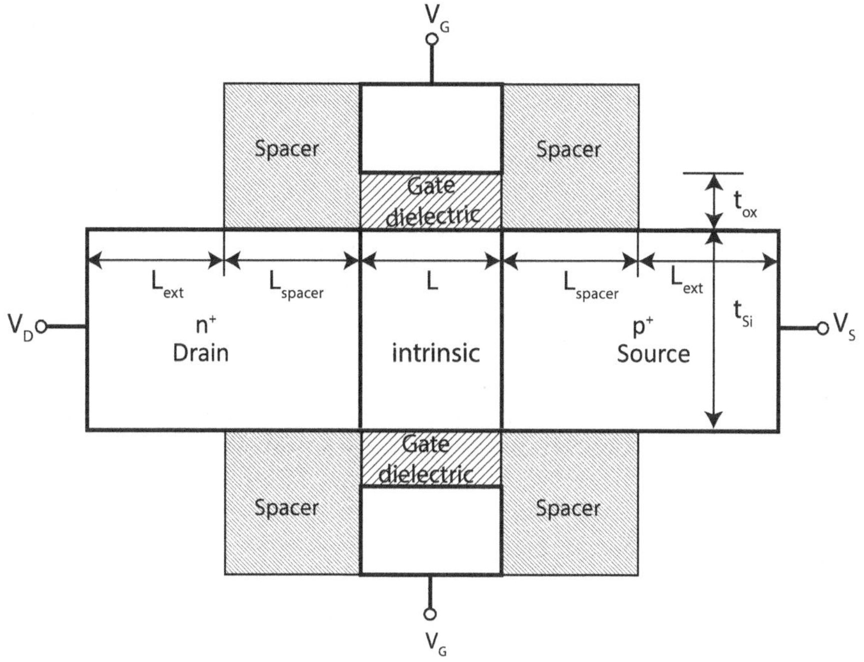

FIGURE 4.1
TFET structure to study the impact of the spacer on the performance of the TFET [15]. The device parameters are $L = 50\ nm$, $L_{spacer} = 50\ nm$, $L_{ext} = 50\ nm$, $t_{ox} = 1\ nm$, $t_{Si} = 10\ nm$.

We illustrate the impact of the spacer on the I_{ON} in a TFET using the device structure shown in Figure 4.1 [15]. The TFET shown in Figure 4.1 is similar to the DGTFET discussed in Chapter 3, except that the gate and the spacer are composed of materials with different dielectric constants. The gate is composed of HfO_2 with an equivalent oxide thickness of 1 nm. The dielectric constant of the spacer material is varied from $\kappa = 2$ to $\kappa = 21$ [15]. The source and the drain regions are of p-type and n-type, respectively, with a doping of $1 \times 10^{20}\ atoms/cm^3$. The channel is of n-type with a doping of $1 \times 10^{17}\ atoms/cm^3$. As the dielectric constant of the spacer is increased, the fringe fields arising out of the spacer increase [15]. Due to the increase in the fringe fields, the position of the maximum electric field and the position of the minimum tunnel width shift away from the semiconductor gate–dielectric interface towards the middle of the semiconductor body [15]. As a result, tunneling occurs at the surface when the spacer dielectric constant is low and it occurs deeper inside the TFET body when the spacer dielectric constant is high [15]. The increase in the fringe fields also results in an increase in the minimum tunnel width in the TFET [15]. Since the I_{ON} in the TFET is determined by the minimum tunnel width, there is a noticeable degradation in the I_{ON} of the TFET as the dielectric constant of the spacer material is increased.

However, from the above description, it should not be concluded that the I_{ON} in the TFETs, in general, decreases with an increase in the dielectric constant of the spacer. The effect of the dielectric constant of the spacer depends on the electrostatics of a given TFET. For example, in a TFET with SiGe source, it has been observed that the I_{ON} improves with the increase in the dielectric constant of the spacer material [14]. In the TFET with SiGe source, the fringe field increases the electric field across the tunneling junction and reduces the minimum tunnel width and hence the I_{ON} improves with the increase in the dielectric constant of the spacer material [14, 15]. Therefore, a high-κ spacer can improve, as well as degrade the performance of the TFET depending on the electrostatics of the TFET and the nature of the tunneling junction. Additionally, the width of the spacer also plays an important role in governing the performance of the TFET [11, 15]. For the TFET shown in Figure 4.1, the I_{ON} in the TFET degrades with the increase in the width of the spacer. In fact, it has been suggested that the spacer can altogether be avoided to improve the I_{ON} in the TFET [11, 15]. The effect of the spacer dielectric material reduces as the silicon body thickness is increased or the gate dielectric constant is decreased [15]. The effect of the spacer on the performance of the TFET also reduces if the source doping is increased or the gate-source overlap is increased [15]. For the TFET shown in Figure 4.1, when a high-κ spacer is employed, a small gate underlap has been found to enhance the I_{ON} [15]. These observations suggest that the properties of the spacer can be adjusted to boost the I_{ON} in the TFETs.

4.3.4 Asymmetric Gate Structures

In a TFET, depending on the gate-source voltage, BTBT can occur on the source side as well as on the drain side, as described in Section 3.11. Therefore, techniques that are applied symmetrically on the TFETs to boost the BTBT, not only increase the BTBT on the source side, but also increase the BTBT on the drain side. As a result, techniques that are applied symmetrically to the TFETs also tend to increase the ambipolar current and the I_{OFF} in the TFETs. Since it is desirable that the ambipolar current and the I_{OFF} in the TFETs are minimized, asymmetrical gate structures have been proposed to suppress them. In an asymmetrical gate design, the properties of the gate on the source and the drain sides are different. These techniques not only suppress ambipolar conduction, but also decrease the I_{OFF} for a given I_{ON} [17–23]. It is noteworthy that the asymmetric gate structure can also be combined with other symmetric current enhancement techniques. Due to the combined effect of the asymmetric gate structure and the symmetric current enhancement technique, a high I_{ON} and a low I_{OFF} can be achieved simultaneously for a given TFET. We illustrate the asymmetric gate structures using a couple of examples from the literature.

1. **Asymmetric gate work function**:

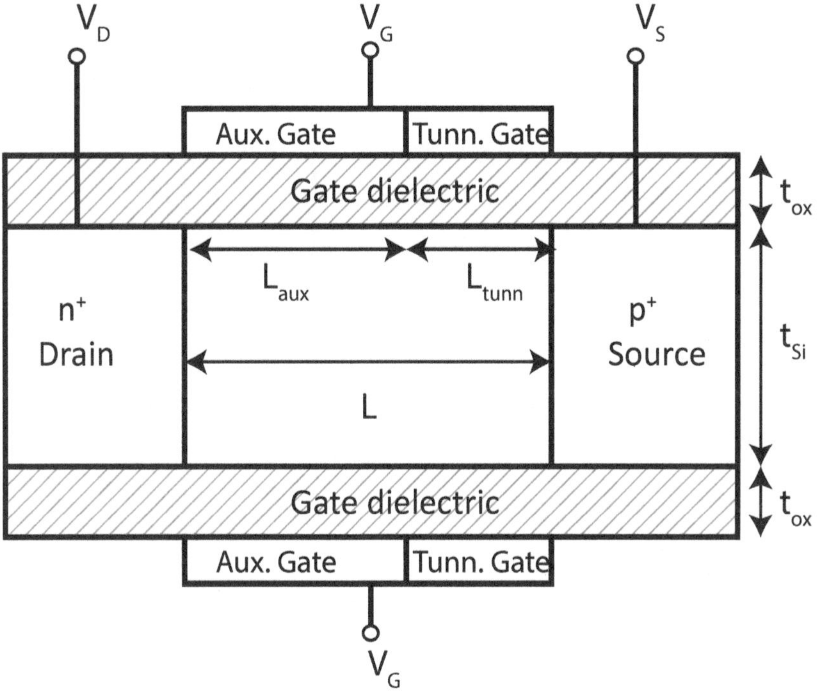

FIGURE 4.2
Dual material gate DGTFET [17].

Figure 4.2 shows a dual material gate DGTFET (DMG-DGTFET) [17]. The structure of the DMG-DGTFET is similar to the DGTFET described in Chapter 3, except that both the top and the bottom gates comprise materials with two different work functions [17]. The gate closer to the source (tunnel gate) is of low work function material and has a greater control over the I_{ON} of the device. The gate closer to the drain (auxiliary gate) is of high work function material and has a greater control over the I_{OFF} and the nature of output characteristics. By choosing optimized work functions for the tunnel gate, the work functions for the auxiliary gate, and the length of the tunnel gate, the subthreshold slope and the I_{ON}/I_{OFF} of the TFET can be improved [17, 24]. Using simulation, it has been demonstrated that the technique of employing a dual material gate (DMG) can be combined with the symmetric current enhancement techniques such as using strained silicon in a DGTFET to achieve a high I_{ON} and a low I_{OFF}, simultaneously, along with the improvement in the subthreshold slope.

2. **Asymmetric gate dielectric material:**

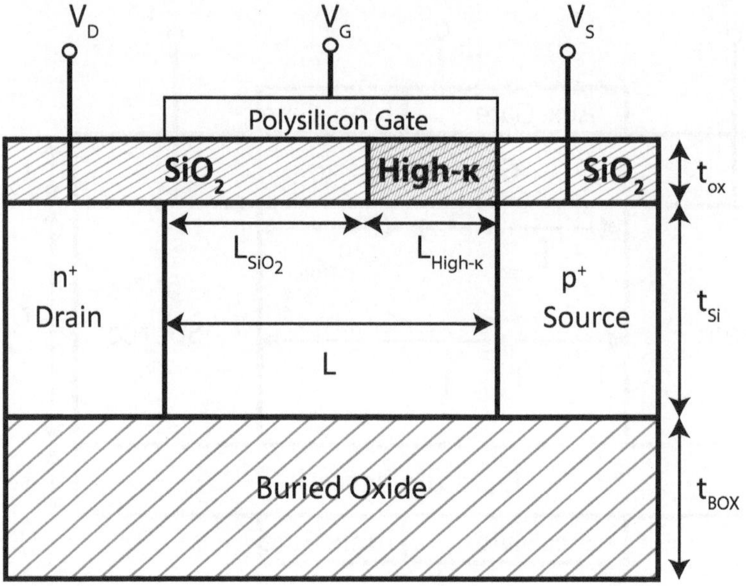

FIGURE 4.3
TFET with hetero-gate structure [18].

Figure 4.3 shows a TFET with a hetero-gate structure [18]. The gate oxide on the source side is composed of a high-κ material such as HfO_2 and the gate oxide on the drain side is composed of a low-κ material such as SiO_2 [18]. This asymmetrical gate design induces local minima at the edge of the conduction band at the tunneling junction which improves the subthreshold swing and enhances the I_{ON} in the TFET compared to the conventional TFET that has the same oxide material all over the gate region. The length of the high-κ gate, $L_{High-\kappa}$, needs to be optimized for improved device performance compared to the conventional TFET.

4.3.5 Quality of Gate Oxide

The quality of the gate dielectric and the quality of the semiconductor gate–dielectric interface plays an important role in determining the subthreshold swing in a TFET. The midgap interface trap density (D_{it}) results in the degradation of the subthreshold swing. The interface traps not only retard the Fermi-level movement caused by the gate potential, but can also lead to trap-assisted tunneling and subsequent thermal emission, which are responsible for degrading the subthreshold swing [5, 6, 25–28]. Therefore, with the improvement in the quality of semiconductor gate–dielectric interface, the subthreshold slope is expected to improve.

4.4 Tunneling Junction Engineering

The I_{ON} in a TFET is strongly dependent on the property of the tunneling junction, particularly on the source side. The property of the tunneling junction on the source side is primarily determined by the following factors, which can be engineered to achieve a high I_{ON}:

1. Source doping
2. Tunneling area
3. Source and channel materials

In this section, we describe some of the tunnel junction engineering techniques that are employed to improve the I_{ON} in TFETs.

4.4.1 Source Doping

The impact of source doping concentration on the I_{ON} of the DGTFET is described in Section 3.10. The I_{ON} in the DGTFET increases as the source doping concentration increases. With an increase in the source doping concentration, the tunnel-barrier width decreases for the following reasons [29]:

1. The built-in field between the source and the channel increases.
2. Due to bandgap narrowing, the energy bandgap decreases.

Therefore, given a TFET, the source doping concentration needs to be optimized to achieve a high I_{ON}. It should also be noted that the source doping concentration, in general, does not affect the I_{OFF} of the TFET.

The source doping profile strongly affects the tunneling barrier at the source–channel junction, which controls the I_{ON} in the TFET [30–41]. With an increase in the abruptness of the source doping, the electric field at the source–

channel junction increases and the tunneling width decreases. This results in an increase in the I_{ON}. It should be noted that the doping profile does not appear in the equation for the tunnel current (Equation (4.1)) because an abrupt doping profile is assumed in deriving the expression for the tunneling current. In practical devices, there is always some smearing out of the lateral doping profile into the channel region, which results in an increase in the tunneling distance [36]. It should also be noted that, in contrast to a conventional MOSFET, the I_{ON} and the subthreshold swing are ultrasensitive to the doping profile in a TFET. Therefore, optimizing the doping profile plays a pivotal role in the TFET [36].

It must be mentioned that attaining an abrupt doping profile is a non-trivial device processing challenge. In fact, many researchers have attributed the inferior subthreshold swing observed in their experimental TFETs, compared to the simulated results, to their inability to attain an abrupt source doping profile, which is often assumed in the simulation. Therefore, a great deal of research is directed toward attaining a sharp doping profile in a TFET. Several techniques that reduce the dopant diffusion into the channel and improve the activation of the dopants have been explored and experimentally demonstrated to improve the abruptness of the source doping profile and boost the I_{ON} in the TFET [33–36, 41]. Avoiding the high-temperature activation of dopants suppresses diffusion of dopants, improves the subthreshold swing, and inhibits drain-induced barrier lowering in the TFETs [41]. Another approach to avoid the high-temperature doping and annealing process and the effect of random dopant fluctuation is to use a doping-less TFET. The source and drain are formed electrostatically in a doping-less TFET by appropriately choosing the work functions of the source and the drain electrodes. A doping-less TFET exhibits electrical behavior similar to the TFETs having abruptly doped source and drain regions, avoiding the high-temperature processes [38, 40, 42, 43].

To attain a higher I_{ON}, a highly doped source pocket can also be added to the TFET [39, 44–48]. The pocket and the source are doped with dopants of opposite types. In this section, we illustrate the source pocket concept using the PNPN TFET.

Figure 4.4 shows the device structure of a PNPN TFET [44]. The most distinguishing feature of the PNPN TFET is the addition of an n^+ pocket on the source side. The width and the doping of an n^+ pocket are adjusted such that it is just fully depleted. The fully depleted pocket is a crucial requirement for the proper operation of the PNPN TFET. Due to the addition of a thin fully depleted n^+ pocket, the tunneling width at the source reduces and the lateral electric field increases. This results in an improvement in the I_{ON} and the subthreshold slope in the PNPN TFET. It should be noted that the doping concentration, doping profiles, and the width of the source pocket must be carefully engineered to attain good electrical characteristics in the PNPN TFET [39, 44]. In an experimental demonstration, using a spike rapid thermal

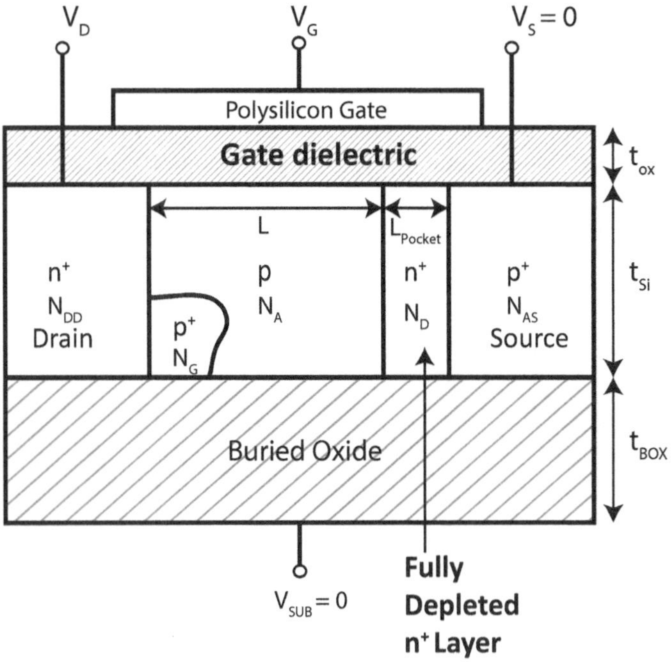

FIGURE 4.4
PNPN TFET with an n^+ pocket on the source side, source/drain doping (N_{DD} and N_{AS}) $= 2 \times 10^{20}$ $atoms/cm^3$, channel doping (N_A) $= 5 \times 10^{17}$ $atoms/cm^3$, pocket doping (N_D) $= 5 \times 10^{19}$ $atoms/cm^3$, $t_{ox} = 2.5\ nm$, $t_{Si} = 60\ nm$, $t_{BOX} = 100\ nm$, $L = 100\ nm$, $L_{Pocket} = 4\ nm$ [44].

annealing technique to obtain a sharp doping profile, the silicon-based PNPN TFET has exhibited an I_{ON} enhancement of around 10× compared to the similar p-i-n TFET [35]. Alternatively, instead of creating a pocket by doping, an n^+ pocket can be created electrostatically in the TFET by employing a metal electrode on the source side and choosing the work function of the electrode appropriately [45]. Using the concept of charge plasma, a part of an n^+ doped source is converted to a p^+ source, leaving behind an in-built n^+ pocket that boosts the I_{ON} and improves the subthreshold swing [45].

4.4.2 Tunneling Area

In a conventional TFET, the direction of tunneling is predominantly parallel to the semiconductor gate–dielectric interface [49]. Therefore, the effective tunneling area in a conventional TFET is a small region around the source–channel junction at the semiconductor gate–dielectric interface. As a result, the effective tunneling area is determined by the thickness of the inversion layer and is very small for a conventional TFET. This observation has motivated researchers to modify the geometry of the TFETs such that the tunneling direction becomes predominantly perpendicular to the semiconductor gate–dielectric interface, and thus a larger tunneling area can be obtained [39, 50–60]. We illustrate this concept using the green transistor or g-FET [50]. The tunnel transistor in [50] is named the green transistor because it exhibits steep subthreshold swing and can be potentially applied in ultra low supply voltage circuits [50]. However, TFETs in general are capable of exhibiting steep subthreshold slope and are promising candidates for future applications in energy-efficient circuits. Therefore, TFETs are in general, widely referred to as "green" transistors.

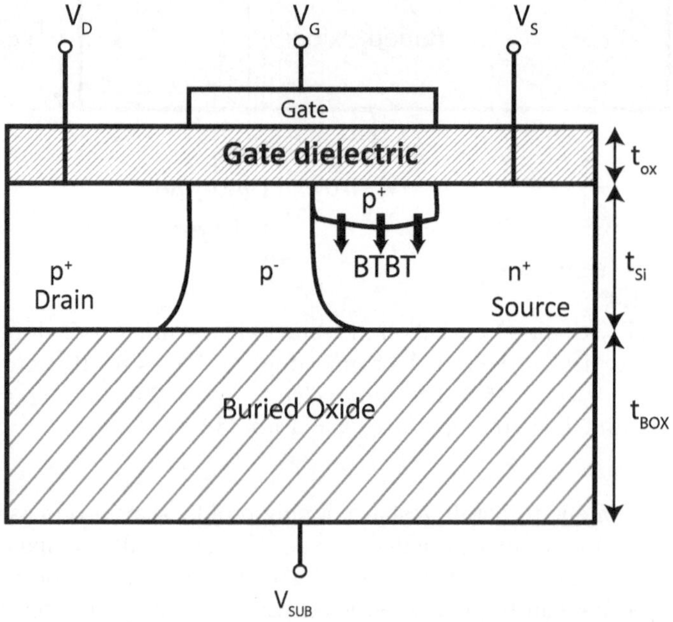

FIGURE 4.5
Schematic representation of the p-channel g-FET with a p^+ pocket in the source. The orientation of the pocket is parallel to the semiconductor gate–dielectric interface. The tunneling direction is perpendicular to the semiconductor gate–dielectric interface as shown. [50].

Figure 4.5 shows the schematic representation of a p-channel g-FET [50]. The distinctive feature of the g-FET is the introduction of a thin highly doped p^+ pocket close to the gate contained within the n^+ source. When a negative potential is applied to the gate, the potential of the floating p^+ pocket is pulled down, enabling BTBT of the electrons from the p^+ pocket to the n^+

source [50]. As a result, the holes are generated in the p^+ pocket and are swept to the drain as the drain current [50]. The doping of n^+ source and p^+ pocket is adjusted such that the tunneling path between the source and the pocket is triggered easily compared to the tunneling path between the n^+ source and the p^+ drain. Therefore, the tunneling is predominantly in the direction perpendicular to the semiconductor gate–dielectric interface as shown in Figure 4.5. The changed orientation of the tunneling has the following advantages compared to the conventional TFET:

1. The tunneling rate is proportional to the pocket length. Therefore, the effective tunneling area can be increased.
2. The gate field assists the internal field of the tunneling junction.

It should be noted that the I_{OFF} in the g-FET could be high due to the lateral source-to-drain leakage far from the gate oxide as demonstrated for an n-channel g-FET [51]. The I_{OFF} can be decreased by pushing up the buried oxide to block the lateral source-to-drain leakage current in the n-channel g-FET [51]. The concept of enhancing the I_{ON} by aligning the gate field in the direction of the tunnel junction internal field has also been demonstrated on TFET structures employing III-V materials and hetero-junctions [52–58, 61]. Along similar lines, there are other geometrical modifications proposed for the TFETs to obtain a high I_{ON}. For example, the channel can be inserted between the gate and the source confining the electric field in a narrow undoped region, thus enhancing the I_{ON} [62, 63]. The source geometry can also be modified, for example, in the form of a wedge such that the electric field can be enhanced and short tunneling lengths are available, thus increasing the I_{ON} [64, 65]. A U-shaped TFET has been proposed to increase the tunneling area without increasing the footprint of the device [66].

4.4.3 Heterojunctions

Equation (4.1) shows that the tunneling current in a TFET can be increased by employing a low-bandgap material. However, in general, when a low-bandgap material is employed throughout the TFET body, the I_{ON} increases and the I_{OFF} also increases. Since the I_{ON} in a TFET is primarily determined by the material property at the source tunneling junction, techniques have been developed to achieve a low effective bandgap only at the source–channel junction by employing heterojunctions, while the high-bandgap elsewhere in the TFET is maintained by using a suitable material.

A semiconductor heterojunction is the interface formed when two semiconductors that have dissimilar bandgaps are placed in conjunction or layered together. The conduction band energy and valence band energy change abruptly at the heterojunction. The magnitude of these changes determines the band alignment and the effective bandgap in the semiconductor heterojunction.

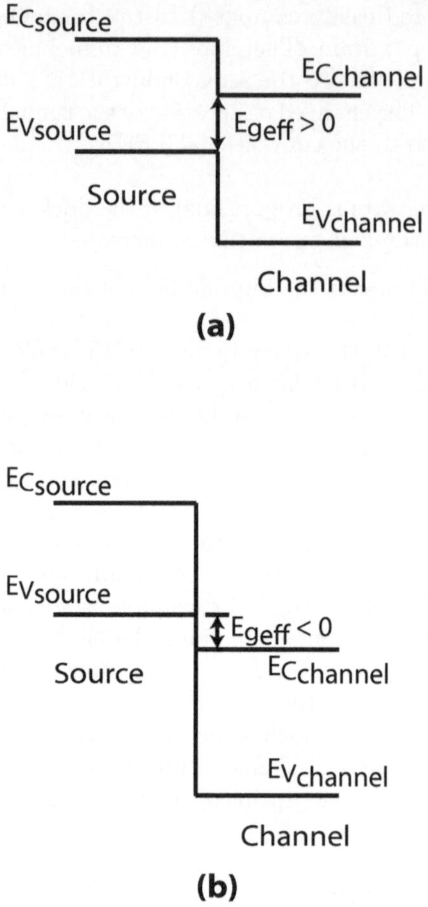

FIGURE 4.6
Band alignment in different types of heterojunctions: (a) staggered type and (b) broken-gap type.

Figure 4.6 shows a typical band alignment in the staggered and the broken-gap type of heterojunction in a TFET. The effective bandgap $E_{g_{eff}}$ for tunneling from the source to the channel in a heterojunction can be computed as

$$E_{g_{eff}} = E_{c_{channel}} - E_{v_{source}}, \tag{4.2}$$

where $E_{c_{channel}}$ is the conduction band energy in the channel and $E_{v_{source}}$ is the valence band energy in the source. In a staggered type of heterojunction,

$E_{g_{eff}}$ is positive. When $E_{c_{channel}}$ is decreased or $E_{v_{source}}$ is increased, the effective bandgap $E_{g_{eff}}$ decreases. When $E_{c_{channel}}$ is sufficiently decreased and/or $E_{v_{source}}$ is sufficiently increased, the effective bandgap $E_{g_{eff}}$ becomes negative and the heterojunction becomes the broken-gap type. An important point that should be noted is that the effective bandgap in a heterojunction can be considerably lower than the bandgap of the constituent materials. Therefore, the I_{ON} in a heterojunction-based TFET can be higher than the I_{ON} in the homojunction TFET based on either of the two materials of the heterojunction. A heterojunction improves the characteristics of a TFET for the following reasons:

1. It reduces the effective bandgap in the TFET and hence a higher I_{ON} can be achieved.
2. It suppresses the ambipolar behavior of the TFET and hence a lower subthreshold swing and a lower I_{OFF} can be achieved.

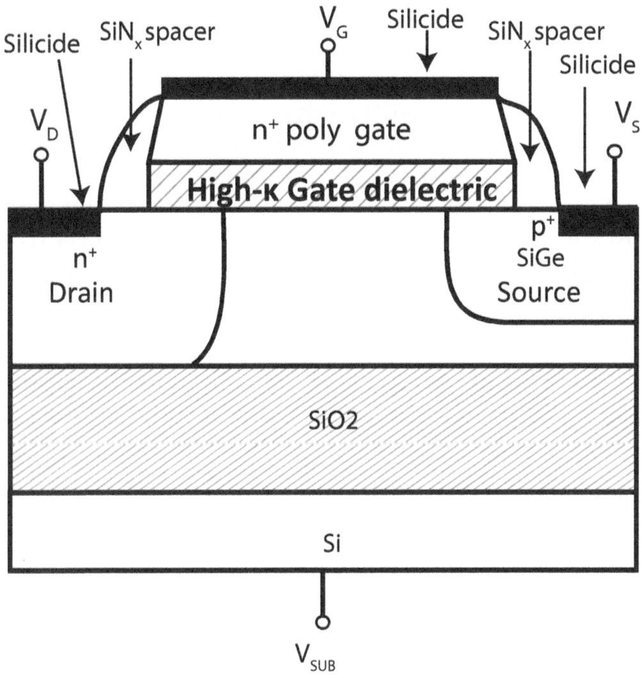

FIGURE 4.7
Heterojunction TFET (HETT) [67].

Heterojunctions have been employed in TFETs using different material systems. Heterojunctions realized using Si/Ge material systems have been widely explored by researchers [67–80]. Figure 4.7 shows a schematic representation of a heterojunction tunneling transistor (HETT) fabricated using an SiGe/Si material system [67]. In the HETT shown in Figure 4.7, the source is composed of SiGe, while the rest of the TFET body is composed of silicon. As a result, a heterojunction is formed at the source–channel junction. The effective bandgap at the SiGe-Si heterojunction is determined by the mole-fraction of Ge in SiGe. As the mole-fraction of Ge in the SiGe source is increased, the effective bandgap at the source–channel heterojunction decreases, thus increasing the I_{ON} and improving the subthreshold slope. It has been experimentally demonstrated that the drain current at a constant subthreshold slope increases by 15× when the Ge concentration is increased from 7% to 25%, indicating the benefits of the heterojunction in the HETT in enhancing the I_{ON}. It should be noted that the relative position of the edge of the gate and the heterojunction plays an important role in determining the I_{ON} and the I_{OFF} in an HETT [69]. Since heterojunctions are employed asymmetrically in the TFET body, the application of a heterojunction in a TFET also helps in ameliorating the ambipolar behavior in the TFET and lowering the I_{OFF} [70]. In general, the I_{ON} in a heterojunction-based TFET can be increased by decreasing the effective bandgap $E_{g_{eff}}$ of the heterojunction by engineering the material systems. However, the minimum $E_{g_{eff}}$ that can be achieved in a heterojunction is limited by the intrinsic band parameters of the corresponding materials and the critical thickness limitations in the lattice mismatched material systems. This has motivated researchers to employ a heterojunction with a negative $E_{g_{eff}}$ or a broken-gap type of heterojunction [67, 71, 72]. With a Si/Ge material system, a broken-gap-type heterojunction cannot be realized. Therefore, researchers have explored novel III-V material systems like InAs and GaSb to realize broken-gap alignment. We shall discuss III-V heterojunction TFETs in greater detail in Chapter 5.

4.5 Materials Engineering

Since the silicon-based fabrication technology has evolved and matured over last several decades, TFETs based on silicon are highly desirable. This will help the TFETs take advantage of the already proven silicon processing technology and will ease the deployment of TFETs into the existing CMOS process flows. However, silicon supports low tunneling current due to its large bandgap and the indirect mechanism of tunneling. As a result, the TFETs that are based purely on silicon fail to meet the I_{ON} requirement of the advanced technology nodes. Therefore, researchers are actively investigating alternative materials for TFETs so that a higher I_{ON} can be achieved. For example, germanium, III-V semiconductors, carbon-based nanomaterials, nanowires, 2D transition metal dichalcogenides, etc., have been employed to realize TFETs

[81–127]. In this section, we present some of the alternative materials that are being actively pursued by researchers for application in TFETs.

4.5.1 Germanium

Germanium is chemically similar to silicon and is largely compatible with the existing CMOS processes. The higher carrier mobility of germanium compared to silicon has attracted the attention of device engineers for the last several decades. However, the bandgap of germanium is 0.67 eV, which is smaller than the bandgap of silicon. Therefore, the leakage current due to band-to-band tunneling is high in a germanium-based MOSFET [81]. However, the smaller direct bandgap of germanium compared to silicon is a beneficial property for employing germanium for TFET applications. Since germanium inherently supports higher band-to-band tunneling compared to silicon, the I_{ON} in the germanium-based TFET is more than an order of magnitude higher than the silicon-based TFET with a similar structure [29]. This has motivated researchers to incorporate germanium in TFET structures. However, a conventional TFET that is solely based on germanium has a high and unacceptable I_{OFF}. Therefore, techniques that can keep the I_{OFF} under control are unavoidably employed in germanium-based TFETs. For example, the I_{OFF} can be reduced by using a short gate configuration or using a lower drain doping [82, 121]. In a short gate configuration, the gate is shortened on the drain side such that the high channel carrier density under the gate does not extend up to the drain [82]. As a result, the tunneling on the drain side is suppressed. Using simulations, it has been shown that the I_{OFF} in a germanium-based TFET can be reduced by three to five orders of magnitude by using a short gate configuration [82]. The ambipolar current can be reduced by two orders of magnitude by reducing the Gaussian doping in the drain by 5× [121].

One of the most commonly employed techniques to utilize the small bandgap of germanium keeping the I_{OFF} within required limits, is to use germanium combined with silicon in a heterojunction. Since the effective bandgap in the Si-Ge heterojunction can be smaller than the bandgap of both silicon and germanium, the I_{ON} obtained in a Si-Ge heterojunction-based TFET is greater than all-Si and all-Ge TFETs. The large bandgap of silicon is exploited to achieve a low I_{OFF} in Si-Ge heterojunction-based TFET. Germanium is also employed in silicon-based TFETs to introduce strain and achieve a higher I_{ON}, as we shall describe in Section 4.6.

4.5.2 III-V Semiconductors

A III-V semiconductor is a compound that is obtained by combining metallic elements of group III of the periodic table (aluminum, gallium and indium) with the non-metallic elements of group V of the periodic table (nitrogen, phosphorous, arsenic and antimony). The III-V semiconductor compounds can be formed as binary, ternary, or quaternary compounds by combining two, three, or four elements of group III and the group V. Some of the III-V compounds that are being widely investigated by researchers for use in TFETs are GaSb, InAs, InGaAs, InAsSb, GaAsSb, InAlAs and AlGaAsSb. In this book, we refer to TFETs that are based on the III-V compound semiconductors as III-V TFETs.

In general, the tunneling rates of III-V semiconductors are higher compared to silicon, which can be utilized to obtain a higher I_{ON} compared to the silicon-based TFETs. The III-V semiconductors offer higher tunneling rates than the tunneling rates in silicon because:

1. A smaller bandgap can be obtained in a III-V compound semiconductor compared to silicon.
2. The mechanism for tunneling is direct in several III-V semiconductors. However, the mechanism of tunneling is indirect in silicon.
3. The tunneling mass of III-V semiconductors is smaller than the tunneling mass in silicon.

It should be noted that, in general, due to the smaller bandgap of the III-V semiconductors, the I_{ON} and I_{OFF} in the III-V TFETs are both high. Therefore, the III-V semiconductors are generally employed as heterojunctions in TFETs such that a high I_{ON} and a low I_{OFF} can be achieved simultaneously.

We illustrate how a high I_{ON} can be realized in a III-V TFET using the device structure proposed in reference [83]. Figure 4.8 shows a schematic representation of a III-V TFET [83]. The source is composed of p$^+$ ($2 \times 10^{19} cm-3$) $Al_{0.5}Ga_{0.5}As_{0.3}Sb_{0.7}$, the channel is n$^-$ ($5 \times 10^{16} cm^{-3}$) $In_{0.8}Ga_{0.2}As$, and the drain is composed of n$^+$ ($5 \times 10^{19} cm^{-3}$) $In_{0.8}Ga_{0.2}As$. The barrier layer beneath the channel is composed of intrinsic $In_{0.8}Al_{0.2}As$. The materials are chosen to avoid lattice mismatch and associated stress and strain. The gate is composed of a high-κ gate dielectric like Al_2O_3 with an equivalent oxide thickness of 0.6 nm.

The band alignment at the source–channel heterojunction is shown in Figure 4.9. The staggered band alignment of the source–channel heterojunction results in an effective bandgap of 0.23 eV. The small effective bandgap of 0.23 eV at the source ensures that a high I_{ON} can be achieved. However, the bandgaps of the individual materials used in the TFET are greater than 0.5 eV. Therefore, the I_{OFF}, limited by the thermal generation of the carriers, is low. Using simulation, it has been shown that the I_{ON} is around 0.4 $mA/\mu m$ and $I_{ON}/I_{OFF} \approx 10^4$, for $V_{DD} = 0.3\ V$ [83]. These results suggest that the III-V TFETs can possibly be utilized for low-voltage applications in future.

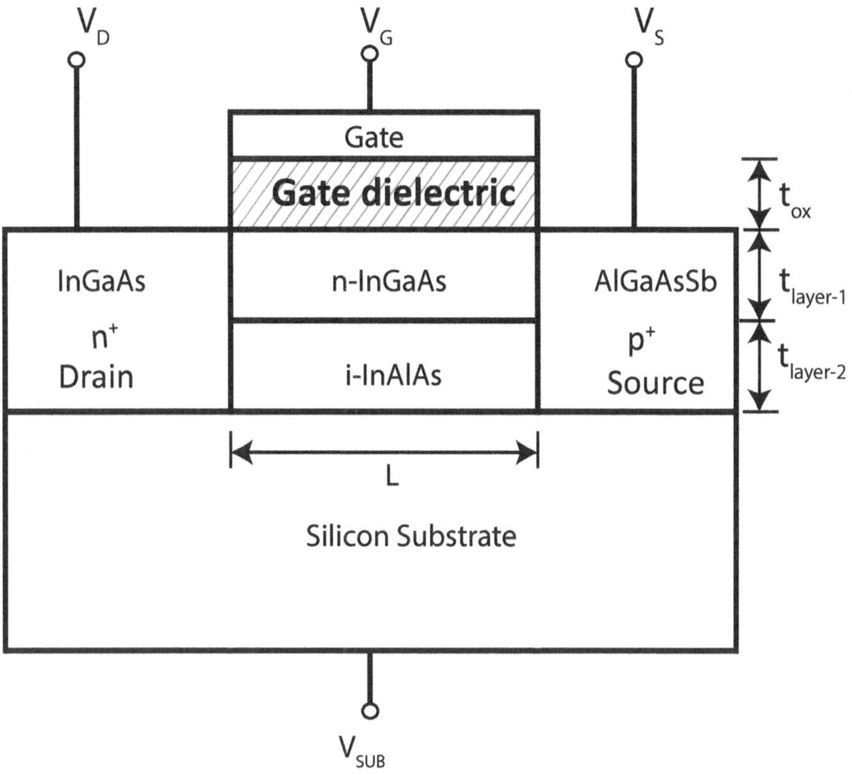

FIGURE 4.8
III-V heterojunction TFET. Device parameters are $L = 100\ nm$, $t_{layer-1} = 10\ nm$, $t_{layer-2} = 10\ nm$ and gate effective oxide thickness $= 0.6\ nm$ [83].

It should be noted that a wide range of heterojunction band alignments can be achieved in III-V semiconductors. Both staggered and broken-gap types of band alignment have been proposed for III-V TFETs [7, 84–87]. Additionally, the electrical properties of the III-V semiconductors can be subtly engineered by changing their material composition. Lattice matching can also be achieved with a wide range of alloy compositions in III-V semiconductors, thus avoiding the problem of stress/strain in the III-V heterojunction TFETs. As a result, researchers are actively pursuing III-V semiconductors for TFET applications. The projection of circuit performance obtained using simulation for InAs-channel TFETs suggests that the TFETs can outperform MOSFETs at low voltages [88]. There have been several experimental demonstrations of high I_{ON} that have been achieved in III-V TFETs [5–7, 84–86].

However, there are significant technological challenges to overcome before III-V compound semiconductors can be deployed in mainstream CMOS technol-

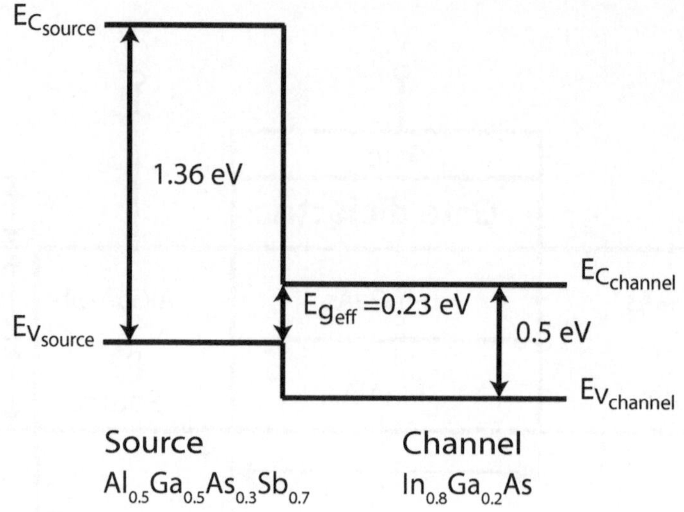

FIGURE 4.9
Band alignment at the source–channel junction for the III-V TFET shown in Figure 4.8.

ogy. The greatest challenge is to grow a high-quality high-κ dielectric over a III-V semiconductor channel. The traps at the semiconductor–dielectric interface limit the field control and degrade the subthreshold slope. These problems of III-V TFETs and possible solutions are described in detail in Chapter 5.

4.5.3 Carbon

Graphene, one of the allotropic forms of carbon, is attracting a great deal of attention from device engineers. Due to the recent advancements in the technology to fabricate graphene, it is being considered as a potential candidate for future electronic applications. Graphene is a one-atom-thick sheet of carbon atoms that are arranged in hexagonal patterns like graphite. The most distinguishing property of graphene is its excellent conductivity, which could possibly be exploited in a MOSFET. However, graphene is a zero-energy bandgap material and the absence of the bandgap restricts its potential application. Therefore, researchers are exploring methods to open a usable bandgap in graphene. One of the successful techniques to achieve a usable bandgap in graphene is to reduce the dimensionality of the graphene using carbon nanotubes (CNT) or graphene nano-ribbons (GNR). Graphenes that are rolled into one-dimensional tubes are known as carbon nanotubes (CNT). Graphene that is fabricated as an ultrathin strip is known as graphene nano-ribbons (GNR). Recently, CNTs and GNRs have attracted the attention of device engineers for application in TFETs because of their unique and tunable properties. These properties could potentially be applied in TFETs to obtain a high I_{ON}, especially at smaller supply voltages. In this section, we briefly describe GNR- and CNT-based TFETs to illustrate that a high I_{ON} can be achieved using these materials. A detailed description of carbon-based TFETs is presented in Chapter 6.

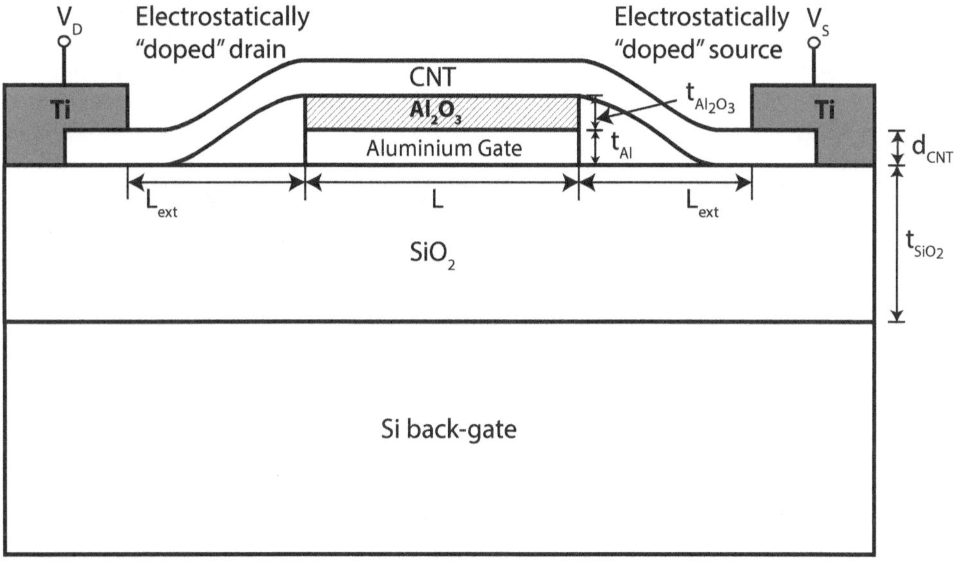

FIGURE 4.10
Schematic representation of the CNT TFET used in the experiments in reference [89]. Device parameters are $L_{ext} = 200\ nm$, $t_{SiO_2} = 10\ nm$, $t_{Al} = 20\ nm$, $t_{Al_2O_3} = 4\ nm$.

CNTs have been investigated for more than a decade for use in a MOSFET because of their geometrical smallness, excellent transport properties, and 1-D effects. Recently, CNTs are being actively pursued for use in TFETs. The first experimental demonstration of the CNT TFET is presented in 2005 [89]. This device structure is shown in Figure 4.10. The device is fabricated on a highly doped Si substrate, which is separated from the CNT by a 10 nm-thick thermally grown SiO_2. The Si substrate acts as a back gate and is used to electrostatically "dope" the source/drain extensions. A 20 nm-thick aluminum layer is deposited over the SiO_2 and serves as the gate for the TFET and also shields the channel from the back gate. A 4 nm thick Al_2O_3 layer is grown over Al and acts as the gate dielectric between the CNT channel and the Al gate. The back-gate is fixed at a negative potential, while a large positive potential is applied to the Al gate. This creates a tunneling junction around the source-channel region and electrons can tunnel from the valence band in the source to the conduction band in the channel. The subthreshold swing experimentally obtained in this device was 40 *mV/decade* [89] and was one of the first experimental demonstrations breaking the barrier of 60 *mV/decade* in a TFET. However, the magnitude of the tunneling current was a few orders of magnitude lower than the conventional MOSFET that was realized in the similar structure [89].

The CNT TFETs can achieve an excellent subthreshold slope and potentially a high I_{ON} because they have a moderate bandgap and a small tunneling mass. Additionally, since CNTs are 1-D systems, a superior gate control can be achieved in the CNT TFETs. The high-κ gate technology can potentially be combined with CNTs and a high I_{ON} can be obtained in the CNT-based TFETs.

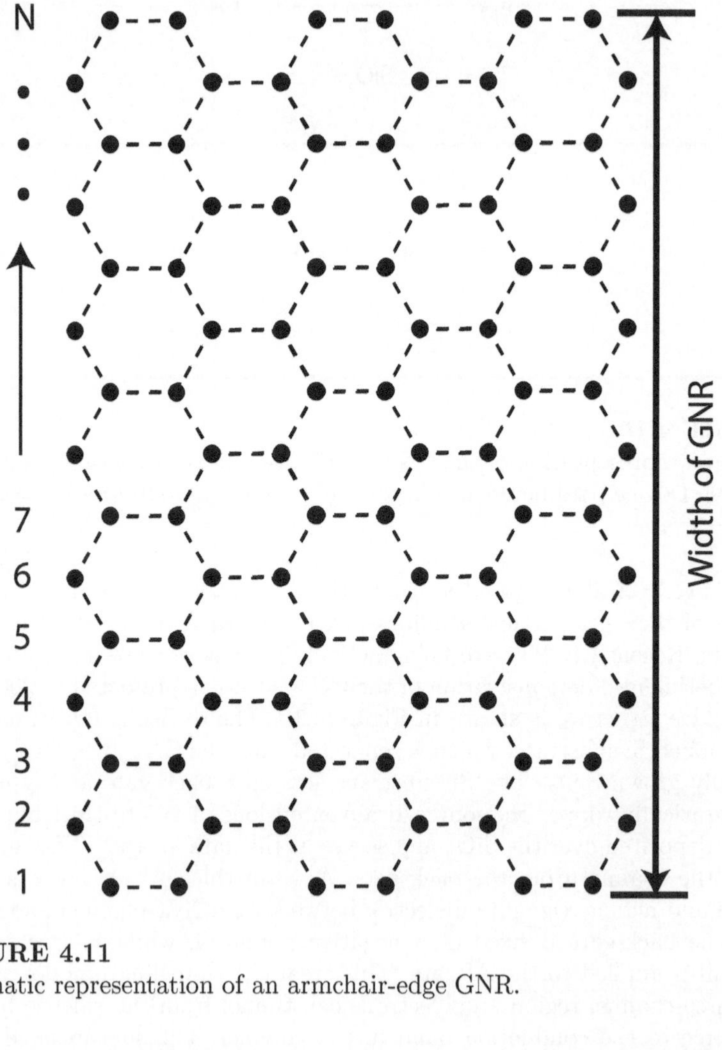

FIGURE 4.11
Schematic representation of an armchair-edge GNR.

GNRs are also being actively investigated for application in TFETs because of their small tunneling mass, direct bandgap, and compatibility with the planar technology. GNRs are classified as zig-zag and armchair types, depending on the shape of the edge in the GNR. An armchair-edge GNR (AGNR) is a geometrically terminated single graphene layer as shown in Figure 4.11. The bandgap in a GNR depends on the width and the external stimuli like strain and the electric field. The flexibility of tuning the bandgap of the GNR makes GNR quite interesting to device engineers. The energy bandgap in the GNR originates from the quantum confinement effect and is heavily dependent on the edge configurations. The characteristics of the carbon bonds at the edges change abruptly in the GNR. The carbon–carbon bonds at the edges are around 3.5% shorter compared to the carbon–carbon bonds inside the GNR [90]. This leads to several interesting effects in GNR TFETs since the bandgap and the electron effective mass are dependent on the edge bond relaxation in GNR. Depending on the number of carbon atoms in the GNR (N in Figure 4.11), three distinct families of GNRs can be identified: $N = 3p$, $N = 3p + 1$ and $N = 3p + 2$, where p is an integer. As an illustration: the GNR with 12 atoms belongs to the family $3p$, the GNR with 13 atoms belong to the family $3p + 1$ and the GNR with 14 atoms belong to the family $3p + 2$. The properties of the GNR for the same family show a similar trend. For example, the edge bond relaxation results in an increase in the bandgap in $3p + 1$ type of GNRs, while the edge bond relaxation results in a decrease in the bandgap in $3p$ type of GNRs [91]. Since the I_{ON} in a GNR TFET is dependent on the bandgap of the material, which in turn depends on the width and the edge bond relaxation, these factors need to be taken into account in optimizing the performance of the GNR TFETs [90, 91].

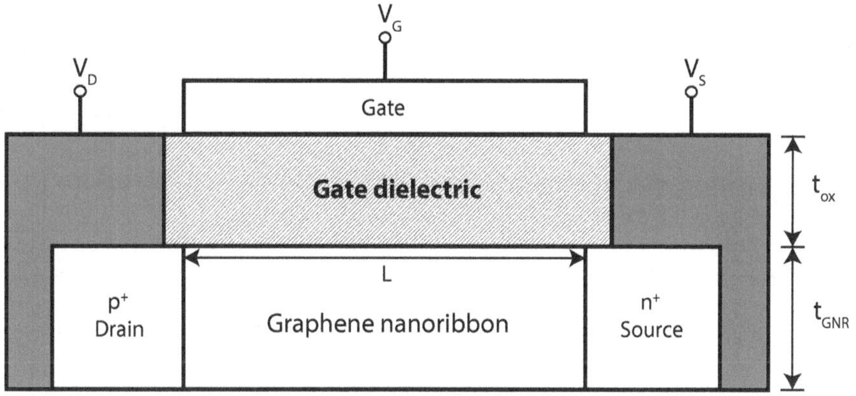

FIGURE 4.12
p-channel GNR TFET [92].

Figure 4.12 shows the p-channel GNR-based TFET [92]. The source is a highly doped n^+ region and the channel and the drain are p^+ doped. The device parameters are as follows: thickness of the body $t_{GNR} = 0.35\ nm$, channel length $L = 20\ nm$, equivalent oxide thickness $t_{ox} = 1\ nm$ and ribbon width $= 5\ nm$ [92]. It has been theoretically predicted that $I_{ON} = 800\ \mu A/\mu m$ can be achieved in the GNR TFET and the subthreshold swing as low as $0.19\ mV/decade$ can be obtained for $V_{DD} = 0.1\ V$. This prediction neglects the gate oxide leakage, interface charge, and parasitic resistances. Although, the theoretical predications of [92] can be considered as an upper bound on the performance of the GNR TFET, it shows that GNR TFETs have a tremendous potential for use in low power applications. We will be discussing GNR TFETs in detail in Chapter 6.

4.5.4 Nanowires

A nanowire is an very thin wire with a diameter on the order of a few nanometers. During recent times, nanowire-based electronic devices have drawn significant attention of researchers. The research on nanowire-based electronics has gained momentum during the last few years due to the following reasons [98]:

1. Techniques have been developed to produce nanowires with a high yield and reproducible electrical properties.
2. At least one critical device dimension can be precisely controlled using bottom-up synthesis of nanowires.
3. There have been recent advances in assembly methods that have enabled the fabrication and testing of circuits realized using nanowires.

A nanowire is a 1-D system where the lateral dimension is constrained to a few nanometers, while the longitudinal dimension is not constrained. As a result, quantum confinement effects that are exhibited in 1-D systems play an important role in determining the electrical properties of nanowires. Several attractive electrical properties of nanowires combined with their geometrical smallness have led researchers to explore conventional FETs, as well as TFETs, that are based on nanowires. It should be noted that the electrostatics of the nanowire-based TFETs (NW-TFET), especially the control of gate over the potential inside the nanowire channel, is similar to the electrostatics of the nanowire-based conventional FET (NW-FET) and fully depleted SOI-MOSFET (FDSOI-MOSFET) [89, 98]. The relationship between the tunneling current and the natural length λ in a nanowire-based TFET (NW-TFET) can be given as [89]

$$I_D \propto \exp\left[-\frac{4\lambda\sqrt{2m^*}E_g^{*\frac{3}{2}}}{3|e|\hbar\left(E_g^* + \Delta\Phi\right)}\right]\Delta\Phi, \tag{4.3}$$

where m^* is the effective carrier mass, E_g is the bandgap of the material, $\Delta\Phi$ is the range of energy over which tunneling can take place, e is the charge of an electron and $\hbar$ is the reduced Planck constant, and λ is given by

$$\lambda = \sqrt{\frac{\varepsilon_{NW}}{\varepsilon_{ox}} t_{ox} d_{NW}}, \tag{4.4}$$

where t_{ox} is the gate oxide thickness, d_{NW} is the diameter of nanowire, ε_{ox} is the dielectric constant of the gate oxide and ε_{NW} is the dielectric constant of the nanowire. Since an extremely small λ can be achieved in a NW-TFET, a high tunneling current can be achieved in a NW-TFET. Another important reason that makes the nanowires interesting for use in TFETs is the high tunneling rates in the 1-D material systems compared to the 2-D and 3-D systems. Moreover, in general, the tunneling effective mass of nanowires is smaller compared to the tunneling effective mass in conventional silicon and a high I_{ON} is expected in NW-TFETs.

It should be noted that there is trade-off between the subthreshold slope and the I_{ON} in a bulk TFET [103]. Due to the $\sqrt{E}$ dependence of the density of states in a 3-D material, a good subthreshold slope and a high I_{ON} cannot be achieved simultaneously in a bulk TFET [103]. However, in a NW-TFET, due to the $\frac{1}{\sqrt{E}}$ dependence of the density of states in the 1-D system, the subthreshold slope and the I_{ON} can be improved simultaneously [103]. This is one of the major advantages of NW-TFETs.

There are nanowires of several different materials that are being actively investigated for use in TFETs. For example, silicon, germanium, carbon, InAs, InSb and GaSb based NW-TFETs are being investigated [3, 34, 36, 71, 72, 89, 102, 103]. Additionally, nanowires composed of different materials can form staggered or broken-gap-type heterojunctions [71, 72]. The reduced effective bandgap in a heterojunction NW-TFET enhances the I_{ON}. The quantization effects in the nanowires result in an increase in the bandgap. This can be utilized in reducing the I_{OFF} in the NW-TFET and turning off the transistor in the broken-gap-type heterojunction NW-TFET. It should be noted that, due to the quantum confinement effects, the bandgap in a nanowire depends on the width of the nanowire. Therefore, nanowires of different widths, but of the same material, can also be combined together to yield a heterojunction and boost the I_{ON}.

It must be mentioned that there are still several hurdles to overcome before NW-TFETs can be deployed for large-scale application. The challenges of processing, synthesizing and assembling the nanowires are non-trivial. Controlling the process-induced variations in the fabrication of NW-TFETs can be one of the biggest hurdles for circuit applications [1]. Additionally, reproducibly fabricating nanowires of a given property and achieving sharp doping profiles at smaller geometries are still concerns. We will look at the interesting properties, problems and possible solutions for NW-TFETs in detail in Chapter 7.

4.6 Strain Engineering

The application of strain in a semiconducting material affects the band structure of the material. As a result, there are subtle changes in the electrical properties of the material when strain is applied. The modulation of the electrical properties due to strain can be utilized to improve the characteristics of the electronic devices. Researchers have studied strained silicon in great detail and have successfully employed strained silicon in the existing CMOS technology. The application of strained silicon instead of normal silicon in a MOSFET results in an improvement of the drive capability of the device. The strained silicon improves the drive capability of a MOSFET primarily because of higher carrier mobility. Recently, researchers are attempting to incorporate strain in different types of TFETs to boost their I_{ON}. However, it should be noted that the mechanism of improvement of the I_{ON} due to strain in a TFET and in a conventional MOSFET are quite different.

The presence of strain in silicon results in a decrease in the bandgap and the effective mass of the carriers and an increase in the electron affinity [104–106]. The mobility of the carriers also improves due to the application of strain in silicon [107]. Equation (4.1) shows that the tunneling current can be increased by decreasing the bandgap and the effective carrier mass of the material of the TFET. Therefore, the application of strain in a silicon-based TFET is expected to increase the I_{ON} [108, 109]. It has been experimentally demonstrated that the application of tensile strain in the silicon-based TFET results in an increase in the drain current [109]. In these experiments, the strain was introduced using wafer bending [109]. The increase in the drain current can be attributed to the combined effect of the decrease in the bandgap, decrease in the effective carrier mass, and increase in the channel mobility. On the other hand, the application of compressive strain has been observed to reduce the drain current [109]. This has been attributed to the decrease in the density of states due to compressive strain. The decrease in the density of states results in a reduction in the available carriers for tunneling on the source side [109].

The strain in silicon-based TFET can also be introduced by employing single-layer strained silicon-on-insulator (s-SOI) instead of the normal silicon [108]. An s-SOI is a SiGe-free material system that retains the merits of strained silicon while improving the scalability of thin-film SOIs [110–112]. The strain in the s-SOI is controlled by changing the mole fraction of Ge in the relaxed SiGe buffer layer that is employed during the fabrication of the s-SOI [108, 110–112]. It has been theoretically shown that an s-SOI-based strained DGTFET (SDGTFET) can achieve an I_{ON} around two orders of magnitude higher than the conventional DGTFET [108].

The strain in a TFET can be applied asymmetrically to the TFET such that the performance is optimized. For example, it has been proposed to apply an asymmetric strain profile using two-local stressor technology for gate-all-around silicon NW-TFETs [113]. The lateral strain profile can be engineered such that the strain is the maximum at the tunneling junction at the source and the strain is the minimum at the drain. Therefore, the bandgap will be the smallest at the source and the largest at the drain. As a result, the I_{ON} of the TFET is increased without sacrificing the I_{OFF} [113].

The strain can also be introduced by employing a lattice mismatched heterostructure in a TFET. For example, heterostructure-induced strain can result in an increase in the I_{ON} by around two orders of magnitude in a nanowire-based germanium source Si-channel TFET [1, 4]. The application of heterostructure-induced strain can also improve the subthreshold slope of the TFET [1, 4].

It has been experimentally shown that $I_{ON} \approx 1\ \mu A/\mu m$ can be achieved in a strained $Si_{0.2}Ge_{0.8}$ on Si TFET [114]. The enhancement in the I_{ON} is due to the increase in the tunneling probability because of the small bandgap of the strained SiGe. It should be noted that the direction of the channel and the crystal orientation play an important role in controlling the I_{ON} in the TFET because the effective masses are different in different directions in the crystal [114].

Strained silicon is also employed as the channel in a TFET to obtain a high I_{ON}. It has been shown experimentally that the compressively strained $Si_{0.5}Ge_{0.5}$ source and tensile strained Si channel heterostructure can achieve a high I_{ON} [74, 115]. The small bandgap of $Si_{0.5}Ge_{0.5}$ ($0.7\ eV$) and large conduction band offset of $Si_{0.5}Ge_{0.5}$ with respect to the strained Si result in a high I_{ON} [115].

Strain has also been proposed to be applied in III-V materials–based TFETs and GNR-based TFETs to boost the I_{ON} [116, 117, 128]. Depending on the material properties and the electrostatics of a given TFET, strain provides another handle to boost the I_{ON} in a wide variety of TFETs.

4.7 Summary

In this chapter, we described various techniques that are being investigated by researchers to solve the problem of low I_{ON} in TFETs. We have seen that there are primarily three techniques to improve the I_{ON} in the TFET: designing an improved gate structure, engineering the tunneling junction and employing novel material systems. These techniques are often combined to further improve the I_{ON} in the TFET. We also highlighted various device and material parameters that can be optimized to boost the I_{ON} in the TFETs. It should be noted that, although the low I_{ON} is the biggest hurdle for TFETs that is blocking their wide-scale electronic applications, there are other important electrical properties of TFETs that need to be optimized. We shall be looking at techniques that improve other electrical properties of the TFETs, for example, subthreshold swing, ambipolar current, saturation characteristics, ON-state resistance, cut-off frequency, gain-bandwidth product, immunity against process variation, etc. in subsequent chapters.

REFERENCES

[1] A. Verhulst, W. Vandenberghe, D. Leonelli, R. Rooyackers, A. Vandooren, J. Zhuge, K.-H. Kao, B. Soree, W. Magnusa, M. Fischetti, G. Pourtois, C. Huyghebaert, R. Huang, Y. Wang, K. D. Meyer, W. Dehaene, and M. H. G. Groeseneken, "Si-based tunnel field-effect transistors for low-power nano-electronics," in *69th Annual Device Research Conference (DRC)*, pp. 193–196, IEEE, 2011.

[2] K. Boucart and A. M. Ionescu, "Double-gate tunnel FET with high-κ gate dielectric," *IEEE Trans. on Electron Devices*, vol. 54, pp. 1725–1733, July 2007.

[3] M. T. Bjork, J. Knoch, H. Schmid, and H. R. W. Riess, "Silicon nanowire tunneling field-effect transistors," *Applied Physics Letters*, no. 193504, 2008.

[4] A. S. Verhulst, W. G. Vandenberghe, D. Leonelli, R. Rooyackers, A. Vandooren, G. Pourtois, S. D. Gendt, M. M. Heyns, and G. Groeseneken, "Boosting the on-current of Si-based tunnel field-effect transistors," *ECS Transactions*, vol. 33, no. 6, pp. 363–372, 2010.

[5] S. Mookerjea, D. Mohata, R. Krishnan, J. Singh, A. Vallett, A. Ali, T. Mayer, V. Narayanan, D. Schlom, and A. L. S. Datta, "Experimental

demonstration of 100 *nm* channel length $In_{0.53}Ga_{0.47}As$-based vertical inter-band tunnel field effect transistors (TFETs) for ultra low-power logic and SRAM applications," in *IEEE International Electron Devices Meeting (IEDM)*, pp. 1–3, IEEE, Dec. 2009.

[6] H. Zhao, Y. Chen, Y. Wang, F. Zhou, and F. X. J. Lee, "$In_{0.7}Ga_{0.3}As$ tunneling field-effect transistors with an ion of 50 $\mu A/\mu m$ and a subthreshold swing of 86 mV/dec using HfO_2 gate oxide," *IEEE Electron Device Letters*, vol. 31, pp. 1392–1394, Dec. 2010.

[7] R. Iida, S.-H. Kim, M. Yokoyama, N. Taoka, S.-H. Lee, and M. T. S. Takagi, "Planar-type $In_{0.53}Ga_{0.47}As$ channel band-to-band tunneling metal-oxide-semiconductor field-effect transistors," *Journal of Applied Physics*, vol. 110, no. 124505, 2011.

[8] E.-H. Toh, G. H. Wang, and G. S. Y.-C. Yeo, "Device physics and design of double-gate tunneling field-effect transistor by silicon film thickness optimization," *Applied Physics Letters*, vol. 90, no. 263507, 2007.

[9] F. Xue, H. Zhao, Y.-T. Chen, Y. Wang, F. Zhou, and J. Lee, "$In_{0.7}Ga_{0.3}As$ tunneling field-effect-transistors with $LaAlO_3$ and ZrO_2 high-κ dielectrics," *ECS Transactions*, vol. 41, no. 3, pp. 249–253, 2011.

[10] Y. Morita, T. Mori, S. Migita, W. Mizubayashi, A. Tanabe, K. Fukuda, T. Matsukawa, K. Endo, S. O'uchi, Y. X. Liu, M. Masahara, and H. Ota, "Performance enhancement of tunnel field-effect transistors by synthetic electric field effect," *IEEE Electron Device Letters*, vol. 35, pp. 792–794, July 2014.

[11] M. Schlosser, K. Bhuwalka, M. Sauter, T. Zilbauer, T. Sulima, and I. Eisele, "Fringing-induced drain current improvement in the tunnel field-effect transistor with high-κ gate dielectrics," *IEEE Trans. on Electron Devices*, vol. 56, pp. 100–108, Jan. 2009.

[12] I. Eisele, H. Lochner, and M. Schlosser, "SiGe tunnel field effect transistors," *ECS Transactions*, vol. 16, no. 10, pp. 961–973, 2008.

[13] C. Anghel, P. Chilagani, A. Amara, and A. Vladimirescu, "Tunnel field effect transistor with increased ON current, low-κ spacer and high-κ dielectric," *Applied Physics Letters*, no. 122104, 2010.

[14] H. G. Virani and A. Kottantharayil, "Optimization of hetero junction n-channel tunnel FET with high-κ spacers," in *Proc. 2nd Int. Workshop Electron Devices Semicond. Technol.*, pp. 1–6, IEEE, 2009.

[15] A. Chattopadhyay and A. Mallik, "Impact of a spacer dielectric and a gate overlap/underlap on the device performance of a tunnel field-effect transistor," *IEEE Trans. on Electron Devices*, vol. 58, pp. 677–683, Mar. 2011.

[16] K. T. Lam, Y. Yang, G. S. Samudra, Y. C. Yeo, and G. Liang, "Electrostatics of ultimately thin-body tunneling FET using graphene nanoribbon," *Electron Device Letters*, vol. 32, pp. 431–433, Apr. 2011.

[17] S. Saurabh and M. J. Kumar, "Novel attributes of a nanoscale dual material gate tunnel field effect transistor," *IEEE Trans. on Electron Devices*, vol. 58, pp. 404–410, Feb. 2011.

[18] W. Y. Choi and W. Lee, "Hetero-gate-dielectric tunneling field-effect transistors," *IEEE Trans. on Electron Devices*, vol. 57, pp. 2317–2319, Sept. 2010.

[19] N. Cui, R. Liang, and J. Xu, "Heteromaterial gate tunnel field effect transistor with lateral energy band profile modulation," *Applied Physics Letters*, vol. 98, no. 142105, 2011.

[20] J. S. Lee, W. Y. Choi, and I. M. Kang, "Characteristics of gate-all-around hetero-gate-dielectric tunneling field-effect transistors," *Japanese Journal of Applied Physics*, vol. 51, 2012.

[21] R. Liang, N. Cui, M. Zhao, J. Wang, and J. Xu, "High performance tunnel field effect transistor with a tri-material-gate structure," in *International Semiconductor Device Research Symposium*, pp. 1–2, IEEE, Dec. 2011.

[22] H. W. Kim, J. H. Lee, W. Kim, M.-C. Sun, J. H. Kim, G. Kim, K.-W. Kim, H. Kim, J. Y. Seo, and B.-G. Park, "A tunneling field-effect transistor using side metal gate/high-κ material for low power application," in *International Semiconductor Device Research Symposium*, pp. 1–2, IEEE, Dec. 2011.

[23] D. B. Abdi and M. J. Kumar, "Controlling ambipolar current in tunneling FETs using overlapping gate-on-drain," *IEEE Journal of the Electron Devices Society*, vol. 2, pp. 187–190, Nov. 2014.

[24] C. Y. Hsu, C. Y. Chang, E. Y. Chang, and C. Hu, "Suppressing non-uniform tunneling in InAs/GaSb TFET with dual-metal gate," *IEEE Journal of the Electron Devices Society*, vol. 4, pp. 60–65, Mar. 2016.

[25] K. E. Moselund, H. Schmid, C. Bessire, M. T. Bjork, H. Ghoneim, and H. Riel, "InAs-Si nanowire heterojunction tunnel fets," *IEEE Electron Device Letters*, vol. 33, pp. 1453–1455, Oct. 2012.

[26] A. Vandooren, D. Leonelli, R. Rooyackers, K. Arstila, G. Groeseneken, and C. Huyghebaert, "Impact of process and geometrical parameters on the electrical characteristics of vertical nanowire silicon n-TFETs," *Solid-State Electronics*, vol. 72, pp. 82–87, 2012.

[27] W. C. Y. Ma and Y.-H. Chen, "Performance improvement of poly-Si tunnel FETs by trap density reduction," *IEEE Transactions on Electron Devices*, vol. 63, pp. 864–868, Feb. 2016.

[28] W. C.-Y. Ma, Y.-H. Chen, Z.-Y. Lin, Y.-S. Huang, B.-S. Huang, and Z.-D. Wu, "Performance improvement of poly-Si tunnel thin-film transistor by NH_3 plasma treatment," *Thin Solid Films*, 2016.

[29] E.-H. Toh, G. H. Wang, G. Samudra, and Y.-C. Yeo, "Device physics and design of germanium tunneling field-effect transistor with source and drain engineering for low power and high performance applications," *Journal of Applied Physics*, vol. 103, no. 104504, 2008.

[30] W. Y. Choi, B.-G. Park, J. D. Lee, and T.-J. K. Liu, "Tunneling field-effect transistors (TFETs) with subthreshold swing (SS) less than 60 mV/dec," *IEEE Electron Device Letters*, vol. 28, pp. 743–745, Aug. 2007.

[31] Y.-C. Yeo, G. Han, Y. Yang, and P. Guo, "Strain engineering and junction design for tunnel field-effect transistor," *ECS Transactions*, vol. 33, no. 6, pp. 77–87, 2010.

[32] C. Hu, P. Patel, A. Bowonder, K. Jeon, S. Kim, W. Loh, C. Kang, J. Oh, P. Majhi, A. Javey, T.-J. Liu, and R. Jammy, "Prospect of tunneling green transistor for 0.1 V CMOS," in *IEEE International Electron Devices Meeting (IEDM)*, pp. 16.1.1–16.1.4, IEEE, Dec. 2010.

[33] W.-Y. Loh, K. Jeon, C. Y. Kang, J. Oh, P. Patel, C. Smith, J. Barnett, C. Park, T.-J. K. Liu, H.-H. Tseng, P. Majhi, R. Jammy, and C. Hu, "Sub-60nm Si tunnel field effect transistors with $Ion > 100\ \mu A/\mu m$," in *Proc. of the European Solid-State Device Research Conference*, pp. 162–165, IEEE, 2010.

[34] J. Nah, E.-S. Liu, K. Varahramyan, D. Dillen, S. MacCoy, J. Chan, and E. Tutuc, "Enhanced-performance germanium nanowire tunneling field-effect transistors using flash-assisted rapid thermal process," *IEEE Electron Device Letters*, vol. 31, pp. 1359–1361, Dec. 2010.

[35] R. Jhaveri, V. Nagavarapu, and J. Woo, "Effect of pocket doping and annealing schemes on the source-pocket tunnel field-effect transistor," *IEEE Trans. on Electron Devices*, vol. 58, pp. 80–86, Jan. 2011.

[36] J. T. Smith, C. Sandow, S. Das, R. A. Minamisawa, S. Mantl, and J. Appenzeller, "Silicon nanowire tunneling field-effect transistor arrays: Improving subthreshold performance using excimer laser annealing," *IEEE Trans. on Electron Devices*, vol. 58, pp. 1822–1829, July 2011.

[37] W. G. Vandenberghe, A. S. Verhulst, K.-H. Kao, K. D. Meyer, B. Sore,

and W. M. G. Groeseneken, "A model determining optimal doping concentration and material's band gap of tunnel field-effect transistors," *Applied Physics Letters*, vol. 100, no. 193509, 2012.

[38] B. Ghosh and M. W. Akram, "Junctionless tunnel field effect transistor," *IEEE Electron Device Letters*, vol. 34, pp. 584–586, May 2013.

[39] D. Verreck, A. S. Verhulst, K.-H. Kao, W. G. Vandenberghe, K. D. Meyer, and G. Groeseneken, "Quantum mechanical performance predictions of p-n-i-n versus pocketed line tunnel field-effect transistors," *IEEE Trans. on Electron Devices*, vol. 60, pp. 2128–2134, July 2013.

[40] M. J. Kumar and S. Janardhanan, "Doping-less tunnel field effect transistor: Design and investigation," *IEEE Trans. on Electron Devices*, vol. 60, pp. 3285–3290, Oct. 2013.

[41] Y.-R. Jhan, Y.-C. Wu, Y.-L. Wang, Y.-J. Lee, M.-F. Hung, H.-Y. Lin, Y.-H. Chen, and M.-S. Yeh, "Low-temperature microwave annealing for tunnel field-effect transistor," *IEEE Electron Device Letters*, vol. 36, pp. 105–107, Feb. 2015.

[42] M. S. Ram and D. B. Abdi, "Single grain boundary dopingless pnpn tunnel FET on recrystallized polysilicon: proposal and theoretical analysis," *Electron Devices Society, IEEE Journal of the*, vol. 3, no. 3, pp. 291–296, 2015.

[43] M. S. Ram and D. B. Abdi, "Dopingless PNPN tunnel FET with improved performance: Design and analysis," *Superlattices and Microstructures*, vol. 82, pp. 430–437, 2015.

[44] V. Nagavarapu, R. Jhaveri, and J. Woo, "The tunnel source (PNPN) n-MOSFET: A novel high performance transistor," *IEEE Trans. on Electron Devices*, vol. 55, pp. 1013–1019, Apr. 2008.

[45] D. B. Abdi and M. J. Kumar, "In-built N^+ pocket p-n-p-n tunnel field-effect transistor," *IEEE Electron Device Letters*, vol. 35, pp. 1170–1172, Dec. 2014.

[46] S. Sahay and M. J. Kumar, "Controlling the drain side tunneling width to reduce ambipolar current in tunnel FETs using heterodielectric BOX," *IEEE Transactions on Electron Devices*, vol. 62, pp. 3882–3886, Nov. 2015.

[47] D. B. Abdi and M. J. Kumar, "PNPN tunnel FET with controllable drain side tunnel barrier width: Proposal and analysis," *Superlattices and Microstructures*, vol. 86, pp. 121–125, 2015.

[48] M. Ram and D. Abdi, "Single grain boundary tunnel field effect transistors on recrystallized polycrystalline silicon: Proposal and investigation," *Electron Device Letters, IEEE*, vol. 35, no. 10, pp. 989–991, 2014.

[49] W. G. Vandenberghe, A. S. Verhulst, G. Groeseneken, B. Soree, and W. Magnus, "Analytical model for a tunnel field-effect transistor," in *Proc. 14th IEEE MELECON*, pp. 923–928, IEEE, 2008.

[50] C. Hu, D. Chou, P. Patel, and A. Bowonder, "Green transistor - a VDD scaling path for future low power ICs," in *International Symposium on VLSI Technology*, pp. 14–15, IEEE, Apr. 2008.

[51] S. Agarwal, G. Klimeck, and M. Luisier, "Leakage-reduction design concepts for low-power vertical tunneling field-effect transistors," *IEEE Electron Device Letters*, vol. 27, pp. 621–623, June 2010.

[52] P. Patel, K. Jeon, A. Bowonder, and C. Hu, "A low voltage steep turn-off tunnel transistor design," in *2009 International Conference on Simulation of Semiconductor Processes and Devices*, pp. 1–4, IEEE, 2009.

[53] R. Li, Y. Lu, S. D. Chae, G. Zhou, Q. Liu, C. Chen, M. Shahriar Rahman, T. Vasen, Q. Zhang, P. Fay, T. Kosel, M. Wistey, H. G. Xing, S. Koswatta, and A. Seabaugh, "InAs/AlGaSb heterojunction tunnel field-effect transistor with tunnelling in-line with the gate field," *Physica Status Solidi (c)*, vol. 9, no. 2, pp. 389–392, 2012.

[54] K. Ganapathi and S. Salahuddin, "Heterojunction vertical band-to-band tunneling transistors for steep subthreshold swing and high on current," *Electron Device Letters*, vol. 32, pp. 689–691, May 2011.

[55] G. Zhou, Y. Lu, R. Li, Q. Zhang, W. S. Hwang, Q. Liu, T. Vasen, C. Chen, H. Zhu, J.-M. Kuo, S. Koswatta, T. Kosel, M. Wistey, P. Fay, A. Seabaugh, and H. Xing, "Vertical InGaAs/InP tunnel FETs with tunneling normal to the gate," *IEEE Electron Device Letters*, vol. 32, pp. 1516–1518, Nov. 2011.

[56] R. Li, Y. Lu, G. Zhou, Q. Liu, S. D. Chae, T. Vasen, W. S. Hwang, Q. Zhang, P. Fay, T. Kosel, M. Wistey, H. Xing, and A. Seabaugh, "AlGaSb/InAs tunnel field-effect transistor with on-current of 78 $\mu A/\mu m$ at $0.5V$," *IEEE Electron Device Letters*, vol. 33, pp. 363–365, Mar. 2012.

[57] Y. Lu, G. Zhou, R. Li, Q. Liu, Q. Zhang, T. Vasen, S. D. Chae, T. Kosel, M. Wistey, H. Xing, A. Seabaugh, and P. Fay, "Performance of AlGaSb/InAs TFETs with gate electric field and tunneling direction aligned," *IEEE Electron Device Letters*, vol. 33, pp. 655–657, May 2012.

[58] I. Fischer, D. Hahnel, H. Isemann, A. Kottantharayil, G. Murali, M. Oehme, and J. Schulze, "Si tunneling field effect transistor with tunnelling in-line with the gate field," in *International Silicon-Germanium Technology and Device Meeting*, pp. 1–2, IEEE, 2012.

[59] D. Verreck, A. S. Verhulst, K.-H. Kao, W. G. Vandenberghe, K. D.

Meyer, and G. Groeseneken, "Quantum mechanical performance predictions of p-n-i-n versus pocketed line tunnel field-effect transistors," *IEEE Trans. on Electron Devices*, vol. 60, pp. 2128–2134, July 2013.

[60] A. M. Walke, A. Vandooren, R. Rooyackers, D. Leonelli, A. Hikavyy, R. Loo, A. S. Verhulst, K.-H. Kao, C. Huyghebaert, G. Groeseneken, V. R. Rao, K. K. Bhuwalka, M. M. Heyns, N. Collaert, and A.-Y. Thean, "Fabrication and analysis of a $Si/Si_{0.55}Ge_{0.45}$ heterojunction line tunnel fet," *IEEE Trans. on Electron Devices*, vol. 61, pp. 707–715, Mar. 2014.

[61] W. Hsu, J. Mantey, L. Register, and S. Banerjee, "On the electrostatic control of gate-normal-tunneling field-effect transistor," *IEEE Trans. on Electron Devices*, vol. 62, pp. 2292–2299, July 2015.

[62] S. W. Kim, W. Y. Choi, M.-C. Sun, H. W. Kim, and B.-G. Park, "Design guideline of Si-based L-shaped tunneling field-effect transistors," *Japanese Journal of Applied Physics*, vol. 51, 2012.

[63] D. Leonelli, A. Vandooren, R. Rooyackers, A. Verhulst, C. Huyghebaert, S. D. Gendt, M. M. Heyns, and G. Groeseneken, "Novel architecture to boost the vertical tunneling in tunnel field effect transistors," in *IEEE International SOI Conference*, pp. 1–2, IEEE, 2011.

[64] K. Jeon, W. Y. Loh, P. Patel, C. Y. Kang, J. Oh, A. Bowonder, C. Park, C. S. Park, C. Smith, P. Majhi, H. H. Tseng, R. Jammy, T. J. K. Liu, and C. Hu, "Si tunnel transistors with a novel silicided source and 46 mV/dec swing," in *Symposium on VLSI Technology (VLSIT)*, pp. 121–122, IEEE, June 2010.

[65] Y. Yang, P. Guo, G. Han, K. L. Low, and C. Zhan, "Simulation of tunneling field-effect transistors with extended source structures," *Journal of Applied Physics*, vol. 111, no. 114514, 2012.

[66] W. Wang, P.-F. Wang, C.-M. Zhang, X. Lin, X.-Y. Liu, Q.-Q. Sun, P. Zhou, and D. W. Zhang, "Design of U-shape channel tunnel FETs with SiGe source regions," *IEEE Trans. on Electron Devices*, vol. 61, pp. 193–197, Jan. 2014.

[67] S. J. Koester, I. Lauer, A. Majumdar, J. Cai, J. Sleight, S. Bedell, P. Solomon, S. Laux, L. Chang, S. Koswatta, W. Haensch, P. Tomasini, and S. Thomas, "Are Si/SiGe tunneling field-effect transistors a good idea?," *ECS Transactions*, vol. 33, no. 6, pp. 357–361, 2010.

[68] K. K. Bhuwalka, J. Schulze, and I. Eisele, "Vertical tunnel field-effect transistor with bandgap modulation and workfunction engineering," in *Proc. of 34th European Solid-State Device Research Conference*, vol. 41, pp. 241–244, IEEE, 2004.

[69] E.-H. Toh, G. H. Wang, L. Chan, G. Samudra, and Y.-C. Yeo, "Device physics and guiding principles for the design of double-gate tunneling field effect transistor with silicon-germanium source heterojunction," *Applied Physics Letters*, vol. 91, no. 243505, 2007.

[70] T. Krishnamohan, D. Kim, S. Raghunathan, and K. Saraswat, "Double-gate strained-Ge heterostructure tunneling FET (TFET) with record high drive currents and $< 60\ mV/dec$ subthreshold slope," in *IEEE International Electron Devices Meeting (IEDM)*, pp. 1–3, IEEE, Dec. 2008.

[71] M. Luisier and G. Klimeck, "Performance comparisons of tunneling field-effect transistors made of InSb, carbon, and GaSb-InAs broken gap heterostructures," in *IEEE International Electron Devices Meeting (IEDM)*, pp. 1–4, IEEE, 2009.

[72] S. O. Koswatta, S. J. Koester, and W. Haensch, "On the possibility of obtaining MOSFET-like performance and sub-60-mV/dec swing in 1-D broken-gap tunnel transistors," *IEEE Trans. on Electron Devices*, vol. 57, pp. 3222–3230, Dec. 2010.

[73] K. Saraswat, C. O. Chui, T. Krishnamohan, D. Kim, A. Nayfeh, and A. Pethe, "High performance germanium MOSFETs," *Materials Science and Engineering*, vol. B 135, pp. 242–249, 2006.

[74] P. K. Asthana, B. Ghosh, Y. Goswami, and B. M. Tripathi, "High-speed and low-power ultradeep-submicrometer III-V heterojunctionless tunnel field-effect transistor," *IEEE Trans. on Electron Devices*, vol. 61, pp. 479–486, Feb. 2014.

[75] M. Schmidt, A. Schafer, R. Minamisawa, D. Buca, S. Trellenkamp, J.-M. Hartmann, Q.-T. Zhao, and S. Mantl, "Line and point tunneling in scaled Si/SiGe heterostructure TFETs," *IEEE Electron Device Letters*, vol. 35, pp. 699–701, July 2014.

[76] P.-Y. Wang and B.-Y. Tsui, "Band engineering to improve average subthreshold swing by suppressing low electric field band-to-band tunneling with epitaxial tunnel layer tunnel FET structure," *IEEE Transactions on Nanotechnology*, vol. 15, pp. 74–79, Jan. 2016.

[77] H. Wang, G. Han, Y. Liu, S. Hu, C. Zhang, J. Zhang, and Y. Hao, "Theoretical investigation of performance enhancement in GeSn/SiGeSn Type-II staggered heterojunction tunneling FET," *IEEE Transactions on Electron Devices*, vol. 63, pp. 303–310, Jan. 2016.

[78] G. Singh, S. I. Amin, S. Anand, and R. Sarin, "Design of $Si_{0.5}Ge_{0.5}$ based tunnel field effect transistor and its performance evaluation," *Superlattices and Microstructures*, vol. 92, pp. 143–156, 2016.

[79] V. Brouzet, B. Salem, P. Periwal, R. Alcotte, F. Chouchane, F. Bassani, T. Baron, and G. Ghibaudo, "Fabrication and electrical characterization of homo- and hetero-structure Si/SiGe nanowire tunnel field effect transistor grown by vapor-liquid-solid mechanism," *Solid-State Electronics*, vol. 118, pp. 26 - 29, 2016.

[80] P. Y. Wang and B. Y. Tsui, "Band engineering to improve average subthreshold swing by suppressing low electric field band-to-band tunneling with epitaxial tunnel layer tunnel FET structure," *IEEE Transactions on Nanotechnology*, vol. 15, pp. 74–79, Jan. 2016.

[81] P. G. D. Agopian, M. D. V. Martino, S. D. dos Santos, F. S. Neves, J. A. Martino, R. Rooyackers, A. Vandooren, E. Simoen, A.-Y. Thean, and C. Claeys, "Influence of the source composition on the analog performance parameters of vertical nanowire-TFETs," *IEEE Trans. on Electron Devices*, vol. 61, pp. 16–22, Jan. 2015.

[82] A. S. Verhulst, W. G. Vandenberghe, K. Maex, and G. Groeseneken, "Tunnel field-effect transistor without gate-drain overlap," *Applied Physics Letters*, vol. 91, no. 053102, 2007.

[83] L. Wang, E. Yu, Y. Taur, and P. Asbeck, "Design of tunneling field-effect transistors based on staggered heterojunctions for ultralow-power applications," *IEEE Electron Device Letters*, vol. 31, pp. 431–433, May 2010.

[84] G. Dewey, B. Chu-Kung, J. Boardman, J. M. Fastenau, J. Kavalieros, R. Kotlyar, W. K. Liu, D. Lubyshev, M. Metz, N. Mukherjee, P. Oakey, R. Pillarisetty, M. Radosavljevic, H. Then, and R. Chau, "Fabrication, characterization, and physics of III-V heterojunction tunneling field effect transistors (H-TFET) for steep sub-threshold swing," in *IEEE International Electron Devices Meeting (IEDM)*, pp. 33.6.1–33.6.4, IEEE, 2011.

[85] H. Zhao, Y. Chen, Y. Wang, F. Zhou, F. Xue, and J. Lee, "InGaAs tunneling field-effect-transistors with atomic-layer deposited gate oxides," *IEEE Trans. on Electron Devices*, vol. 58, pp. 2990–2995, Sept. 2011.

[86] D. Mohata, S. Mookerjea, A. Agrawal, Y. Li, T. Mayer, V. Narayanan, A. Liu, D. Loubychev, J. Fastenau, and S. Datta, "Experimental staggered-source and N^+ pocket-doped channel III-V tunnel field-effect transistors and their scalabilities," *Applied Phys. Express*, vol. 4, pp. 24105(1)–024105(3), 2011.

[87] L. F. Register, M. M. Hasan, and S. K. Banerjee, "Stepped broken-gap heterobarrier tunneling field-effect transistor for ultralow power and high speed," *IEEE Electron Device Letters*, vol. 32, pp. 743–745, June 2011.

[88] A. Seabaugh, S. D. Chae, P. Fay, W. S. Hwang, T. Kosel, R. Li, Q. Liu, Y. Lu, T. Vasen, M. Wistey, G. Xing, G. Zhou, and Q. Zhang, "III-V tunnel field-effect transistors," *ECS Transactions*, vol. 41, no. 7, pp. 227–229, 2011.

[89] J. Knoch and J. Appenzeller, "A novel concept for field-effect transistors: The tunneling carbon nanotube FET," pp. 153–156, IEEE, June 2005.

[90] R. Sako, H. Hosokawa, and H. Tsuchiya, "Computational study of edge configuration and quantum confinement effects on graphene nanoribbon transport," *IEEE Electron Device Letters*, vol. 32, pp. 6–8, Jan. 2011.

[91] P. Zhao, J. Chauhan, and J. Guo, "Computational study of tunneling transistor based on graphene nanoribbon," *Nano Letters*, pp. 684–688, 2009.

[92] Q. Zhang, T. Fang, H. Xing, A. Seabaugh, and D. Jena, "Graphene nanoribbon tunnel transistors," *IEEE Electron Device Letters*, vol. 29, pp. 1344–1346, Dec. 2008.

[93] K.-T. Lam, D. Seah, S.-K. Chin, S. B. Kumar, G. Samudra, Y.-C. Yeo, and G. Liang, "A simulation study of graphene-nanoribbon tunneling FET with heterojunction channel," *IEEE Electron Device Letters*, vol. 31, pp. 555–557, June 2010.

[94] H. Da, K.-T. Lam, G. Samudra, S.-K. Chin, and G. Liang, "Graphene nanoribbon tunneling field-effect transistors with a semiconducting and a semimetallic heterojunction channel," *IEEE Trans. on Electron Devices*, vol. 59, pp. 1454–1461, May 2012.

[95] Y. Yoon, S. H. Kim, and S. Salahuddin, "Performance analysis of carbon-based tunnel field-effect transistors for high frequency and ultralow power applications," *Applied Physics Letters*, vol. 97, no. 233504, 2010.

[96] Y. Yoon and S. Salahuddin, "Barrier-free tunneling in a carbon heterojunction transistor," *Applied Physics Letters*, vol. 97, no. 033102, 2010.

[97] L. Leem, A. Srivastava, S. Li, B. Magyari-Kope, G. Iannaccone, J. S. Harris, and G. Fiori, "Multi-scale simulations of partially unzipped CNT hetero-junction tunneling field effect transistor," in *IEEE International Electron Devices Meeting*, pp. 32.5.1–32.5.4, IEEE, 2010.

[98] W. Lu, P. Xie, and C. M. Lieber, "Nanowire transistor performance limits and applications," *IEEE Trans. on Electron Devices*, vol. 55, pp. 2859–2876, Nov. 2008.

[99] K. K. Young, "Short-channel effect in fully depleted SOI MOSFETs," *IEEE Trans. on Electron Devices*, vol. 36, pp. 399–402, Feb. 1989.

[100] R.-H. Yan, A. Ourmazd, and K. F. Lee, "Scaling the Si MOSFET: From bulk to SOI to bulk," *IEEE Trans. on Electron Devices*, vol. 39, pp. 1704–1710, July 1992.

[101] S. Chuang, Q. Gao, R. Kapadia, A. C. Ford, J. Guo, and A. Javey, "Ballistic InAs nanowire transistors," *NANO Letters*, pp. 555–558, 2013.

[102] M. Luisier and G. Klimeck, "Atomistic full-band design study of InAs band-to-band tunneling field-effect transistors," *IEEE Electron Device Letters*, vol. 30, pp. 602–604, June 2009.

[103] J. Knoch, S. Mantl, and J. Appenzeller, "Impact of the dimensionality on the performance of tunneling FETs: Bulk versus one-dimensional devices," *Solid-State Electronics*, vol. 51, pp. 572–578, 2007.

[104] T. Numata, T. Mizuno, T. Tezuka, J. Koga, and S. Takagi, "Control of threshold-voltage and short-channel effects in ultrathin strained-SOI CMOS devices," *IEEE Trans. on Electron Devices*, vol. 52, pp. 1780–1786, Aug. 2005.

[105] J. S. Lim, S. E. Thompson, and J. G. Fossum, "Comparison of threshold voltage shifts for uniaxial and biaxial tensile-stressed n-MOSFETs," *IEEE Electron Device Letters*, vol. 25, pp. 731–733, Nov. 2004.

[106] W. Zhang and J. G. Fossum, "On the threshold voltage of strained-Si-$Si_{1-x}Ge_x$ MOSFETs," *IEEE Trans. on Electron Devices*, vol. 52, pp. 263–268, Feb. 2005.

[107] S. Dhar, H. Kosina, V. Palankovski, S. E. Ungersboeck, and S. Selberherr, "Electron mobility model for strained-Si devices," *IEEE Trans. on Electron Devices*, vol. 52, pp. 527–533, Apr. 2005.

[108] S. Saurabh and M. J. Kumar, "Impact of strain on drain current and threshold voltage of nanoscale double gate tunnel field effect transistor: Theoretical investigation and analysis," *Japanese Journal of Applied Physics*, vol. 48, June 2009.

[109] P. F. Guo, L. T. Yang, Y. Yang, L. Fan, G. Q. Han, G. S. Samudra, and Y. C. Yeo, "Tunneling field-effect transistor: Effect of strain and temperature on tunneling current," *IEEE Electron Device Letters*, vol. 30, pp. 81–983, Sept. 2009.

[110] H. Yin, K. Hobart, R. L. Peterson, F.J.Kub, S. R. Shieh, T. S. Duffy, and J. C. Sturm, "Fully-depleted strained-Si on insulator NMOSFETs without relaxed SiGe buffers," in *IEDM Tech. Digest*, pp. 3.2.1–3.2.4, IEEE, 2003.

[111] K. Rim, K. Chan, L. Shi, D. Boyd, J. Ott, N. Klymko, F. Cardone, L. Tai, S. Koester, M. Cobb, D. Canaperi, B. To, E. Duch, I. Babich,

R. Carruthers, P. Saunders, G. Walker, Y. Zhang, M. Steen, and M. Ieong, "Fabrication and mobility characteristics of ultra-thin strained Si directly on insulator (SSDOI) MOSFETs," in *IEDM Tech. Digest*, pp. 3.1.1–3.1.4, IEEE, 2003.

[112] S.Takagi, T. Mizuno, T. Tezuka, N. Sugiyama, T. Numata, K. Usuda, Y. Moriyama, S. Nakaharai, J. Koga, A. Tanabe, and T. Maeda, "Fabrication and device characteristics of strained-Si-on-insulator (strained SOI) CMOS," *Applied Surface Science*, vol. 224, pp. 241–247, Mar. 2004.

[113] M. Najmzadeh, K. Boucart, W. Riess, and A. Ionescu, "Asymmetrically strained all-silicon multi-gate n-Tunnel FETs," *Solid-State Electronics*, vol. 54, pp. 35–941, 2010.

[114] M. H. Lee, S. T. Chang, T.-H.Wu, and W.-N. Tseng, "Driving current enhancement of strained Ge (110) p-type tunnel FETs and anisotropic effect," *IEEE Electron Device Letters*, vol. 32, pp. 1355–1357, Oct. 2011.

[115] Q. T. Zhao, W. J. Yu, B. Zhang, M. Schmidt, S. Richter, D. Buca, J.-M. Hartmann, R. Luptak, A. Fox, K. K. Bourdelle, and S. Mantl, "Tunneling field-effect transistor with a strained Si channel and a $Si_{0.5}Ge_{0.5}$ source," in *Proc. of European Solid-State Device Research Conference*, pp. 251–254, IEEE, 2011.

[116] Y. Lu and J. Guo, "Local strain in tunneling transistors based on graphene nanoribbons," *Applied Physics Letters*, vol. 97, no. 073105, 2010.

[117] J. Kang, Y. He, J. Zhang, X. Yu, X. Guan, and Z. Yu, "Modeling and simulation of uniaxial strain effects in armchair graphene nanoribbon tunneling field effect transistors," *Applied Physics Letters*, vol. 96, no. 252105, 2010.

[118] H. Lu and A. Seabaugh, "Tunnel field-effect transistors: State-of-the-art," *Journal of the Electron Devices Society*, pp. 44–49, May 2014.

[119] R. Ghosh and M. B. S. Mahapatra, "Germanane: A low effective mass and high bandgap 2-D channel material for future FETs," *IEEE Trans. on Electron Devices*, vol. 61, pp. 2309–2315, July 2014.

[120] G. Fiori, F. Bonaccorso, G. Iannaccone, T. Palacios, D. Neumaier, A. Seabaugh, S. Banerjee, and L. Colombo, "Electronics based on two-dimensional materials," *Nature Nanotechnology*, pp. 768–779, Oct. 2014.

[121] K. Alam, S. Takagi, and M. Takenaka, "A Ge ultrathin-body n-channel tunnel FET: Effects of surface orientation," *IEEE Trans. on Electron Devices*, vol. 61, pp. 3594–3600, Nov. 2014.

[122] H. Ilatikhameneh, Y. Tan, B. Novakovic, G. Klimeck, R. Rahman, and J. Appenzeller, "Tunnel field-effect transistors in 2-D transition metal dichalcogenide materials," *IEEE Journal on Exploratory Solid-State Computational Devices and Circuits*, vol. 1, pp. 12–15, Apr. 2015.

[123] W. Li, S. Sharmin, H. Ilatikhameneh, R. Rahman, Y. Lu, J. Wang, X. Yan, A. Seabaugh, G. Klimeck, and D. Jena, "Polarization-engineered III-Nitride heterojunction tunnel field-effect transistors," *IEEE Journal on Exploratory Solid-State Computational Devices and Circuits*, pp. 28–34, Apr. 2015.

[124] M. Li, D. Esseni, J. J. Nahas, D. Jena, and H. Xing, "Two-dimensional heterojunction interlayer tunneling field effect transistors (thin-TFETs)," *IEEE Journal of the Electron Devices Society*, vol. 3, May 2015.

[125] A. Szabo, S. J. Koester, and M. Luisier, "Ab-initio simulation of van der waals $MoTe_2-SnS_2$ heterotunneling fets for low-power electronics," *IEEE Electron Device Letters*, vol. 36, pp. 514–516, May 2015.

[126] S.-C. Lu, M. Mohamed, and W. Zhu, "Novel vertical hetero- and homo-junction tunnel field-effect transistors based on multi-layer 2D crystals," *2D Materials*, vol. 3, no. 1, p. 011010, 2016.

[127] T. Roy, M. Tosun, M. Hettick, G. H. Ahn, C. Hu, and A. Javey, "2D-2D tunneling field-effect transistors using $WSe_2/SnSe_2$ heterostructures," *Applied Physics Letters*, vol. 108, no. 8, 2016.

[128] D. Verreck, A. S. Verhulst, M. L. V. de Put, B. Sore, N. Collaert, A. Mocuta, A. Thean, and G. Groeseneken, "Uniform strain in heterostructure tunnel field-effect transistors," *IEEE Electron Device Letters*, vol. 37, pp. 337–340, Mar. 2016.

5

III-V Tunnel Field-Effect Transistor

5.1 Introduction

III-V semiconductors have attracted the attention of device engineers for application in MOSFETs since 1960s. However, until recently, the attempts to employ III-V semiconductors in MOSFET technology have been largely unsuccessful, primarily because of the difficulty in obtaining a high-quality III-V semiconductor–dielectric interface. During the last decade, researchers have been successful in obtaining reasonably good-quality III-V semiconductor–dielectric interfaces using novel processing techniques and these successes have reignited the interest of device researchers in III-V MOSFETs.

Since TFETs are structurally similar to MOSFETs, TFETs employing III-V semiconductors derive direct benefit from the advancements in III-V MOSFET technology. Additionally, some of the material properties of the III-V semiconductors can be exploited in TFETs to obtain better characteristics compared to TFETs based on group IV semiconductors like silicon and germanium. Therefore, several research groups are actively working on III-V semiconductor-based TFETs and have already demonstrated that TFETs realized using III-V semiconductors could potentially outperform conventional MOSFETs in low-voltage CMOS applications [1–15].

In this chapter, we describe TFETs that are composed of III-V semiconductors. We describe the advantages of employing III-V semiconductors in TFETs and illustrate some of the novel techniques that can improve the performance of III-V TFETs. We also highlight distinct challenges involved in employing III-V semiconductors in TFETs and describe some of the solutions that can address these challenges.

At the end of this chapter, readers will be able to appreciate the distinctive characteristics of III-V TFETs and their advantages and their disadvantages compared to silicon-based TFETs.

5.2 III-V Semiconductors

The binary III-V compound semiconductors are of the form $A^{III}Z^{V}$ where A^{III} is an element of column III of the periodic table and Z^{V} is an element of column V of the periodic table. For electronics applications, the commonly employed III-V compound semiconductors are the arsenide, phosphide and antimonide of aluminum, gallium and indium. These compound semiconductors generally form a zinc blend type of crystal structure. The atoms in III-V compound semiconductors have covalent bonds similar to the group IV elements like silicon. However, since group III elements are more electropositive and group V elements are more electronegative compared to group IV elements, III-V compound semiconductors have greater ionic characteristics compared to group IV elements.

Several binary III-V compounds have identical structures, similar lattice constants, and high solid solubility in each other. These binary III-V compounds can be combined to obtain ternary and quaternary compounds. The binary compounds $A^{III}Z^{V}$ and $B^{III}Z^{V}$ can be combined in the ratio $x : (1 - x)$ to obtain a ternary compound $A^{III}_{x}B^{III}_{(1\text{-}x)}Z^{V}$. The properties of $A^{III}_{x}B^{III}_{(1\text{-}x)}Z^{V}$ is intermediate between the properties of $A^{III}Z^{V}$ and $B^{III}Z^{V}$ and can be varied by changing the mole ratio x from 0 to 1.

A comparison of the properties of some of the III-V semiconductors with the corresponding properties of silicon is shown in Table 5.1 [16]. The following inferences can be made from Table 5.1:

1. The mobility of carriers in III-V semiconductors is appreciably higher compared to the mobility of carriers in silicon. For example, the mobility of electrons in InAs is more than an order of magnitude higher than the mobility of electrons in silicon. The outstanding mobility of the III-V semiconductors can be utilized in transistors to obtain a high current and an excellent frequency response. For example, in InGaAs-based high-electron-mobility transistors (HEMTs), power-gain cutoff frequencies exceeding 1 THz can be achieved [17]. The excellent transport properties of III-V semiconductors have motivated device engineers to explore the possibility of employing these materials in mainstream CMOS technology since 1960s.

2. The bandgap of III-V semiconductors is generally smaller than the bandgap of silicon. For example, the bandgap of InAs is 0.35 eV while the bandgap of silicon is 1.12 eV. The smaller bandgap of III-V semiconductors can be utilized in a TFET to obtain a high ON-state current (I_{ON}).

TABLE 5.1
Properties of III-V semiconductors compared with silicon [16].

	Silicon	**GaAs**	$\mathbf{Ga_{0.47}In_{0.53}As}$	**InAs**	**InP**
Structure	Diamond	Zinc Blende	Zinc Blende	Zinc Blende	Zinc Blende
Lattice constant ($\AA$)	5.43	5.65	5.87	6.06	5.87
Electron mobility (cm^2/Vs)	≤ 1400	≤ 8500	< 12000	≤ 40000	≤ 5400
Hole mobility (cm^2/Vs)	≤ 450	≤ 400	< 300	≤ 500	≤ 200
Band gap (eV)	1.12	1.42	0.74	0.35	1.34
Nature of bandgap	Indirect	Direct	Direct	Direct	Direct
Donors	As, P, Sb	S, Se, Si, Ge, Sn, Te	Sn, Ge, Si, C	Se, S, Te, Ge, Si, Sn, Cu	S, Si, Sn, Ge
Acceptors	Al, B, Ga, In	C, Si, Ge, Sn	Mg, Zn, Cd, Mn, Fe	Sn, Ge, Si, Cd, Zn	C, Hg, Zn, Cd, Si, Cu, Be, Mg, Ge, Mn

3. The bandgap nature of III-V materials is direct, while bandgap nature of silicon is indirect.

   In direct bandgap material the momentum corresponding to the conduction band minima and the valence band maxima are the same as shown in Figure 5.1(a). Therefore, an electron can jump from the conduction band to the valence band and vice versa by directly emitting or absorbing the photon. However, in an indirect bandgap material, like silicon and germanium, the momentum corresponding to the conduction band minima and the valence band maxima are different, as shown in Figure 5.1(b). As a result, making a transition from one band to another requires a change in the momentum and the electrons must also absorb some extra energy from the vibrations in the crystal lattice to attain the required change in momentum. This extra obstacle in making the transition

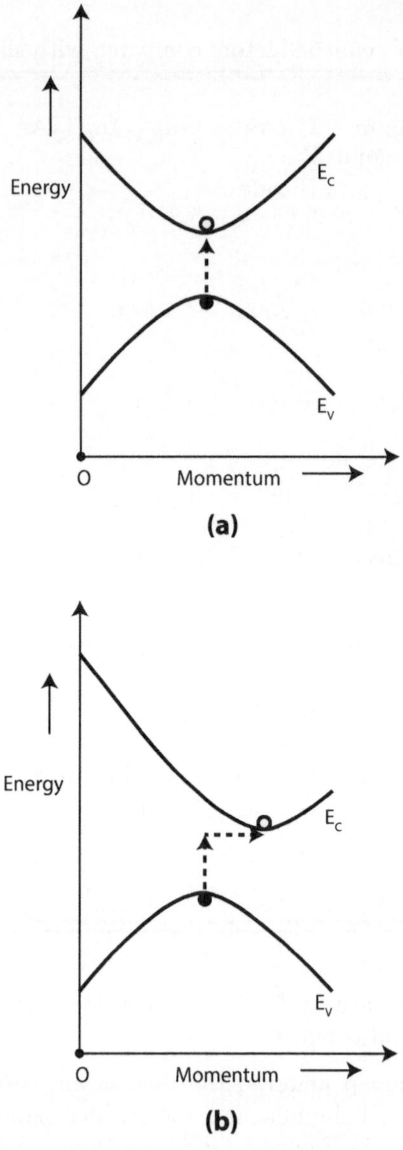

FIGURE 5.1
Energy vs. momentum of the carrier in (a) direct bandgap material and (b) indirect bandgap material.

significantly lowers the probability of the carriers for band-to-band tunneling (BTBT). As a result, III-V-based TFETs are expected to support a higher I_{ON} compared to the silicon- and germanium-based TFETs.

4. The bandgap of ternary III-V semiconductors can be engineered by varying the mole fraction of the constituents. For example, the bandgap of $Ga_{0.47}In_{0.53}As$ is intermediate between the bandgap of GaAs and InAs. In fact, the bandgap E_g of $Ga_xIn_{1-x}As$ can be smoothly varied by changing x as given by the following relationship [16]:

$$E_g = (0.36 + 0.63x + 0.43x^2)\ eV. \tag{5.1}$$

The ease of tuning the properties of the III-V semiconductors by changing the mole fraction of the constituents makes III-V semiconductors extremely versatile for application in TFETs, as we shall see in the following sections.

5.3 Challenges in III-V Materials–Based MOSFETs

The outstanding properties of III-V semiconductors have motivated device engineers to realize III-V MOSFETs and incorporate these materials into the standard CMOS technology. However, these efforts have been severely hampered due to the poor quality of the interface of the III-V semiconductors. In this section, we describe some of the challenges in realizing III-V MOSFETs and the recent advancements made in the processing technology that can make the fabrication of high-performance III-V MOSFETs possible and commercially exploitable. It is important to appreciate the challenges involved in realizing III-V MOSFETs, since similar challenges are faced in realizing III-V TFETs, as we shall see in the following sections.

The key component in a MOSFET is the capacitor composed of a metal gate and the semiconductor channel separated by a dielectric layer. For a MOSFET, to have good electrical characteristics and function reliably, the dielectric layer must contain minimum trapped charges and other defects, the dielectric–semiconductor interface must be smooth with minimum imperfections, and the dielectric should be thermodynamically stable. For silicon MOSFETs, SiO_2 meets these requirements of the dielectric and has been the key enabler of the large-scale application of silicon into the mainstream CMOS technology. However, no such native oxides meet the above requirements for III-V semiconductors. For example, the oxidation of GaAs results in the formation of various types of oxides and suboxides of Ga and As, as well as elemental arsenic and Ga dangling bonds [17]. The existence of these species at the interface of GaAs results in a high density of interface states in GaAs. Additionally, in silicon-based MOSFETs, the surface states of Si–SiO_2 can easily be passivated using hydrogen. However, such techniques do not work for the surface states of III-V semiconductors [18].

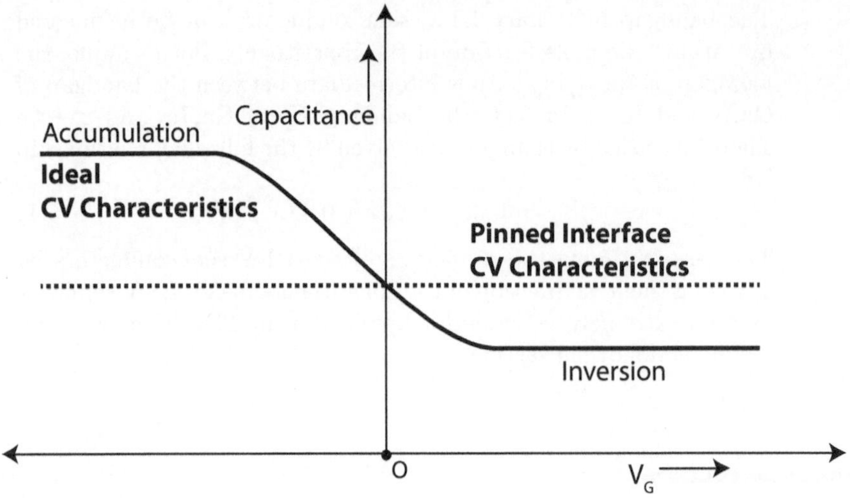

FIGURE 5.2
Comparison of ideal high-frequency CV characteristics of a MOSFET with the CV characteristics of the same MOSFET when there is a high interface state density D_{it}.

The consequences of high density of interface states at the III-V semiconductor surface is a phenomenon known as Fermi-level pinning [18]. When a metal and a semiconductor are in contact in an ideal interface, the charge distribution occurs such that the Fermi level is equalized in the two materials. As a result, the Schottky barrier in an ideal metal–semiconductor interface will be the difference of the work function of the metal and the electron affinity of the semiconductor. However, when a metal is in contact with a semiconductor that has a high density of interface states, the charge required to equalize the Fermi level is provided by the interface states rather than from the semiconductor, effectively shielding the semiconductor from the details of the metal. As a result, the bands inside the semiconductor align to the interface states, which are in turn pinned to the Fermi level due to their high density. Therefore, the work function of the metal would have little effect on the bands of the semiconductor with a high density of interface states. For example, the Schottky barrier height observed for almost all the metals and the air-exposed GaAs, which is rich in interface states, is $0.8 \pm 0.1\ eV$ [18].

The phenomenon of Fermi-level pinning in the III-V semiconductor interface poses a serious challenge in the realization of III-V MOSFETs. The surface potential in the MOSFET that is realized using a semiconductor with a high density of interface states cannot be modulated effectively due to Fermi-level pinning. Figure 5.2 shows an ideal high-frequency capacitance-voltage (CV) characteristic of a MOSFET. Figure 5.2 also shows the CV characteristics of the same MOSFET when there is a high interface state density D_{it}. The capacitance of the MOSFET that has a high interface state density cannot be modulated by the gate voltage, thus rendering the MOSFET useless for any practical purpose. In general, depending on the nature of the III-V semiconductor, the energy of the interface states can be at the band edges or within the bandgap of the material [18, 19]. In a MOSFET, the interface states below the edge of the conduction band result in a degradation in the subthreshold swing, whereas the interface states inside the conduction band trap electrons, leading to a reduced I_{ON}/I_{OFF} ratio. The interface states can also lead to other problems in a MOSFET like a shift in the threshold voltage and degraded mobility of carriers in the channel. As a result, attempts to fabricate MOSFETs using III-V semiconductors like GaAs, which has a high interface state density D_{it}, yielded devices with poor performance. Therefore, achieving a GaAs–oxide interface without Fermi-level pinning was the goal of researchers for several decades since 1960s. Early attempts to obtain a GaAs interface without Fermi-level pinning were directed at employing native oxides on GaAs using techniques like thermal oxidation, wet chemical anodization, laser-assisted oxidation, and photo-washing procedures [18, 20, 21]. Though these techniques greatly helped in understanding the physics and the chemistry of the III-V interfaces, these techniques did not lead to commercially feasible GaAs MOSFET technology. However, a major breakthrough was achieved in the last decade when a GaAs MOSFET was realized using Al_2O_3 deposited by atomic layer deposition (ALD) [21–23].

ALD is a technique which is widely used in silicon manufacturing at advanced processing nodes for depositing atomic-scale high-κ gate dielectrics. An ALD involves layer-by-layer adsorption and reaction of reactant species on the surface. The chemistry of ALD is similar to chemical vapor deposition (CVD). However, the ALD reaction decomposes the CVD reaction into two half-reactions. A cycle of ALD consists of exposure to the first precursor, followed by purging, and then exposure to the second precursor followed by purging. The growth of film by ALD is self-limited and based on surface reactions. Therefore, the amount of material deposited in each cycle is constant and atomic-scale control of the thickness of the film can be achieved. The precise control of the thickness of the deposited films, the conformal step coverage, and the possibility of low temperature deposition make ALD attractive for depositing high-κ gate dielectrics in CMOS technology.

The ALD technique is suitable for depositing dielectric in III-V MOSFETs too. The ALD technique enables obtaining a good-quality dielectric–III-V interface and unpinning of the Fermi levels on most III-V semiconductors [21]. It should be noted that the uniqueness of the ALD in III-V dielectric processing lies in a phenomenon called "self-clean effect." In the early stages of ALD over III-V materials, a kind of "self-clean effect" takes place in which surface oxides are greatly reduced and these oxides do not re-grow, even after exposure to the ALD oxidants [21, 23]. Therefore, even when the starting material is a

III-V semiconductor like GaAs with native oxides, a good-quality dielectric–III-V interface can be obtained by ALD [21, 23]. The good-quality interfaces obtained using ALD have made III-V MOSFETs commercially feasible and has largely been responsible for the renewed interest in III-V MOSFETs. Consequently, the ALD technique is widely used by researchers in experiments to realize III-V TFETs, as we shall see in the following sections.

5.4 A III-V TFET Prototype

In this section, we illustrate a simple implementation of a III-V TFET from [1]. Though the TFET in [1] exhibits a poor subthreshold swing and a small I_{ON}/I_{OFF} ratio, it is one of the earliest experimental demonstrations of a III-V TFET and highlights the challenges in realizing III-V TFETs.

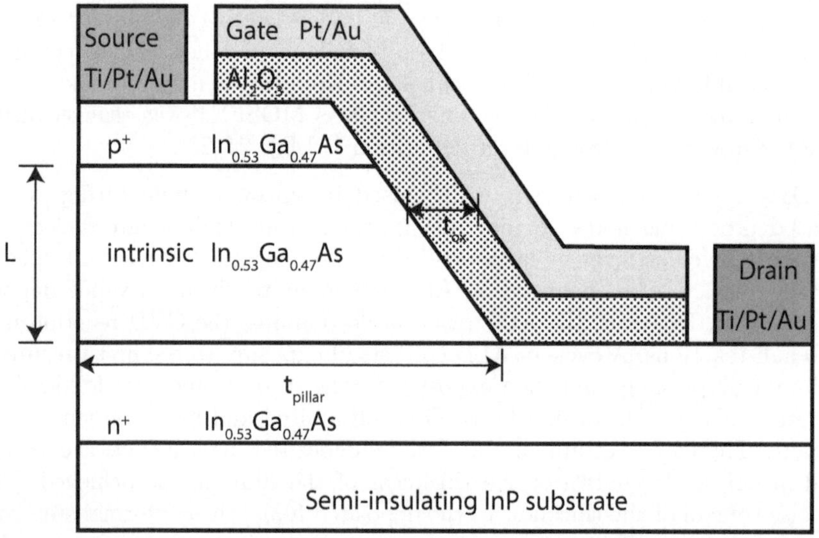

FIGURE 5.3
Schematic representation of an n-type $In_{0.53}Ga_{0.47}As$ TFET [1]. Device parameters are $L = 100\ nm$, $t_{pillar} = 20\mu m$, $t_{ox} = 10\ nm$.

Figure 5.3 shows the schematic representation of the III-V TFET [1]. The TFET consists of an n-channel $In_{0.53}Ga_{0.47}As$ on a semi-insulating InP substrate. The drain was n$^+$ doped with Si doping $N_D = 5\times10^{19}\ atoms/cm^3$ and the source was p$^+$ doped with C doping of $N_A = 1 \times 10^{20}\ atoms/cm^3$. The device was fabricated using molecular beam epitaxy (MBE) and the vertical device architecture along with *in situ* doped layers ensures that the tunnel junctions are highly abrupt. A mesa sidewall etch was performed using citric acid. On the mesa sidewall, the poor-quality native oxide was removed using NH_4OH and a $10 - nm$-thick conformal layer of Al_2O_3 was deposited using ALD at $300°C$.

The ON-current (I_{ON}) obtained in the TFET was 20 $\mu A/\mu m$ at $V_{DS} = 0.75\ V$ and 0.5 $\mu A/\mu m$ at $V_{DS} = 50\ mV$. When $V_{GS} < 1.2\ V$, the drain current primarily depends on BTBT. The smaller bandgap of $In_{0.53}Ga_{0.47}As$ compared to silicon results in a relatively higher I_{ON} compared to Si-based TFETs at similar voltages. When $V_{GS} > 1.2\ V$, the lateral electric field in the TFET is small, leading to the diffusion of carriers becoming the dominant mechanism of the current transport. Therefore, at high gate voltages, the mobility of the carriers also plays an important role in determining the drain current. The OFF-state current (I_{OFF}) realized in the TFET was 6 $nA/\mu m$ at $V_{DS} = 0.75\ V$ and 40 $pA/\mu m$ at $V_{DS} = 50\ mV$. The Shockley–Read–Hall (SRH) generation-recombination current of the reverse-biased p^+-i-n^+ diode dominates the I_{OFF} in the TFET. The SRH generation-recombination current is proportional to the intrinsic carrier concentration, which in turn is proportional to $\exp(-E_g/2kT)$ where E_g is the bandgap, T is the absolute temperature and k is the Boltzmann constant. The smaller bandgap $E_g \approx 0.72\ eV$ of $In_{0.53}Ga_{0.47}As$ results in a high I_{OFF} in the TFET.

The average subthreshold swing obtained in this TFET was 260 $mV/decade$ at $300K$. The current transport in the subthreshold region can be modeled using the Poole–Frenkel (PF) effect. The current transport model using the PF effect shows a good match with the experimental data [24]. The PF effect arises due to the traps which are charged when empty and are neutral when the traps are filled. These traps provide an intermediate state in the carrier transport. The PF effect manifests itself in the semiconductor or the insulator by the exponential increase in the conductivity of the material proportional to the square root of the electric field. The applied electric field assists the transition of the carriers from the trap states into the conduction band, thereby increasing the conductivity of the material. The experimental data for the TFET yields the trap to be located at 0.4 eV from the conduction band edge and within the bandgap of $In_{0.53}Ga_{0.47}As$ [24]. The tunneling of carriers from the valence band of the p^+ source region to the mid-gap trap levels followed by thermal emission of the carrier into the conduction band, explains the poor average subthreshold swing obtained in this TFET.

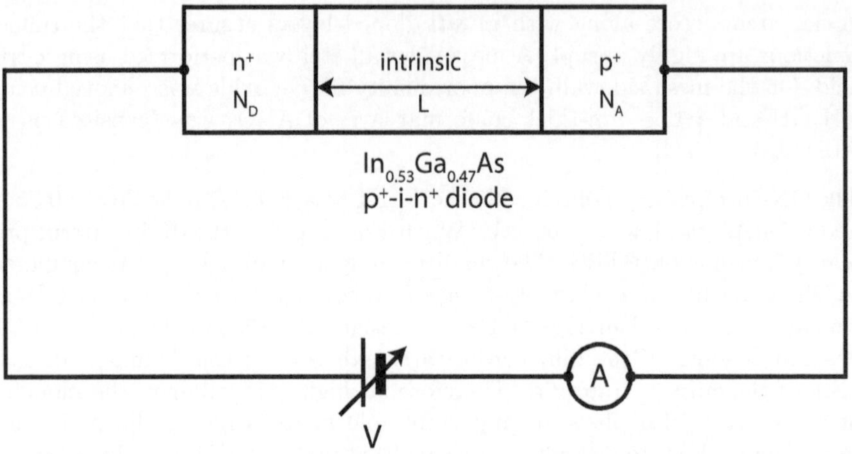

FIGURE 5.4
Schematic representation of the p^+–i–n^+ diode used in the study of the origin of the trap states. Device parameters are $N_D = 5 \times 10^{19}\ /cm^3$, $N_A = 1 \times 10^{20}\ /cm^3$ and $L = 100\ nm$ [1].

The origin of the trap states is the poor quality of the $Al_2O_3/In_{0.53}Ga_{0.47}As$ gate–channel interface. This was confirmed by characterizing a p^+-i-n^+ diode that is similar to the experimental TFET, but without the gate [1]. The schematic representation of the p^+-i-n^+ diode is shown in Figure 5.4. It was found that the reverse bias characteristics of the diode in Figure 5.4 can be well explained by SRH recombination/generation current at low bias voltages and by BTBT at higher bias voltages. Moreover, it was found that the contribution of the mid-gap trap in the reverse-biased characteristics of the diode was negligible. Therefore, the origin of the mid-gap traps in the TFET can be attributed to the $Al_2O_3/In_{0.53}Ga_{0.47}As$ gate–channel interface. Additionally, the interface state density extracted for an $In_{0.53}Ga_{0.47}As$ MOSFET similar to the experimental TFET exhibited a peak near the mid-gap level and this corroborates that the origin of the trap is the oxide–semiconductor interface.

The TFET shown in Figure 5.3 is composed of a homojunction and exhibits a poor I_{ON}/I_{OFF} ratio. In a homojunction TFET, it is difficult to simultaneously obtain a high I_{ON} and a low I_{OFF}. To obtain a high I_{ON}, the bandgap of the material needs to be decreased. However, the I_{OFF} would also increase with the decrease in the bandgap since the I_{OFF} is often determined by the SRH generation-recombination mechanism in the III-V TFETs. These conflicting requirements of the I_{ON} and the I_{OFF} for the bandgap can be addressed by employing a heterojunction in a III-V TFET as described in the following section.

5.5 Heterojunction III-V TFETs

III-V semiconductors are often employed in heterojunctions for the following reasons [25]:

1. Properties like the bandgap and the alignment of bands in the heterojunction can easily be tuned in III-V ternary semiconductors by changing the mole-fraction of the constituent elements, thus obtaining staggered and broken-gap heterojunction using the same material.
2. Lattice matching between materials realizing the heterojunction can be achieved with several III-V semiconductors. The lattice matching of the materials is required to avoid unnecessary stress and strain and the associated defects. However, in certain devices, the strain due to lattice mismatch can be exploited to improve the electrical characteristics of the device.

The heterojunction composed of InAs/$Al_xGa_{1\text{-}x}Sb$ has been explored by several researchers for application in TFETs [2, 8, 11, 12, 15]. One such p-type heterojunction TFET is shown in shown in Figure 5.5 [2, 15]. The channel is composed of an $Al_xGa_{1\text{-}x}Sb$ nanowire (NW). The nanowire is grown over a highly doped n-type InAs, which acts as the source. The area of the source contact is kept large to avoid vertical quantization in the source and ensure that the position of the heterointerface is fixed. The gate underlap is kept to ensure that there is no short circuit between the gate and the substrate. Since there is a small lattice mismatch between InAs/$Al_xGa_{1\text{-}x}Sb$, the heterojunction TFET can be realized by vertically growing a nanowire or by etching out a nanowire from a 2-D heterostructure [2]. The bandgap in $Al_xGa_{1\text{-}x}Sb$ can be adjusted by varying the mole fraction x of Al as follows [2]:

$$E_g = (0.73 + 1.1x + 0.47x^2)\ eV. \tag{5.2}$$

Therefore, the band alignment in the InAs/$Al_xGa_{1\text{-}x}Sb$ heterojunction can be changed from staggered to broken-gap type by changing the mole fraction of Al in $Al_xGa_{1\text{-}x}Sb$. When the mole fraction of Al is small, the alignment is of broken-gap type, and when the mole fraction of Al is high, the alignment is of staggered type. Due to the alignment of the bands in the source and the channel, the effective bandgap in the heterojunction TFET is significantly lower than the corresponding homojunction TFET, thus resulting in a high I_{ON}. However, due to the greater bandgap in the channel, the I_{OFF} of the TFET is kept low, thus simultaneously optimizing both the I_{ON} and the I_{OFF}.

It should be noted that, when the source doping is low, the large screening

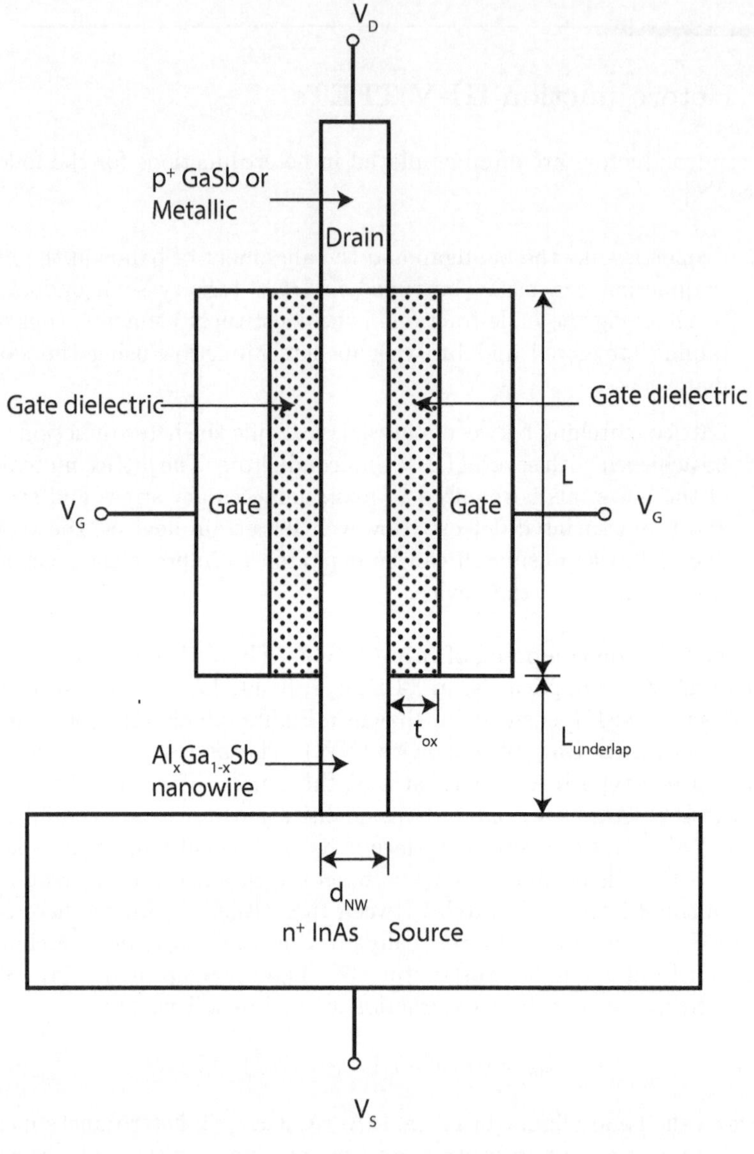

FIGURE 5.5
Schematic representation of a p-type heterojunction III-V TFET. Device parameters are $d_{NW} = 4\ nm$, $L = 10\ nm$, $L_{underlap} = 1\ nm$ and $t_{ox} = 1\ nm$ [2].

length within the source leads to a small band-to-band tunneling resulting in a low I_{ON}. However, when the doping density in the source is too high, the band-pass filter operation of cutting the low- and high-energy tails of the

Fermi distribution in the TFET becomes ineffective, leading to a subthreshold swing greater than 60 $mV/decade$. Therefore, for a given mole fraction x of the Al, the doping density in the source can be optimized to maximize the I_{ON}/I_{OFF} ratio. For the TFET shown in Figure 5.5, for $x = 0.6$, the optimum subthreshold slope of 35 $mV/decade$ can be obtained when the source doping is 1×10^{19} $atoms/cm^3$. Further, the quantization induced by the thin body will also have an impact on the band alignment of the heterojunction TFETs. In fact, for a given thickness of the TFET body, an optimum mole-fraction can be found that results in a small effective bandgap, without the device entering into the broken gap regime [15, 26].

The application of heterojunction in a III-V TFET not only improves the I_{ON} and the subthreshold slope, but also improves the scalability of the device. When the junction is highly staggered, the contribution of the channel in the tunnel junction width is smaller. This results in the peak of the tunneling generation being close to the source–channel junction away from the drain potential. Therefore, drain-induced barrier thinning (DIBT) is observed to be smaller in heterojunction III-V TFETs [14, 27].

Similar to the p-type TFET shown in Figure 5.5, an n-type TFET can be realized using a GaSb/$In_xGa_{1-x}As$ heterojunction or an $Al_xGa_{1-x}Sb$/InAs heterojunction [2, 15]. Other III-V semiconductor heterojunctions like $Al_{0.5}Ga_{0.5}As_{0.3}Sb_{0.7}/In_{0.8}Ga_{0.2}As$ and $GaAs_xSb_{1-x}/In_yGa_{1-y}As$ have also been explored for heterojunction TFETs and have shown better performance than their corresponding homojunction TFETs [25–28]. Additionally, the heterojunctions between silicon/germanium and III-V materials have been explored by several researchers [29–35]. However, it should be pointed out that there is an appreciable mismatch in the lattice constant of the Si and III-V compound semiconductors, as is evident from Table 5.1. For example, the mismatch in the lattice constant of InAs and Si is as large as 11% [30]. Therefore, traditional techniques of growing Si/III-V compound heterojunctions are expected to yield a large defect density in the heterojunction. Si/III-V heterojunction TFETs have been fabricated using the selective area growth of an n^+-InAs/undoped-InAs axial NW on a silicon substrate and could achieve a subthreshold swing of 104 $mV/decade$ [29]. The poor subthreshold swing obtained in the Si/InAs heterojunction TFET is attributed to the misfit dislocation in the heterojunction [29]. The technique to obtain coherent growth could potentially improve the electrical characteristics of III-V/Si heterojunction TFETs [29, 31]. It has been shown that the template-assisted growth of vertical III-V NW on a Si substrate holds good promise in integrating III-V heterostructures on a Si platform [35].

As we have already seen, a broken-gap type of heterojunction alignment can also be obtained in a III-V heterojunction TFET by choosing a suitable composition of the heterojunction materials. A broken-gap heterojunction is expected to yield a high I_{ON} because of the reduced tunneling barrier [33]. However, due to the transitions of the carriers caused by optical phonon scattering at the heterointerface, the subthreshold swing in a broken-gap heterojunction III-V TFET can be degraded [2, 15, 33]. For example, when an InAs source is employed, which has a considerably small effective mass of the carrier, there would be a larger density of states in the potential notch in the channel created due to band bending. As a result, the inelastic scattering at the heterointerface significantly contributes to the leakage current and degradation in the subthreshold swing is observed [2, 15, 33]. Therefore, the broken-gap alignment is generally avoided in InAs/$Al_xGa_{1-x}Sb$ heterojunction TFETs. Also, the broken-gap heterojunction can potentially be employed in other types of TFETs like CNT-based TFETs as we shall show in Chapter 6.

A novel TFET architecture has been recently proposed that utilizes multiple reflectors in the channel to improve the I_{ON} and the subthreshold swing [36]. At certain energies, the waves from the source–channel junction interfere destructively with the waves from the reflectors, yielding a high net transmission co-efficient and improving the overall characteristics of the device [36].

5.6 Device Optimization in III-V TFETs

The performance of the III-V TFETs can be improved by intelligent device modifications. For example, the tunneling direction can be aligned with the direction of the gate field to boost the I_{ON} and improve the subthreshold swing [3, 4, 8, 11, 12, 37–40]. TFETs in which the dominant tunneling occurs in the direction normal to the semiconductor–dielectric interface are also referred to as vertical TFETs. Figure 5.6 shows an InAs vertical TFET in which a highly doped n^+ pocket in the gate–source overlap region is introduced [3]. A back gate underneath the channel is added to improve the electrostatic control of the channel [3]. The doping of the source, channel, and drain are 5×10^{19} $atoms/cm^3$, 1×10^{15} $atoms/cm^3$ and 5×10^{18} $atoms/cm^3$, respectively. The gate is 1.2 nm thick with a dielectric constant $\kappa = 15.4$. The doping of the pocket is 5×10^{18} $atoms/cm^3$. The addition of the pocket on the source side introduces an additional tunneling barrier on the source in the OFF-state, resulting in a lower I_{OFF}. The heavily doped n^+ pocket leads to a thinner tunneling barrier in the ON-state leading to a higher I_{ON}. The pocket also creates a vertical tunneling component, in addition to the lateral tunneling component, which results in a higher I_{ON} and a steeper subthreshold swing. However, the device shown in Figure 5.6 exhibits poor scalability because direct source-to-drain tunneling becomes more severe as the channel length is reduced. The direct source-to-drain tunneling can be inhibited by pushing up

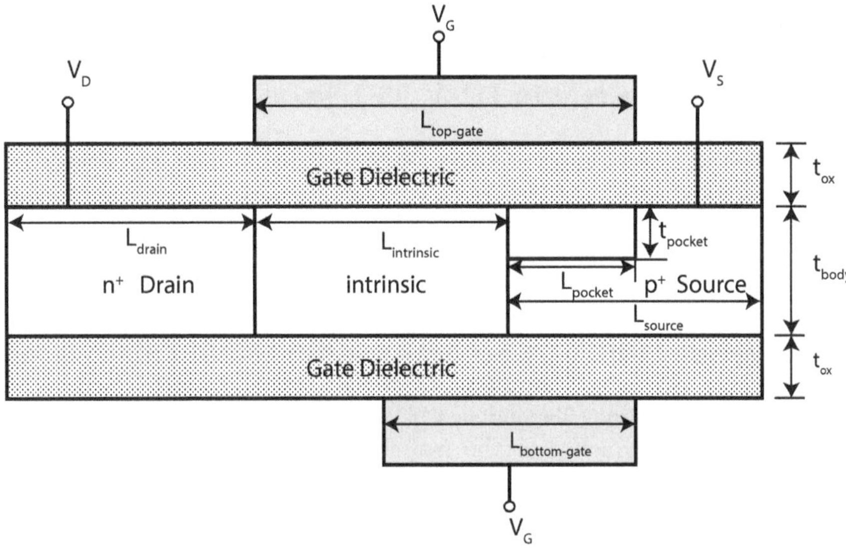

FIGURE 5.6
Schematic representation of an InAs vertical tunnel field-effect transistor [3]. Device parameters are $L_{drain} = L_{source} = L_{intrinsic} = 20\ nm$, $L_{top-gate} = 30\ nm$, $L_{bottom-gate} = 20\ nm$, $t_{body} = 10\ nm$, $t_{pocket} = 3.6\ nm$, $L_{pocket} = 10\ nm$, $t_{ox} = 1.2\ nm$.

the buried oxide on the drain side such that the path of direct source-to-drain tunneling is blocked [4].

The performance of a vertical TFET can be further improved by employing a heterojunction [8, 11, 12, 37]. The subthreshold swing in a heterojunction vertical TFET can be made steeper by tuning the mole-fraction in the ternary III-V semiconductor such that the lateral tunneling and the vertical tunneling are turned on simultaneously [8]. The I_{OFF} in a heterojunction vertical TFET can be kept low by choosing material such that there is a larger tunneling barrier height in the OFF-state and the effective mass of the carrier in the channel is larger. The wide range of heterojunction alignment possible with the III-V compound semiconductors makes III-V semiconductors ideal candidates for realizing high-performance heterojunction-based vertical TFETs.

A thin delta-doping layer can also be introduced in a III-V TFET to reduce the tunneling width to obtain a higher I_{ON} and smaller subthreshold swing, and also suppress the ambipolar transport [9, 12, 27, 41, 42]. The ambipolar transport can be suppressed in a III-V TFET, along with obtaining a steep subthreshold swing, by employing a dual material gate (DMG) and introducing an intrinsic drain region [43–45]. The I_{ON} can also be enhanced by the application of uniform strain in III-V heterostructure TFETs [46].

5.7 Gate Dielectric in III-V TFETs

As described in the previous sections, obtaining a gate dielectric over a III-V semiconductor with a low interface state density D_{it} is one of the biggest challenges in fabricating III-V TFETs. The presence of the interface states is normally confirmed by capacitance-voltage (CV) measurements in the TFETs [7, 11, 24, 26, 39]. The interface trap density D_{it} at the semiconductor–dielectric interface can be estimated by making the low-frequency (LF) and the high-frequency (HF) CV measurements for the MOS structure and then computing the D_{it} using the difference of the LF and the HF capacitances [47]. However, the strong dependence between the measured capacitance on the measurement frequency or the frequency dispersion effect, observed in highly scaled devices in the accumulation mode, can complicate the estimation of D_{it} [19, 48, 49]. In MOSFETs, the dispersion effect could be 50% or more in the frequency range of $1\ kHz - 1\ MHz$ [48]. A small frequency dispersion observed in the CV measurement is an indication of a good dielectric–semiconductor interface [7, 26]. Figure 5.7 shows a comparison of the mea-

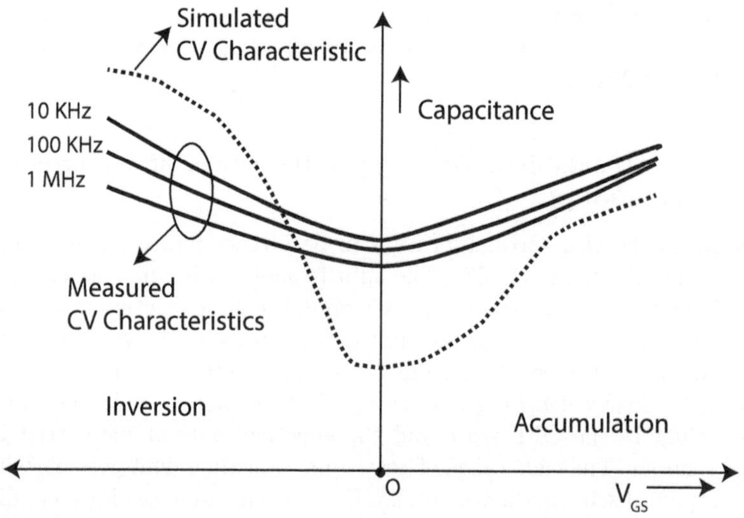

FIGURE 5.7
Comparison of the measured CV characteristics of a III-V TFET with the characteristics of the same TFET obtained using numerical simulations [11].

sured CV characteristics of a III-V TFET and the characteristics of the same TFET obtained using numerical simulations assuming that there are no interface traps. The measured CV characteristics show a much gentler transition from the accumulation mode to the inversion mode and the measured characteristics are stretched out compared to the simulated characteristics. These observations are indicative of the presence of the interface states in the dielectric–semiconductor interface. Additionally, there is little frequency dispersion effect observed in the CV characteristics.

The interface traps retard the Fermi-level movement of the carriers in the channel and result in trap-assisted tunneling followed by thermal emission. These effects lead to degradation in the subthreshold swing. The existence of the interface traps explains the experimentally observed poorer subthreshold swing in several III-V TFETs compared to the subthreshold swing obtained by numerical simulation [9, 11, 14, 24, 39, 50]. As a result, a great deal of ongoing research on III-V TFETs is targeted at obtaining a good-quality high-κ gate dielectric in III-V TFETs. The ALD technique has been successful in obtaining a reasonably good-quality gate dielectric in III-V TFETs [7, 9, 11, 14, 15, 28, 31, 37, 50]. However, analysis of several recent experimental results on III-V TFETs that employed ALD for deposition of dielectric suggest that further improvement in the dielectric–channel interface is the key to the III-V TFETs achieving an average subthreshold swing smaller than 60 $mV/decade$. In the rest of this section, we look at some of the techniques to improve the gate dielectric–semiconductor interface in III-V TFETs.

5.7.1 Processing Techniques

Pre-deposition cleaning treatments of the interfaces, the use of interfacial layers, the deposition chemistry, and post-deposition treatments can be employed to reduce the interface state density D_{it} [17]. Processing techniques like post-deposition annealing (PDA) has been shown to improve the quality of the interface [37, 39]. For example, in the $In_{0.53}Ga_{0.47}As$ TFET the interface state density D_{it} was found to decrease by 30% near the midgap after PDA [14, 37, 51]. Due to the reduced interface state density D_{it}, the TFET with PDA exhibited a subthreshold swing of 130 $mV/decade$, while the TFET without PDA exhibited a subthreshold swing of 240 $mV/decade$ [37].

In order to prevent the semiconductor from oxidizing in air, a conformal layer of non-reactive material is coated over the device using techniques like CVD, ALD, etc. The process of coating non-reactive material on the semiconductor is referred to as passivation. Normally, passivation does not improve the electrical characteristics of the device but improves the reliability of the device [37, 39]. However, the mesa passivation of TFET with 10 nm SiN_x at 120°C using plasma enhanced chemical vapor deposition (PECVD) instead of ALD, can potentially improve the performance of the TFET [39]. For vertical InGaAs/InP heterojunction TFETs, it is found that PECVD of SiN_x improves the minimum subthreshold slope of the TFET by 40% [37, 39]. On the other hand, no improvement in performance is observed in the ALD passivated device [37, 39]. The improvement in subthreshold slope with SiN_x passivation is somewhat attributed to the decrease in the D_{it} within the bandgap [39]. It has also been shown that if appropriate gate metal is deposited using thermal evaporation rather than electron-beam evaporation, the damage to the underlying oxide-semiconductor interface is avoided and an un-intentional oxide layer does not form [28]. This improves the I_{ON} and the subthreshold swing by reducing effective oxide thickness and mid-gap D_{it} [28].

5.7.2 Interface Material Engineering

By using a suitable combination of an interface material and a high-κ gate dielectric, improved electrical characteristics of III-V TFETs can be obtained [7, 9]. For example, in the InGaAs TFET, for the same effective oxide thickness, the subthreshold swing is found to be better when an Al_2O_3–HfO_2 bilayer is used instead of a single HfO_2 gate oxide [9]. The reduced interface trap density in the InGaAs/Al_2O_3 interface compared to that in the InGaAs/HfO_2 interface is believed to be the underlying reason for the better performance of the bilayer compared to a single layer [9]. Alternative materials, like $TaSiO_x$, have also been explored and $TaSiO_x$-$In_{0.53}Ga_{0.47}As$ obtained using ALD has been found to be of fairly good quality [7].

5.7.3 Hybrid III-V/Silicon TFET

Recently InAs/Si heterojunction TFETs have been explored to obtain higher I_{ON} and a smaller subthreshold swing due to the smaller effective bandgap of the heterojunction [29, 31, 52]. However, incorporating Si as the channel material opens avenues for utilizing a high-quality Si–dielectric interface [29, 52]. In the future, hybrid III-V/Silicon TFETs could provide a solution to the poor quality of the III-V–dielectric interface. However, it should be noted that realizing a defect-free III-V/Si heterojunction can itself be a serious challenge [29].

5.8 Tunneling Junction

The abruptness of the source doping profile plays a crucial role in determining the electrical characteristics of a TFET. To obtain a sharp subthreshold slope, the source doping profile must be as abrupt as possible. The abruptness of the tunneling junction is strongly dependent on the technique that is employed for doping the source. Besides ion implantation, the tunneling junction can be formed by re-growth of III-V materials with *in situ* impurity doping. The *in situ* doping of the tunneling junction can lead to the formation of a junction without introducing a significant number of defects and can result in a steep doping profile [9, 10, 53]. The defects introduced while forming the junction by ion implantation can be eliminated by advanced annealing techniques such as laser annealing or flash lamp annealing [10]. In addition to the fabrication techniques, the process parameters play a significant role in determining the abruptness of the junction. For example, in the $In_{0.53}Ga_{0.47}As$ TFET, lower ion implantation energy and the resulting steeper junction profile have been shown to improve the subthreshold slope [10]. The properties of the dopants affect the abruptness of the doping profile as well. The dopants with lighter mass and smaller thermal diffusion coefficients would be more suitable for p^+/n source formation [9, 10]. For example, in the InGaAs TFET, the implantation of Be is found to result in a better junction compared to the implantation of Zn since the defects introduced by Be implantation can be recovered easily compared to the defects introduced by Zn implantation [10]. The device structure and the geometry of the tunneling junction can be engineered to optimize the electrical characteristics of a TFET. As an illustration, two heterojunction $In_{0.7}Ga_{0.3}As$ TFETs with different structures of the tunneling junction are shown in Figure 5.8 [9]. Both these devices were fabricated using molecular beam epitaxy (MBE) and the high-κ gate dielectric were deposited using ALD. In TFET-1, the p^+ $In_{0.7}Ga_{0.3}As$ region is 6 nm and the undoped $In_{0.7}Ga_{0.3}As$ region is 6 nm. However, in TFET-2, the p^+ $In_{0.7}Ga_{0.3}As$ region is 4 nm, while the undoped $In_{0.7}Ga_{0.3}As$ region is 8 nm. The p^+ region is doped with Be with acceptor density $N_A = 2 \times 10^{19}$ $atoms/cm^3$. It is experimentally found that TFET-2 exhibits 80% higher I_{ON} compared to TFET-1 [9]. This is attributed to the thicker undoped layer in TFET-2, which allows a thicker effective undoped region taking into account the diffusion of Be from the p^+ region. However, TFET-1 exhibits a better subthreshold swing compared to TFET-2. It should be noted that the Be of the p^+ region diffuses into the undoped $In_{0.7}Ga_{0.3}As$ region in both TFET-1 and TFET-2. Using secondary ion mass spectrometry (SIMS), it was found that TFET-1 exhibits a more abrupt doping profile compared to TFET-2, which might have resulted in a better subthreshold swing observed in TFET-1 compared to TFET-2 [9]. These results illustrate that the geometry of the tunneling junction can be optimized to improve the performance of the TFET.

The problem of obtaining an abrupt tunneling junction is further complicated in heterojunction TFETs [9, 13, 32, 53–55]. The lattice mismatch between the constituents of the heterojunction can give rise to defects at the heterojunction interface leading to a formation of the trap states in the bandgap. For example, obtaining a defect-free $Ge/In_{0.53}Ga_{0.47}As$ heterojunction is a challenge, since there is a 3.7% mismatch in the lattice of Ge and InGaAs [32, 54]. When there are large defects in the heterojunction, the electrons can tunnel from the source to the channel via these trap states leading to degradation in the subthreshold slope [32, 53]. The defect density in the heterojunction can be

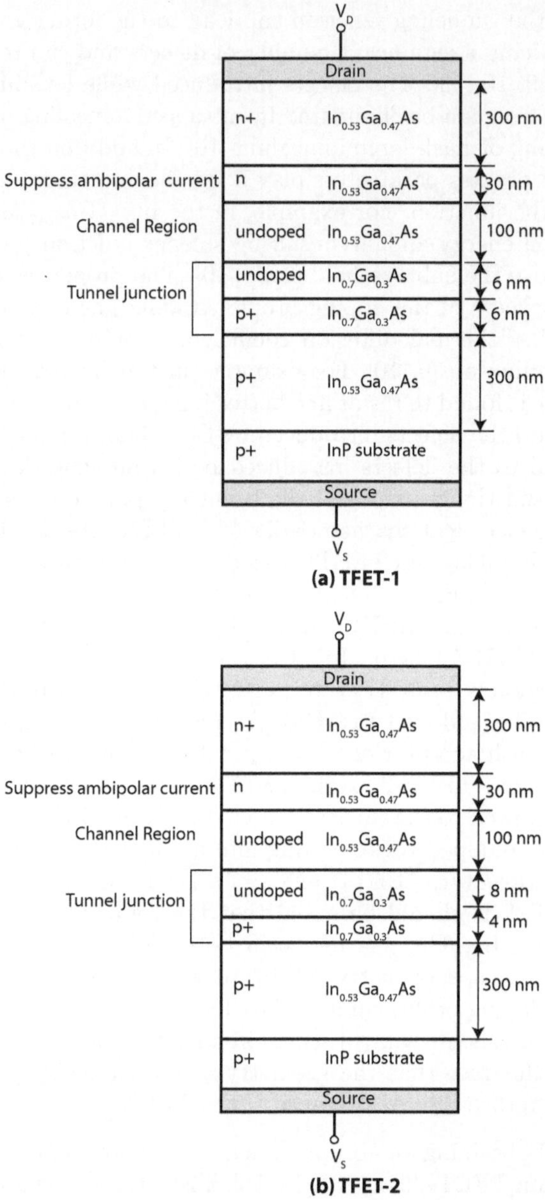

FIGURE 5.8
Heterojunction $In_{0.7}Ga_{0.3}As$ TFETs with different structure of the tunneling junction [9].

reduced by adjusting the process parameters like temperature or choosing the right heterointerface [32, 53]. For example, in the $GaAs_{0.35}Sb_{0.65}/In_{0.7}Ga_{0.3}As$

heterojunction TFET under different growth conditions, the heterojunction can be GaAs terminated or InAs terminated. It is found that the device layers with heterointerface terminated by GaAs are strain relaxed and are vulnerable to the formation of defects, whereas the device layers that are terminated by InAs are pseudomorphic to the underlying layer and are likely to be defect-free. As a result, in $GaAs_{0.35}Sb_{0.65}/In_{0.7}Ga_{0.3}As$ heterojunction TFETs with an InAs-terminated device layer has more than three orders of magnitude improvement in the I_{ON}/I_{OFF} ratio compared to a GaAs-terminated device layer due to the reduction in defect-assisted conduction [53].

5.9 Impact of Temperature

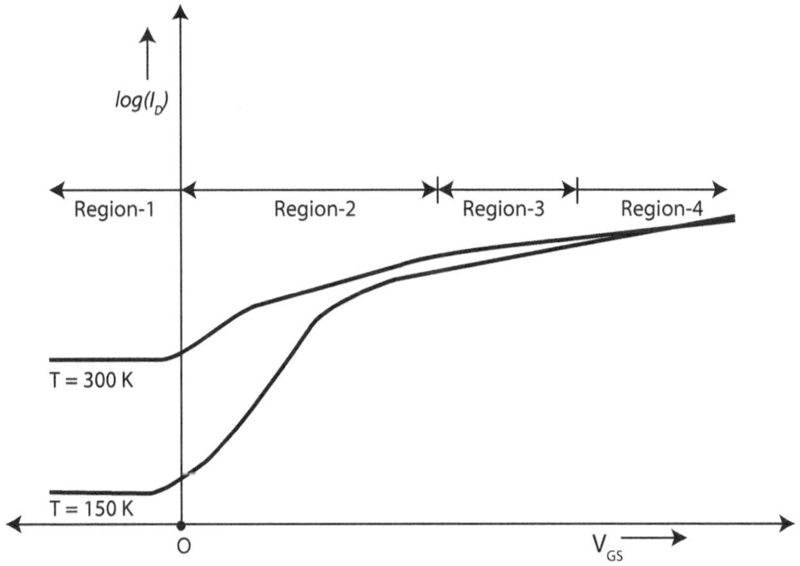

FIGURE 5.9
Qualitative comparison of transfer characteristics of a TFET at $T = 150\ K$ and $T = 300\ K$ [24].

The impact of temperature on the drain current in a TFET depends on the region in which the TFET is operating [24]. Since different mechanisms of current transport are dominant in different regions of operation, the transfer characteristics exhibit different impacts of temperature in four regions of operation shown in Figure 5.9.

In Region-1, the TFET is in the OFF-state. In this region of operation, the drain current is independent of the gate voltage and is determined by the SRH generation-recombination current in the reverse-biased diode. Therefore, in Region-1 of operation, the drain current increases exponentially with temperature, since the SRH generation-recombination current is directly proportional to $\exp(-E_g/2kT)$ where E_g is the bandgap, T is the absolute temperature, and k is the Boltzmann constant [1, 24, 27, 50].

When the gate voltage is increased beyond the OFF-state, the TFET enters the subthreshold region of operation, which is shown as Region-2 in Figure 5.9. The average subthreshold swing in III-V TFETs normally increases as the temperature is increased and exhibits strong temperature dependence [9, 24, 27]. The primary reason for this behavior is the presence of the interface traps in the III-V TFETs. The tunneling of carriers from the valence band of the p^+ source to the mid-gap trap levels followed by thermal emission of the carrier into the conduction band can explain the increase of the average subthreshold swing with temperature [9, 24].

When the gate voltage is increased further, the TFET enters Region-3 of operation. In this region of operation, the dependence of the drain current on temperature becomes weaker, since the drain current is primarily governed by BTBT [24, 50]. The weak temperature dependence is due to the dependence of the tunneling barrier on the bandgap $E_g(T)$, which in turn is dependent on the temperature as follows [56]:

$$E_g\left(T\right) = E_g\left(0\right) - \frac{\alpha T^2}{\beta + T}, \tag{5.3}$$

where $E_g(0)$ is the bandgap at $T = 0$ and α and β are the material constants.

When the gate voltage is increased further, the TFET enters Region-4. In this region of operation, the drain current is primarily determined by the diffusion current, since the lateral electric field is small. The diffusion current is dependent on the mobility of the electrons in the channel, which shows negative dependence on temperature due to phonon scattering [1, 24, 50]. Therefore, the drain current decreases with an increase in temperature at high gate voltages.

5.10 Challenges in p-Type III-V TFETs

For CMOS applications, it is essential that both n-type and p-type TFETs can be realized on the same material system. For good performance of a p-type TFET, it is critical to achieve proper band alignment, particularly the valence band offset [57]. However, there are additional challenges involved in realizing a p-type TFET using III-V semiconductors. Additionally, in the optimization of the heterostructure III-V TFETs, there are conflicting requirements for the p-type and the n-type TFETs [58]. Therefore, simultaneously obtaining optimized p-type and n-type TFETs on the same source/channel material system, is not possible [58]. In III-V semiconductors, the effective density of states in the conduction band is low. For example, in $In_{0.53}Ga_{0.47}As$ the conduction band effective density of state (N_C) is 2.1×10^{17} cm^{-3} and the valence band effective density of state (N_V) is 7.7×10^{18} cm^{-3} [59]. As a result, when the source of a p-type III-V TFET is heavily doped, the Fermi level in the source moves deep into the conduction band [59]. For example, when a $In_{0.53}Ga_{0.47}As$ source is doped with n^+ dopants with a density $N_D = 8 \times 10^{19}$ $atoms/cm^3$, the Fermi level moves 1.442 eV above the conduction band edge in a p-type TFET [59]. In contrast, when the $In_{0.53}Ga_{0.47}As$ source is doped with p^+ dopant with density $N_A = 8 \times 10^{19}$ $atoms/cm^3$, the Fermi level moves only 0.14 eV below the valence band edge in an n-type TFET. It should be noted that the source is heavily doped in TFETs to obtain a high I_{ON}. Therefore, due to the Fermi level lying deep in the conduction band in a p-type III-V TFET, the carriers with energies in the exponential tail of the source Fermi–Dirac distribution function participate in BTBT. This results in the minimum subthreshold swing being restricted to 60 $mV/decade$ at room temperature in p-type III-V TFETs. Therefore, it is more difficult to simultaneously obtain high drive currents and a steep subthreshold swing in a p-type III-V TFET compared to an n-type III-V TFET, and it poses a serious challenge in employing III-V TFETs for future low-power CMOS applications [18, 53, 58, 59].

A source underlapped GaSb-InAs TFET has been shown to perform better than the Si-FinFET for low-power applications providing higher throughput, especially in near-threshold computing [60]. Similarly, inverters realized using $InAs/Al_{0.05}Ga_{0.95}Sb$ TFETs are reported to be faster than the corresponding CMOS logic based on multigate architecture in low operating power and low standby power environments [61].

5.11 Current State and Future Perspective

The research on III-V TFETs is advancing at a rapid pace. The focus area for researchers has been to demonstrate a high I_{ON} in III-V TFETs, along with obtaining a steep average subthreshold swing at low supply voltages ($V_{DD} < 0.5$ V). To achieve this, novel III-V TFET device structures have been proposed, advanced processing techniques have been incorporated into the experiments with III-V TFETs and the material properties like effective bandgap have been tuned to achieve the best possible results.

Researchers have achieved considerable success in obtaining a high I_{ON} in III-V TFETs. Using heterojunction and small effective oxide thickness, $I_{ON} = 135\mu A/\mu m$ at $V_{DS} = 0.5\ V$ has been reported for III-V TFETs [13, 14]. The I_{ON} obtained in III-V TFETs is fairly high and suggests that III-V TFETs are a good candidate for future low-power CMOS applications. However, to clearly demonstrate that III-V TFETs are superior to the conventional MOSFET, it is essential that the III-V TFETs exhibit an average subthreshold swing smaller than 60 $mV/decade$ for at least 4 decades of drain current [62, 63]. The average subthreshold swing demonstrated for experimental III-V TFETs do not meet this criterion and researchers are actively working to improve the subthreshold swing of the III-V TFETs. Since the poor subthreshold swing obtained in experimental III-V TFETs has been mostly attributed to the interface states at the dielectric–channel interface, obtaining a good-quality dielectric–semiconductor interface continues to be the biggest challenge to overcome for researchers working on III-V TFETs.

In addition to the processing challenges involved in fabricating III-V TFETs, the integration of III-V TFETs into the mainstream CMOS technology would pose additional hurdles. The p-type III-V TFETs with good electrical characteristics need to be realized along with n-type III-V TFETs on the same material system. The integration of III-V semiconductors into the mainstream Si-based CMOS technology would require a major change in the fabrication methodology and would pose its own challenges.

Nevertheless, it must be noted that numerous simulation results and theoretical studies have shown that III-V TFETs hold great promise for future CMOS applications. Therefore, research on III-V TFETs should be pursued aggressively in the quest for transistors that are feasible for commercial applications and replace the conventional MOSFETs at advanced technology nodes.

5.12 Summary

In this chapter, we described recent advancements in III-V TFETs. We looked at the advantages of employing III-V semiconductors in TFETs. We highlighted the challenges that are involved in realizing III-V TFETs and the techniques that are used to tackle these challenges. In the next chapter, we will look at the TFETs that are realized using different forms of carbon like carbon nanotubes and graphene. Carbon-based TFETs offer distinct advantages over other types of TFETs and are worth exploring for future CMOS applications.

REFERENCES

[1] S. Mookerjea, D. Mohata, R. Krishnan, J. Singh, A. Vallett, A. Ali, T. Mayer, V. Narayanan, D. Schlom, A. Liu, and S. Datta, "Experimental demonstration of 100 nm channel length $In_{0.53}Ga_{0.47}As$-based vertical inter-band tunnel field effect transistors (TFETs) for ultra low-power logic and SRAM applications,," in *IEEE International Electron Devices Meeting (IEDM)*, pp. 1–3, IEEE, Dec. 2009.

[2] J. Knoch and J. Appenzeller, "Modeling of high-performance p-type III-V heterojunction tunnel FETs," *IEEE Electron Device Letters*, vol. 31, pp. 305–307, Apr. 2010.

[3] K. Ganapathi, Y. Yoon, and S. Salahuddin, "Analysis of InAs vertical and lateral band-to-band tunneling transistors: Leveraging vertical tunneling for improved performance," *Applied Physics Letters*, vol. 97, no. 033504, 2010.

[4] S. Agarwal, G. Klimeck, and M. Luisier, "Leakage-reduction design concepts for low-power vertical tunneling field-effect transistors," *IEEE Electron Device Letters*, vol. 27, pp. 621–623, June 2010.

[5] A. Verhulst, W. Vandenberghe, D. Leonelli, R. Rooyackers, A. Vandooren, J. Zhuge, K.-H. Kao, B. Soree, W. Magnusa, M. Fischetti, G. Pourtois, C. Huyghebaert, R. Huang, Y. Wang, K. D. Meyer, W. Dehaene, M. Heynsa, and G. Groeseneken, "Si-based tunnel field-effect transistors for low-power nano-electronics," in *69th Annual Device Research Conference (DRC)*, pp. 193–196, IEEE, 2011.

[6] A. Seabaugh, S. D. Chae, P. Fay, W. S. Hwang, T. Kosel, R. Li, Q. Liu, Y. Lu, T. Vasen, M. Wistey, G. Xing, G. Zhou, and Q. Zhang, "III-V tunnel field-effect transistors," *ECS Transactions*, vol. 41.

[7] G. Dewey, B. Chu-Kung, J. Boardman, J. M. Fastenau, J. Kavalieros, R. Kotlyar, W. K. Liu, D. Lubyshev, M. Metz, N. Mukherjee, P. Oakey, R. Pillarisetty, M. Radosavljevic, H. Then, and R. Chau, "Fabrication, characterization, and physics of III-V heterojunction tunneling field effect transistors (H-TFET) for steep sub-threshold swing," in *IEEE International Electron Devices Meeting (IEDM)*, pp. 33.6.1–33.6.4, IEEE, 2011.

[8] K. Ganapathi and S. Salahuddin, "Heterojunction vertical band-to-band

tunneling transistors for steep subthreshold swing and high ON current," *Electron Device Letters*, vol. 32, pp. 689–691, May 2011.

[9] H. Zhao, Y. Chen, Y. Wang, F. Zhou, F. Xue, and J. Lee, "InGaAs tunneling field-effect-transistors with atomic-layer deposited gate oxides," *IEEE Trans. on Electron Devices*, vol. 58, pp. 2990–2995, Sept. 2011.

[10] R. Iida, S.-H. Kim, M. Yokoyama, N. Taoka, S.-H. Lee, M. Takenaka, and S. Takagi, "Planar-type $In_{0.53}Ga_{0.47}As$ channel band-to-band tunneling metal-oxide-semiconductor field-effect transistors," *Journal of Applied Physics*, vol. 110, no. 124505, 2011.

[11] R. Li, Y. Lu, G. Zhou, Q. Liu, S. D. Chae, T. Vasen, W. S. Hwang, Q. Zhang, P. Fay, T. Kosel, M. Wistey, H. Xing, and A. Seabaugh, "AlGaSb/InAs tunnel field-effect transistor with on-current of 78 $\mu a/\mu m$ at 0.5 V," *IEEE Electron Device Letters*, vol. 33, pp. 363–365, Mar. 2012.

[12] Y. Lu, G. Zhou, R. Li, Q. Liu, Q. Zhang, T. Vasen, S. D. Chae, T. Kosel, M. Wistey, H. Xing, A. Seabaugh, and P. Fay, "Performance of AlGaSb/InAs TFETs with gate electric field and tunneling direction aligned," *IEEE Electron Device Letters*, vol. 33, pp. 655–657, May 2012.

[13] D. Mohata, R. Bijesh, Y. Zhu, M. Hudait, R. Southwick, Z. Chbili, D. Gundlach, J. Suehle, J. M. Fastenau, D. Loubychev, A. K. Liu, T. S. Mayer, V. Narayanan, and S. Datta, "Demonstration of improved heteroepitaxy, scaled gate stack and reduced interface states enabling heterojunction tunnel FETs with high drive current and high On-Off ratio," in *Symposium on VLSI Technology*, pp. 53–54, IEEE, June 2012.

[14] D. Mohata, B. Rajamohanan, T. Mayer, M. Hudait, J. Fastenau, D. Loubychev, A. W. K. Liu, and S. Datta, "Barrier-engineered arsenide-antimonide heterojunction tunnel FETs with enhanced drive current," *IEEE Electron Device Letters*, vol. 33, pp. 1568–1570, Nov. 2012.

[15] E. Baravelli, E. Gnani, R. Grassi, A. Gnudi, S. Reggiani, and G. Baccarani, "Optimization of n- and p-type TFETs integrated on the same $InAs/Al_xGa_{1-x}Sb$ technology platform," *IEEE Trans. on Electron Devices*, vol. 61, pp. 178–185, Jan. 2014.

[16] Ioffe Physico-Technical Institute, "New semiconductor materials: Characteristics and properties." http://www.ioffe.ru/SVA/NSM/Semicond/.

[17] J. A. del Alamo, "Nanometre-scale electronics with III-V compound semiconductor," *Nature*, vol. 479, pp. 317–323, Nov. 2011.

[18] J. M. Woodall, *Fundamentals of III-V Semiconductor MOSFETs*, ch. 1. Springer, USA, 2010.

[19] S. Stemmer, V. Chobpattana, and S. Rajan, "Frequency dispersion in III-V metal-oxide-semiconductor capacitors," *Applied Physics Letters*, vol. 100, no. 233510, 2012.

[20] S. D. Offsey, J. M. Woodall, A. C. Warren, P. D. Kirchner, T. I. Chappell, and G. D. Pettit, "Unpinned (100) GaAs surfaces in air using photochemistry," *Applied Physics Letters*, vol. 48, Feb. 1986.

[21] P. D. Ye, Y. Xuan, Y. Wu, and M. Xu, *Fundamentals of III-V Semiconductor MOSFETs*, ch. 7. Springer, USA, 2010.

[22] P. Ye, G. Wilk, J. Kwo, B. Yang, H.-J. Gossmann, M. Frei, S. Chu, J. Mannaerts, M. Sergent, M. Hong, K. Ng, and J. Bude, "GaAs MOSFET with oxide gate dielectric grown by atomic layer deposition," *Electron Device Letters*, vol. 24, pp. 209–211, Apr. 2003.

[23] P. D. Ye, "InGaAs revolutionizes III-V MOSFETs," *Compound Semiconductor*, pp. 29–31, Apr. 2008.

[24] S. Mookerjea, D. Mohata, T. Mayer, V. Narayanan, and S. Datta, "Temperature-dependent I-V characteristics of a vertical $In_{0.53}Ga_{0.47}As$ tunnel FET," *IEEE Electron Device Letters*, vol. 31, pp. 564–566, June 2010.

[25] L. Wang, E. Yu, Y. Taur, and P. Asbeck, "Design of tunneling field-effect transistors based on staggered heterojunctions for ultralow-power applications," *IEEE Electron Device Letters*, vol. 31, pp. 431–433, May 2010.

[26] T. Yu, J. T. Teherani, D. A. Antoniadis, and J. L. Hoyt, "$In_{0.53}Ga_{0.47}As/GaAs_{0.5}Sb_{0.5}$ quantum-well tunnel-FETs with tunable backward diode characteristics," *IEEE Electron Device Letters*, vol. 34, pp. 1503–1505, Dec. 2013.

[27] D. Mohata, S. Mookerjea, A. Agrawal, Y. Li, T. Mayer, V. Narayanan, A. Liu, D. Loubychev, J. Fastenau, and S. Datta, "Experimental staggered-source and n+ pocket-doped channel III-V tunnel field-effect transistors and their scalabilities," *Appl. Phys. Express*, vol. 4, pp. 24105(1)–024105(3), 2011.

[28] B. Rajamohanan, R. Pandey, V. Chobpattana, C. Vaz, D. Gundlach, K. P. Cheung, J. Suehle, S. Stemmer, and S. Datta, "0.5 V supply voltage operation of $In_{0.65}Ga_{0.35}As/GaAs_{0.4}Sb_{0.6}$ tunnel FET," *IEEE Electron Device Letters*, vol. 36, pp. 20–23, Jan. 2015.

[29] K. Tomioka and T. Fukui, "Tunnel field-effect transistor using InAs nanowire/Si heterojunction," *Applied Physics Letters*, vol. 98, no. 083114, 2011.

[30] H. Schmid, K. E. Moselund, M. T. Bjork, M. Richter, H. Ghoneim, and C. D. Bessire, "Fabrication of vertical InAs-Si heterojunction tunnel field effect transistors," in *DRC*, pp. 181–182, IEEE, 2011.

[31] K. Tomioka, M. Yoshimura, and T. Fukui, "Steep-slope tunnel field-effect transistors using III-V nanowire/Si heterojunction," in *Symposium on VLSI Technology (VLSIT)*, pp. 47–48, IEEE, 2012.

[32] P. Guo, Y. Yang, Y. Cheng, G. Han, J. Pan, Ivana, Z. Zhang, H. Hu, Z. X. Shen, C. K. Chia, and Y.-C. Yeo, "Tunneling field-effect transistor with Ge/$In_{0.53}Ga_{0.47}As$ heterostructure as tunneling junction," *Journal of Applied Physics*, vol. 113, no. 094502, 2013.

[33] M. Luisier and G. Klimeck, "Performance comparisons of tunneling field-effect transistors made of InSb, carbon, and GaSb-InAs broken gap heterostructures," in *IEEE International Electron Devices Meeting (IEDM)*, pp. 1–4, IEEE, 2009.

[34] P. K. Asthana, B. Ghosh, Y. Goswami, and B. M. Tripathi, "High-speed and low-power ultradeep-submicrometer III-V heterojunctionless tunnel field-effect transistor," *IEEE Trans. on Electron Devices*, vol. 61, pp. 479–486, Feb. 2014.

[35] D. Cutaia, K. E. Moselund, M. Borg, H. Schmid, L. Gignac, C. M. Breslin, S. Karg, E. Uccelli, and H. Riel, "Vertical InAs-Si gate-all-around tunnel FETs integrated on Si using selective epitaxy in nanotube templates," *IEEE Journal of the Electron Devices Society*, vol. 3, pp. 176–183, May 2015.

[36] P. Long, E. Wilson, J. Z. Huang, G. Klimeck, M. J. W. Rodwell, and M. Povolotskyi, "Design and simulation of GaSb/InAs 2D transmission-enhanced tunneling FETs," *IEEE Electron Device Letters*, vol. 37, pp. 107–110, Jan. 2016.

[37] G. Zhou, Y. Lu, R. Li, Q. Zhang, W. S. Hwang, Q. Liu, T. Vasen, C. Chen, H. Zhu, J.-M. Kuo, S. Koswatta, T. Kosel, M. Wistey, P. Fay, A. Seabaugh, and H. Xing, "Vertical InGaAs/InP tunnel FETs with tunneling normal to the gate," *IEEE Electron Device Letters*, vol. 32, pp. 1516–1518, Nov. 2011.

[38] R. Li, Y. Lu, S. D. Chae, G. Zhou, Q. Liu, C. Chen, M. S. Rahman, T. Vasen, Q. Zhang, P. Fay, T. Kosel, M.Wistey, H. Xing, S. Koswatta, and A. Seabaugh, "InAs/AlGaSb heterojunction tunnel field-effect transistor with tunnelling in-line with the gate field,," *Phys. Stat. Solidi (c)*, vol. 9, no. 2, pp. 389–392, 2012.

[39] G. Zhou, Y. Lu, R. Li, Q. Zhang, Q. Liu, T. Vasen, H. Zhu, J.-M. Kuo, T. Kosel, M. Wistey, P. Fay, A. Seabaugh, and H. Xing, "InGaAs/InP

tunnel FETs with a subthreshold swing of 93 mV/dec and ION/IOFF ratio near 10^6," *IEEE Electron Device Letters*, vol. 33, pp. 782–784, June 2012.

[40] G. Zhou, R. Li, T. Vasen, M. Qi, S. Chae, Y. Lu, Q. Zhang, H. Zhu, J.-M. Kuo, T. Kosel, M. Wistey, P. Fay, A. Seabaugh, and H. Xing, "Novel gate-recessed vertical InAs/GaSb TFET with record high ION of $180\mu a/\mu m$ at $V_{DS} = 0.5V$," in *IEDM*, pp. 32.6.1–32.6.4, IEEE, Dec. 2012.

[41] D. Mohata, D. Pawlik, L. Liu, S. Mookerjea, V. Saripalli, S. Rommel, and S. Datta, "Implications of record peak current density $In_{0.53}Ga_{0.47}As$ esaki tunnel diode on tunnel FET logic applications," in *Device Research Conference*, pp. 103–104, IEEE, June 2010.

[42] R. Kapadia, K. Takei, A. C. Ford, H. Fang, S. Chuang, M. Madsen, S. Krishna, and A. Javey, "Ultra-thin compound semiconductor on insulator (XOI) for MOSFETs and TFETs," in *Device Research Conference*, pp. 13–16, IEEE, June 2011.

[43] S. Saurabh and M. J. Kumar, "Novel attributes of a nanoscale dual material gate tunnel field effect transistor,," *IEEE Trans. on Electron Devices*, vol. 58, pp. 404–410, Feb. 2011.

[44] G. B. Beneventi, E. Gnani, A. Gnudi, S. Reggiani, and G. Baccarani, "Can interface traps suppress TFET ambipolarity?," *IEEE Electron Device Letters*, vol. 34, pp. 1557–1559, Dec. 2013.

[45] C. Y. Hsu, C. Y. Chang, E. Y. Chang, and C. Hu, "Suppressing non-uniform tunneling in InAs/GaSb TFET with dual-metal gate," *IEEE Journal of the Electron Devices Society*, vol. 4, pp. 60–65, Mar. 2016.

[46] D. Verreck, A. S. Verhulst, M. L. V. de Put, B. Sore, N. Collaert, A. Mocuta, A. Thean, and G. Groeseneken, "Uniform strain in heterostructure tunnel field-effect transistors," *IEEE Electron Device Letters*, vol. 37, pp. 337–340, Mar. 2016.

[47] J. D. Plummer, M. Deal, and P. B. Griffin, *Silicon VLSI Technology: Fundamentals, Practice, and Modeling.* Prentice Hall, 2000.

[48] Y. Xuan, H. C. Lin, and P. D. Ye, "Capacitance-voltage studies on enhancement-mode InGaAs metal-oxide-semiconductor field-effect transistor using atomic-layer-deposited Al_2O_3 gate dielectric," *Applied Physics Letters*, vol. 88, no. 263518, 2006.

[49] J. Tao, C. Z. Zhao, C. Zhao, P. Taechakumput, M. Werner, S. Taylor, and P. R. Chalker, "Extrinsic and intrinsic frequency dispersion of high-k materials in capacitance-voltage measurements," *Materials*, pp. 1005–1032, 2012.

[50] H. Zhao, Y. Chen, Y. Wang, F. Zhou, F. Xue, and J. Lee, "$In_{0.7}Ga_{0.3}As$ tunneling field-effect transistors with an Ion of 50 $\mu a/\mu m$ and a subthreshold swing of 86 mV/dec using HfO_2 gate oxide," *IEEE Electron Device Letters*, vol. 31, pp. 1392–1394, Dec. 2010.

[51] Q. Z. andG. Zhou, H. G. Xing, A. C. Seabaugh, K. Xu, O. A. Kirillov, C. A. Richter, and N. V. Nguyen, "Band alignment of TFET heterojunctions and post deposition annealing effects by internal photoemission spectroscopy," in *ISDRS*, IEEE, Dec. 2011.

[52] C. Kshirsagar and S. J. Koester, "InAs/SiGe on Si nanowire tunneling field effect transistors," in *69th Annual Device Research Conference*, pp. 151–152, IEEE, 2011.

[53] S. Datta, R. Bijesh, H. Liu, D. Mohata, and V. Narayanan, "Tunnel transistors for low power logic," in *IEEE Compound Semiconductor Integrated Circuit Symposium (CSICS)*, pp. 1–4, IEEE, Oct. 2013.

[54] M. Clavel, P. Goley, N. Jain, Y. Zhu, and M. K. Hudait, "Strain-engineered biaxial tensile epitaxial germanium for high-performance Ge/InGaAs tunnel field-effect transistors," *IEEE Journal of the Electron Devices Society*, vol. 3, pp. 184–193, May 2015.

[55] E. Lind, E. Memisevic, A. Dey, and L.-E. Wernersson, "III-V heterostructure nanowire tunnel FETs," *Journal of the Electron Devices Society*, vol. 3, pp. 96–102, May 2015.

[56] P. F. Guo, L. T. Yang, Y. Yang, L. Fan, G. Q. Han, G. S. Samudra, and Y. C. Yeo, "Tunneling field-effect transistor: Effect of strain and temperature on tunneling current," *IEEE Electron Device Letters*, vol. 30, pp. 981–983, Sept. 2009.

[57] Y. Zhu, N. Jain, D. K. Mohata, S. Datta, D. Lubyshev, J. M. Fastenau, A. K. Liu, and M. K. Hudait, "Structural properties and band offset determination of p-channel mixed As/Sb type-II staggered gap tunnel field-effect transistor structure," *Applied Physics Letters*, vol. 101, no. 112106, 2012.

[58] C. Wu, R. Huang, Q. Huang, J. Wang, and Y. Wang, "Design guideline for complementary heterostructure tunnel FETs with steep slope and improved output behavior," *IEEE Electron Device Letters*, vol. 37, pp. 20–23, Jan. 2016.

[59] S. Datta, R. Bijesh, H. Liu, D. Mohata, and V.Narayanan, "Tunnel transistors for energy efficient computing," in *IEEE International Reliability Physics Symposium (IRPS)*, pp. 6A.3.1–6A.3.7, IEEE, 2013.

[60] A. Sharma, A. A. Goud, and K. Roy, "Sub-10 nm FinFETs and tunnel-FETs: From devices to systems," in *Design, Automation and Test in*

Europe Conference and Exhibition (DATE), pp. 1443–1448, IEEE, Mar. 2015.

[61] E. Baravelli, E. Gnani, A. Gnudi, S. Reggiani, and G. Baccarani, "TFET inverters with n-/p-devices on the same technology platform for low-voltage/low-power applications," *IEEE Trans. on Electron Devices*, vol. 61, pp. 473–478, Feb. 2014.

[62] H. Lu and A. Seabaugh, "Tunnel field-effect transistors: State-of-the-art," *Journal of the Electron Devices Society*, pp. 44–49, May 2014.

[63] International Technology Roadmap for Semiconductors (ITRS), "Emerging research devices, 2013 edition." http://www.itrs.net, 2013.

6

Carbon-Based Tunnel Field-Effect Transistor

6.1 Introduction

Since Iijima first reported the formation of needle-like tubular carbon structures in the early 1990s, the research on carbon-based electronics, driven by some ground-breaking discoveries and pioneering processing techniques, has advanced a long way [1–6]. Recently, there has been demonstration of a first computer built entirely using carbon nanotube (CNT)-based transistors [6]. The allotropic forms of carbon such as CNT and graphene possess unique physical properties and exhibit outstanding carrier mobilities. These properties of CNT and graphene, coupled with small geometric size, provide opportunities to be exploited in field-effect transistors (FETs). It is predicted that the transistors fabricated using CNTs can outperform silicon-based FETs in terms of energy efficiency by more than an order of magnitude [6]. As a result, researchers are actively exploring the possibility of employing CNT and graphene in mainstream CMOS technology and are making important breakthroughs.

Since the research on Tunnel Field-Effect Transistors (TFETs) is still in the exploratory phase, researchers are experimenting with different material systems to realize TFETs, including various allotropes of carbon. The bandgaps of CNT and graphene are tunable and sufficiently small for use in a TFET. The outstanding transport property of CNTs and graphene, excellent electrostatic control of the gate over the channel potential, and capability of operating at ultralow voltages make carbon-based TFETs especially attractive for future energy-efficient circuits [7–9]. Additionally, due to the structural similarity of a TFET and a MOSFET, advancements in carbon-based FETs are directly helping research on carbon-based TFETs.

In this chapter, we will first look at CNT-based TFETs, followed by graphene-based TFETs. We will start with an overview of the material properties of CNT/graphene and then look at some of the implementation of TFETs based on them, highlighting their unique characteristics. We will discuss some of the device optimization techniques that are specific to carbon-based TFETs.

Finally, we will look at some of the biggest challenges in deploying carbon-based TFETs for future applications and possible solutions to tackle these challenges.

6.2 Carbon Nanotubes (CNTs)

Carbon nanotubes (CNTs) are hollow cylindrical nanostructures of carbon in which the length is normally several orders of magnitude greater than the diameter. A CNT that is made up of a single shell of carbon atom is called a single-walled CNT (SWCNT) and has a diameter of an order of nanometer. A CNT that is made up of multiple concentric shells of carbon atoms is called multi-walled CNT (MWCNT).

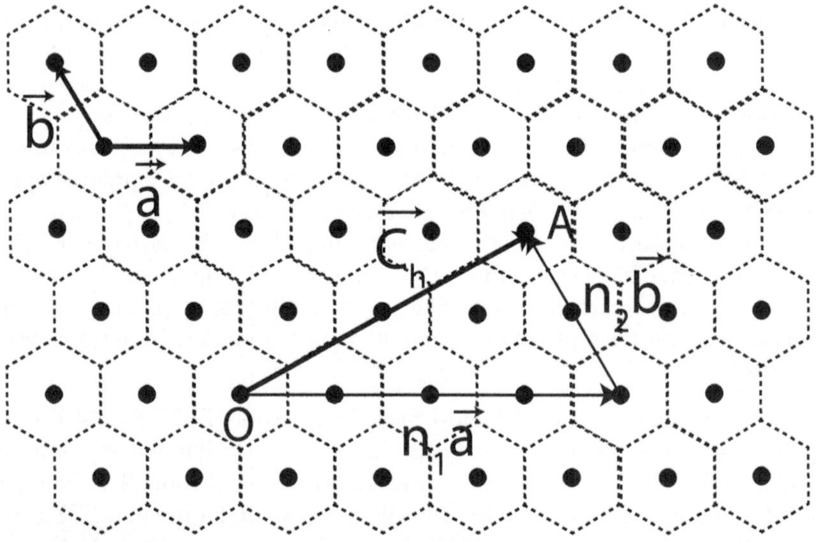

FIGURE 6.1
Hexagonal lattice in 2-D graphene layer. In this figure $n_1 = 4$, $n_2 = 2$ and $\vec{C}_h = 4\vec{a} + 2\vec{b}$. The point A can be represented as $A(4, 2)$.

A SWCNT can be viewed as a structure obtained by rolling-up a monoatomic 2-D graphene layer into a seamless cylinder. A graphene layer consists of a 2-D, atomic-scale, hexagonal lattice made up of carbon atoms as shown in Figure 6.1. It forms the basic building element of other allotropes of carbon such as graphite, charcoal, CNT, and fullerenes. There are an infinite number of different arrangements of carbon atoms that can be obtained by rolling-up a graphene sheet to form a CNT, as described below.

For a 2-D graphene, a lattice point O is designated as the origin as shown in Figure 6.1. Any other lattice point A can be represented as $A(n_1, n_2)$, where n_1 and n_2 are integers representing the position of A in the lattice in terms of the unit vectors $\vec{a}$ and $\vec{b}$. Here, the chiral vector ($\vec{C_h}$) representing the position of A and the chiral angle θ can be written as

$$\vec{C_h} = n_1\vec{a} + n_2\vec{b} \tag{6.1}$$

$$\tan\theta = \frac{\sqrt{3}n_2}{2n_1 + n_2}. \tag{6.2}$$

A CNT formed by rolling up the graphene sheet to form a seamless cylinder such that the lattice point $A(n_1, n_2)$ is superposed to the origin O is designated as (n_1, n_2) CNT. The structure of any given CNT corresponds to one of the points $A(n_1, n_2)$ in the 2-D graphene lattice. The chiral angle θ, representing the "twist" in the cylindrical tube, is the angle between the unit vector $\vec{a}$ and $\vec{C_h}$ and lies between 0° and 30°. The CNT obtained when $n_2 = 0$ or $\theta = 0$ is called a zigzag CNT. The CNT obtained when $n_1 = n_2$ or $\theta = 30°$ is called an armchair CNT. The other types of CNTs with $0 < \theta < 30°$ are called chiral CNTs.

Different arrangements of carbon atoms in CNTs, as represented by the indices (n_1, n_2), give rise to different band structures in CNTs. Unless the diameter of a CNT is very small, the band-structure of the CNT would resemble that of graphene [1]. However, in contrast to graphene, due to the tubular structure of a CNT, a periodic boundary condition is imposed along the circumferential direction, leading to quantization of electronic states in the circumferential direction [1]. Since different boundary conditions are imposed for CNTs with different chiral vectors, the band structure and transport properties of CNTs vary widely with the diameter and chirality. Additionally, the properties such as the bandgap of CNTs do not vary monotonically with diameter and chirality but show a periodic trend. For instance, a CNT is metallic when $n_1 = n_2$, it is semi-conducting with a very small bandgap when $(n_1 - n_2)$ is a multiple of 3, and it is generally semi-conducting in all other cases.

Some of the commonly used techniques to grow SWCNTs are arc discharge, laser ablation, and chemical vapor deposition (CVD) [10]. The methodology employing CVD to grow SWCNTs has the advantage of a low temperature requirement, simple process, good controllability, and low cost [10]. However, as-synthesized CNTs show wide variation in diameter, chirality, bandgap, and transport properties [11]. For electronic applications, it is required that the semiconductor-type CNTs are separated from the metallic CNTs. The post-synthesis sorting of CNTs or controlled breakdown of metallic CNTs with Joule heating can be used to isolate the semiconducting CNTs with the required properties [6, 10, 11].

CNTs exhibit the following properties, which make them particularly attractive for applications in FETs:

1. Due to the long mean free paths in carbon nanotubes, quasi-ballistic transport can be observed in CNTs even at comparatively long channel lengths.
2. The wrap-around gate architecture naturally fits into a CNT and a very good electrostatic control of the gate over the channel potential can be obtained. Additionally, CNTs are amenable to high-κ gate dielectric, which further helps in obtaining better gate control.
3. Due to the inert nature of CNTs, high electric current can be sustained in the CNTs.

As a result, CNTs are being widely investigated for use as a channel material in FETs for future CMOS applications. Experimentally, quasi-ballistic transport and ON-state current (I_{ON}) exceeding silicon-based MOSFETs have already been reported for CNT FETs.

6.3 CNT TFETs

The properties of CNTs that are advantageous for conventional FETs are beneficial for TFETs also. However, some of the properties of CNTs that are especially suitable for implementing TFETs are as follows [12, 13]:

1. CNT is a direct bandgap material with a small effective carrier mass. This results in higher band-to-band tunneling (BTBT) in a CNT TFET compared to a TFET based on an indirect bandgap material such as silicon. In indirect bandgap materials, some other processes such as phonon emission or absorption are required to conserve the transverse momentum of the carriers, which decreases the BTBT rate [7].
2. The bandgap of CNTs is moderate and it can be varied over a wide range based on their diameter and chirality.
3. The one-dimensional electronic transport in a CNT allows tight control of the gate voltage over the bands in the CNT channel. Therefore, the valence band in the channel of a CNT TFET can be raised above the conduction band in the source with the application of a small gate voltage.

4. The small diameter of a CNT boosts BTBT resulting in a larger ON-state current.

Therefore, CNT TFETs have attracted a great deal of attention. In fact, one of the earliest experimental demonstrations of a TFET exhibiting subthreshold swing smaller than 60 $mV/decade$ was for a CNT TFET in 2004 [14]. In this section, we will look at a simple implementation of CNT TFET presented in [15] and discuss the distinct properties of CNT TFETs.

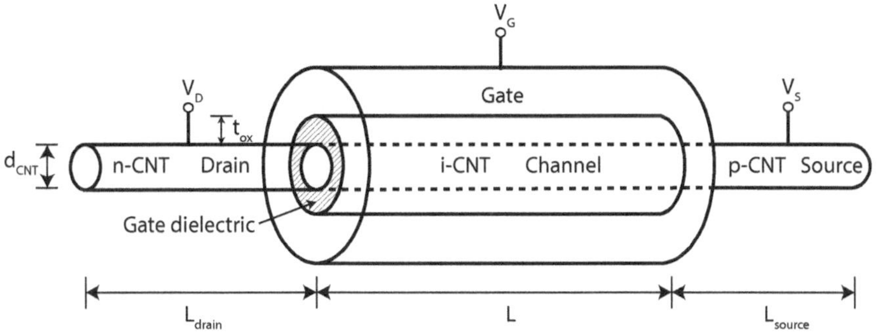

FIGURE 6.2
Schematic representation of CNT TFET [15].

Figure 6.2 shows the schematic representation of the CNT TFET that has been investigated in detail using simulation in [15]. The channel consists of (13, 0) zigzag CNTs with diameter $d_{CNT} = 1\ nm$. The TFET has a gate-all-around architecture consisting of $t_{ox} = 2\ nm$ HfO_2 ($\kappa = 16$). The source and drain regions have a linear doping concentration of 0.8 $dopants/nm$. An important point to note is that the source and drain regions are not covered with high-κ gate oxide, since high-κ gate oxide over source increases the fringe field, which is found to result in a lower tunneling current [15]. On the other hand, increasing the gate-drain fringing can suppress the ambipolar conduction, though it could degrade the performance at high frequency due to larger fringe capacitances [7].

The simulation results in [15] show that, with supply voltage $V_{DD} = 0.3\ V$, the minimum attainable OFF-state current is $I_{OFF} \approx 5 \times 10^{-6}\ \mu A$ at $300K$. However, under these conditions, the ON-state current is extremely small, $I_{ON} \approx 1\ \mu A$.

The phenomena of ballistic transport and quantum capacitance effect are particularly important in a CNT TFET. In the rest of this section, these phenomena are described in detail.

6.3.1 Ballistic Transport

The large current carrying capability, very low intrinsic resistivity, and quantized conductance in a CNT have been experimentally demonstrated and theoretically established [16, 17]. The conduction mechanism in a CNT can be either ballistic or diffusive. A ballistic transport occurs when the electrons get transmitted without any impurity, defect, or phonon scattering. When the length of a CNT is less than the mean free path length in the material, ballistic transport is possible. The mean free path length in a CNT can be quite long (more than a few hundred nanometer). Therefore, ballistic transport can be observed at room temperature in a CNT even for lengths close to 100 nm. The ballistic transport is important in CNT transistors since the conduction mechanism plays an important role in determining the electrical characteristics such as I_{ON} and subthreshold swing of the device. Additionally, if the transport mechanism is purely ballistic in a transistor, then there is no energy dissipation in the channel.

For the TFET shown in Figure 6.2, simulation results in [15] show that the subthreshold swing is lower than 60 $mV/decade$ at $300K$, both under ballistic mode of operation and when phonon scattering is taken into account. However, the presence of phonon scattering leads to degradation of the subthreshold swing since phonon-absorption-assisted transport plays a significant role in the OFF-state [15, 18]. As a result, in the presence of phonon scattering, subthreshold swing shows stronger dependence on the temperature. In the ON-state, properties of the tunneling barrier in a TFET have a greater role in determining the drive current of the TFET and the effect of phonon scattering in the channel is rather limited.

6.3.2 Quantum Capacitance

The concept of quantum capacitance has been used to describe the charge–voltage relationship in one-dimensional (1-D) systems such as CNTs and two-dimensional (2-D) systems such as highly conducting 2-D electron gas and graphene [19–21]. Quantum capacitance becomes important in a MOS capacitor when at least one electrode has low density of states (DOS). In such cases, total gate capacitance can be modeled by adding another capacitor C_Q in series with the normal gate-oxide parallel-plate capacitor C_{ox}. The gate-oxide parallel-plate capacitance is given as

$$C_{ox} = \frac{\varepsilon W L}{t_{ox}}, \tag{6.3}$$

where ε and t_{ox} are the dielectric constant and thickness of the gate oxide, respectively, and W and L are the width and the length of the MOSFET, respectively. The total gate capacitance C_G is given as

$$C_G = \frac{C_{ox} C_Q}{C_{ox} + C_Q}. \tag{6.4}$$

When the parallel-plate capacitance C_{ox} increases, for example, due to reduction in the thickness of the gate-dielectric or increase in the dielectric constant of the gate material, the total gate capacitance C_G becomes strongly dependent on the capacitance C_Q. Additionally, the capacitance C_Q is dependent on the average density of states near the Fermi level and can be given as follows for a MOSFET [22]:

$$C_Q = q \frac{\partial Q_{ch}}{\partial \Phi_f^0}, \tag{6.5}$$

where Q_{ch} is the charge in the channel, Φ_f^0 is the maximum surface potential, and q is the charge of an electron. In the ON-state, the capacitance can be given as [22]

$$C_Q = q^2 D \left(E_f^s - \Phi_f^0 \right), \tag{6.6}$$

where E_f^s is the source Fermi energy level and D is the density of state function. The capacitance C_Q, related to the density of states of the electrodes (gate and channel), is called the quantum capacitance. For a 1-D system, where the density of states decreases with increasing energy, quantum capacitance C_Q can become very small. In a CNT TFET, being a 1-D system, the total capacitance can be given as

$$C_G \approx C_Q. \tag{6.7}$$

When this condition C_G approaches C_Q it is called the quantum capacitance limit (QCL). When a TFET works in the QCL, the charge induced in the channel is quite low, improving the intrinsic delay of the transistor [15]. Additionally, almost perfect gate control is obtained in CNT TFETs due to operation under QCL [12].

6.4 Device Optimization of CNT TFETs

In a CNT TFET, some of the device optimization techniques that can improve the electrical characteristics are as follows:

1. **Diameter and chirality of CNT:** The bandgap of CNTs varies within a wide range with diameter and chirality. For example, the bandgap of (10,0), (13,0), and (16,0) CNTs are 1.053, 0.817, and

0.667 eV, respectively [15]. The simulation results presented in [15] show that the I_{ON} of the CNT TFET increases with a decrease in the bandgap of the material. The decreasing carrier effective mass in (10,0), (13,0), and (16,0) CNTs reinforces the increasing trend of the I_{ON} [15]. It should be noted that the ambipolar current also significantly increases with a decrease in the bandgap of the CNTs [15]. Therefore, techniques that can reduce the ambipolar conduction in a CNT TFET are important, especially in CNT TFETs with smaller bandgaps.

2. **Source doping:** With an increase in the source doping, the thickness of the tunneling barrier reduces, resulting in an increase in the I_{ON} of the TFET [7, 23]. However, in a CNT TFET, the maximum source doping concentration should be realistic in terms of the ratio of the dopant ions and carbon atoms per unit length [23]. For example, in a (13, 0) CNT, the number of carbon atoms per nanometer is 122 [7, 15]. Therefore, the number of dopants per unit length should be substantially lower than 122 $dopant/nm$. Additionally, the impact of increasing the source doping on the subthreshold swing should also be considered.

   It should be noted that ion implantation techniques that are commonly employed to obtain the required doping profile cannot be applied to CNTs since substituting a carbon atom with a dopant atom in a CNT destroys the properties of the nanotube [12, 24]. As a result, obtaining sophisticated doping profiles, as in conventional CMOS technology, is difficult in CNT-based devices. Alternative techniques of obtaining source and drain contacts such as controlling the electrostatics of the CNTs using molecules, additional gates or metal ions are often applied [13, 24].

3. **Drain doping:** In order to reduce the ambipolar conduction, the drain doping can be reduced, which increases the tunneling barrier width on the drain side [7, 23]. However, decreasing the drain doping results in an undesirable increase in the series resistance. Therefore, a lowly doped spacer region combined with a highly doped extended region can be employed in the drain region [7].

4. **Gate source/drain underlap/overlap:** The gate electrode can be underlapped or overlapped with respect to the channel to optimize the characteristics of a CNT TFET. An underlap on the source side results in a higher I_{ON} and a smaller subthreshold swing, while an overlap on the source side results in a lower I_{ON} [15]. The optimal value of the gate underlap on the source side that would result in a maximum I_{ON} is dependent on the thickness of the gate dielectric. On the drain side, a slight overlap is beneficial in obtaining a gentler transition in the potential and decreasing the ambipolar

conduction [23]. However, the overlap should be such that the parasitic capacitances do not increase appreciably [23].

5. **Strain:** The application of axial tensile strain in a CNT TFET results in a considerable decrease in the I_{OFF} and an improvement in the subthreshold swing of the device due to bandgap modulation [25]. However, the application of strain is accompanied by an undesirable decrease in the I_{ON} of the TFET [25, 26]. Therefore, strain can be optimized to obtain the required trade off between the I_{ON} and I_{OFF} in a CNT TFET.

6. **Multiple CNTs:** The I_{ON} in a CNT TFET can be increased by employing multiple single-walled CNTs. For a multiple-CNT TFET, the diameter and the position of the individual CNTs can be optimized to increase the I_{ON}/I_{OFF} ratio [27]. Additionally, from a reliability perspective, employing multiple CNTs is better, since even when some tubes fail, the transistor will keep operating.

6.5 Challenges and Future Perspectives of CNT TFETs

Some of the technological challenges in employing CNT TFETs for future applications are similar to challenges in employing CNT FETs for future applications. These challenges are as follows:

1. **Mass production:** For future wide-scale nanoelectronics applications, billions of CNTs must be fabricated at certain positions on an integrated circuit (IC). These CNTs are expected to have the same electronic properties. Additionally, the cost of mass production of CNT-based ICs needs to be kept low in order to make them competitive with the current CMOS technology. These are very difficult technological challenges to overcome [28].

2. **Doping in CNTs:** Doping in a CNT is a difficult technological challenge since the electronic properties of CNT get destroyed if carbon atoms in the tube are replaced using conventional techniques such as ion implantation.

3. **Reliability issues:** The breakdown of CNT at high electric fields and temperature gradients is a serious problem. Additionally, CNTs show degradation when exposed to oxygen and passivation techniques for CNTs need to be developed.

Some of the biggest challenges that are specific to CNT TFETs are:

1. **Low ON-state current:** The ON-state current, OFF-state current, delay, and energy efficiency of a CNT TFET and a similar CNT FET have been compared in detail in [15]. It is demonstrated that the I_{ON} of a CNT TFET is about 3× smaller than the I_{ON} of a CNT FET. Nevertheless, the intrinsic delay, which is computed as
$$\tau = \frac{Q_{ON} - Q_{OFF}}{I_{ON}}, \tag{6.8}$$
where Q_{ON} and Q_{OFF} are the total charge induced in the transistors under the ON-state and the OFF-state conditions, respectively, is found to be similar for the CNT TFET and the comparable CNT FET. This is because CNT TFETs work close to QCL and the charges involved are low resulting in a smaller intrinsic delay of the CNT TFET [15]. However, when the TFET is employed in a circuit, the low I_{ON} of the CNT TFET will result in a higher delay due to higher capacitive loading of the interconnects [15, 29]. Further, it is demonstrated in [15] that a CNT TFET exhibits a lower power delay product (PDP) compared to a CNT FET. Therefore, CNT TFETs hold great promise for future low-power energy-efficient circuits.

2. **Channel length scalability:** When the channel length of a CNT TFET is reduced below 10 nm, there is a sharp increase in the direct source-to-drain tunneling. As a result, the I_{OFF} and the subthreshold swing of a CNT TFET increase with aggressive channel length scaling [7, 23]. Additionally, the total capacitance of the transistor that is dominated by the source/drain fringe capacitance does not decrease appreciably due to scaling [23]. The increase in the I_{ON} due to scaling is not appreciable since the I_{ON} is mainly determined by the characteristics of the source–channel tunneling junction rather than the nature of the channel [23]. The combined result of these effects is that intrinsic delay (τ) of the TFET that meets reasonable I_{OFF} requirements does not decrease with the channel length. A near-optimum τ is obtained at a channel length of 16 nm by scaling the gate oxide thickness and the supply voltage [23]. Therefore, TFETs can outperform conventional MOSFETs in future applications at smaller device dimensions only when the problem of degradation in the I_{OFF} and the subthreshold swing with channel length scaling is solved.

6.6 Graphene

The groundbreaking experiments of Geim and Novoselov in 2004, which isolated and demonstrated the field-effect in graphene, created great excitement among device engineers and spurred hectic research activity on possible electronic applications of graphene [30]. Since 2004, there has been tremendous progress in understanding of the properties of graphene and the techniques to exploit graphene in FETs and as interconnects for integrated circuits have been explored.

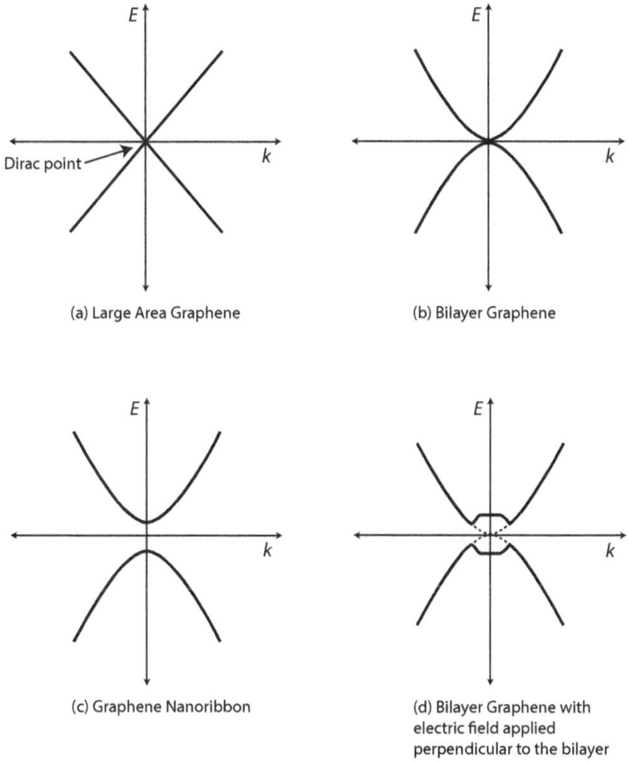

FIGURE 6.3
Schematic representation of bandgap in (a) large-area graphene, (b) bilayer graphene, (c) graphene nanoribbon, and (d) bilayer graphene with transverse electric field [31–35].

Graphene is a two-dimensional (2-D) atomic layer of carbon atoms that is sufficiently isolated from the environment to be considered free-standing [31].

The carbon atoms are arranged in a hexagonal lattice. Among the four valence electrons of the carbon atoms, three electrons are part of sp^2 hybrid orbitals combining the orbitals of s, p_x and p_y that constitute the s-bonds. The fourth p_z electron forms the p-bond. The electrons in the sp^2 orbitals are strongly localized, therefore, the electronic properties of the graphene are primarily determined by the p-electrons. The interaction between the p_z-orbitals in graphene results in delocalization of the p-electrons forming bonding π-bands and antibonding π^*-bands. The band structure of graphene has a linear dispersion relationship with momentum proportional to the energy. In a large-area graphene, the conduction band and the valence bands in graphene are conical and meet at K points of the Brillouin zone, resulting in zero bandgap (as shown in Figure 6.3). As a result, large-area graphene exhibits semi-metallic properties. It has a very high carrier mobility at room temperature, high thermal conductivity, and a very high tensile strength.

The primary difficulty in isolating graphene arises due to the instability of the lower dimension crystals and the impossibility of growing 2-D crystals [31]. However, after Geim and Novoselov's pioneering work, several methods of obtaining graphene have been successfully demonstrated. Graphene can be prepared using mechanical exfoliation, obtaining epitaxial graphene using thermal decomposition of silicon carbide on top of SiC wafers, or using chemical vapor deposition (CVD) over metallic surfaces such as nickel [31, 32]. The mechanical exfoliation method, though quite simple, is limited to laboratory setups. The methods of obtaining epitaxial graphene and CVD can be employed in wide-scale production. In order to fabricate transistors on a graphene sheet, conventional techniques used in the current CMOS technology can be employed, unlike CNTs. This is one of the advantages of graphene that makes it suitable for future CMOS technology.

Graphene has been studied as a possible channel material in FETs for more than a decade. The atomic-layer thickness of graphene provides excellent gate control, suppresses short-channel effects, and allows scaling to smaller device dimensions. The high carrier mobility of graphene is another attraction to employ graphene in future transistors [32]. However, the absence of bandgap in graphene implies that the FETs realized using graphene cannot be effectively switched off and a very low I_{ON}/I_{OFF} ratio is achievable in a graphene FET. This is a serious handicap of graphene FETs in logic circuit applications.

In order to tackle the problem of the absence of bandgap in graphene, several techniques to open a useful bandgap have been explored. The most commonly employed techniques to open a bandgap in graphene are making use of quantum confinement effects in graphene nanoribbons (GNR) and applying an appropriate electric field in bilayer graphene (BLG).

1. **Quantum confinement:** GNR is a narrow strip of graphene. GNR can be viewed as an unrolled single-walled CNT and exhibits proper-

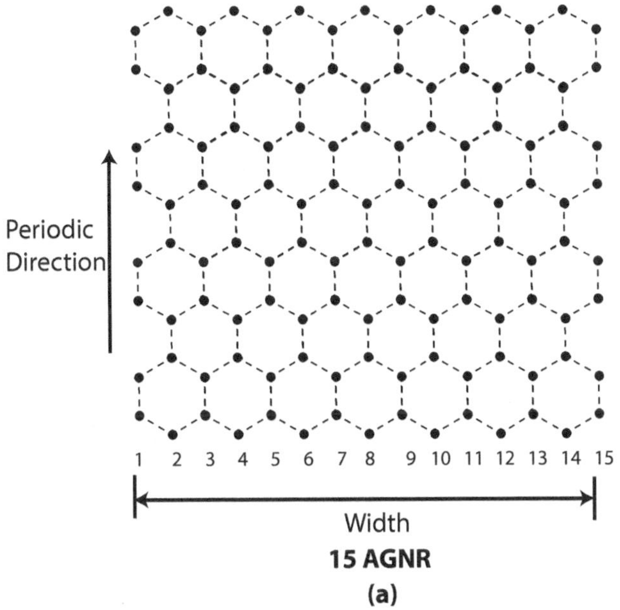

FIGURE 6.4
Schematic representation of armchair GNR (AGNR).

ties of a quasi-one-dimensional material. Due to the boundary conditions imposed on the wave functions in the direction perpendicular to the axis of a GNR, a quantum confinement–induced bandgap opens up in the GNR (as shown in Figure 6.3). The bandgap in a GNR strongly depends on the width and the crystallographic orientation. This dependence can be exploited in transistors to optimize the device characteristics by tuning the electronic structure of the GNRs.

Two common types of GNRs are the armchair GNR (AGNR) and zigzag GNR (ZGNR), named after the shape of the edges. An AGNR with N_a C-C dimer lines along the width of the GNR is known as N_a-AGNR. Figure 6.4 shows a 15-AGNR. A ZGNR with N_z zigzag lines across the GNR is known as a N_z-ZGNR. Figure 6.5 shows an 8-ZGNR. The geometric structure for both AGNR and ZGNR repeats in the direction perpendicular to the width. Since the carbon atoms at the edge of the GNR are not saturated, the dangling s-bonds at the edges are passivated, for example, by hydrogen atoms.

Both theoretical predictions and experimental results show that AGNR and ZGNR possess bandgaps that scale inversely with the

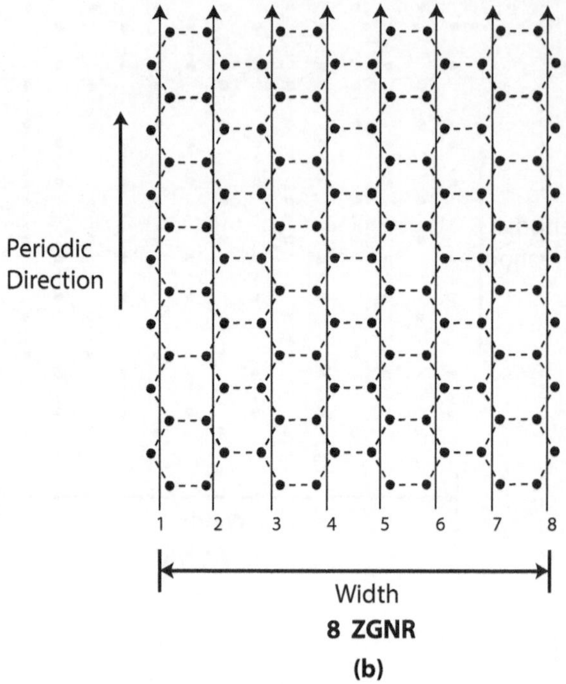

FIGURE 6.5
Schematic representation of zigzag GNR (ZGNR).

width of the GNR [33, 36, 37]. However, the origin of the bandgaps for both types of GNR are different [33]. The bandgaps in AGNRs originate from quantum confinement while bandgaps in ZGNR originate from a staggered sublattice potential [33]. Experimentally, a bandgap of 0.18 eV for a width of 10 nm and a bandgap of 0.5 eV for a width of 2.5 nm have been reported for AGNR [37]. The variations in bandgap in AGNR exhibit three distinct family behaviors. A given N_a-AGNR belongs to one of the families $3p$, $3p+1$ or $3p+2$ based on if N_a can be represented as $3p$, $3p+1$ or $3p+2$ where p is a positive integer.

2. **Biasing bilayer graphene:** Bilayer graphene (BLG) consists of two layers of graphene and has zero bandgap like monolayer graphene. The valence band and the conduction bands have a parabolic shape near the K point (as shown in Figure 6.3(b)). However, when an electric field is applied perpendicular to the bilayer, a bandgap opens up as shown in Figure 6.3(d) taking a Mexican-hat shape [32, 34, 38–40]. The opening of a bandgap in a BLG was predicted by theory and has been experimentally verified. The

bandgap depends on the strength of the electric field perpendicular to the bilayer graphene. The bandgap can be $0.2 - 0.25\ eV$ for a high electric field of the order of $1 \times 10^7\ V/cm$ [32, 34, 38]. A bandgap in BLG can also be opened using top-bottom asymmetric chemical doping [41, 42]. It has been shown that a bandgap can be obtained in BLG due to asymmetric charge distribution between the two layers, for example, by selective adsorption of n- or p-type dopant molecules on one of the layers [42]. The bandgap obtained by chemical doping scales linearly with induced carrier density where the bandgap varies as $47\ meV/10^{13}\ cm^{-2}$ for n-type dopants and $40\ meV/10^{13}\ cm^{-2}$ for p-type dopants [42].

There are other techniques to open a bandgap in graphene, such as applying uniaxial strain to a large-area graphene, using epitaxial graphene grown on SiC or using epitaxial bilayer graphene [32, 43]. It is worth mentioning that although several methods to open a bandgap in graphene have been explored, obtaining a usable bandgap in graphene in a controlled way is still the biggest challenge for graphene transistors [32].

6.7 Graphene TFETs

There are certain properties of graphene that can be exploited in TFETs to obtain superior electrical characteristics:

1. A small bandgap can be opened up in graphene using various techniques. This enables obtaining a high I_{ON} in graphene-based TFETs.
2. The atomic-layer thickness of graphene allows superior gate control and diminished short-channel effects.
3. Graphene has excellent carrier mobility and small effective mass. This can be exploited in TFETs to obtain higher I_{ON}.
4. Graphene is suitable for planar processing and large-scale integration of graphene transistors is possible, in contrast to CNT-based transistors.
5. Graphene can form different types of heterostructures and a small effective bandgap can be obtained in graphene-based heterostructures.

Due to these interesting properties of graphene, several research groups have explored utilizing graphene in TFETs. In this section, we will look at some of the graphene-based TFETs in which different techniques to obtain a usable bandgap have been applied.

6.7.1 GNR-Based TFET

Successful demonstration of GNR FETs with smooth edges and a high I_{ON}/I_{OFF} ratio have encouraged researchers to investigate TFETs based on GNRs [31, 32, 44–47]. The most distinguishing feature of a GNR TFET is the important role of the edge bond relaxation in determining the electrical characteristics of the device [46]. The edge bond relaxation originates due to an abrupt change in the characteristics of the carbon–carbon bonds at the edges of the GNR. For example, the carbon–carbon bonds at the edges are approximately 3.5% shorter compared to the length of the carbon–carbon bonds inside the GNR [48]. This is in contrast to a CNT TFET where no such effect can be observed.

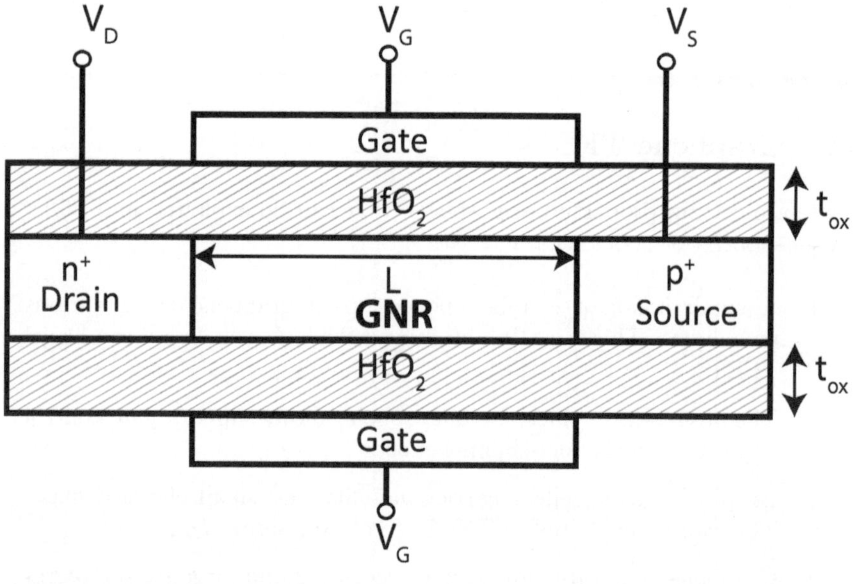

FIGURE 6.6
Schematic representation of a GNR TFET from [46]. Device parameters are $L = 30\ nm$ and $t_{ox} = 1.5\ nm$.

Figure 6.6 shows a schematic representation of a GNR TFET [46]. The electrical characteristics obtained using an atomistic quantum transport simulator for the GNR TFET shown in Figure 6.6 have been presented in [46]. The doping density of the source and the drain extension regions are assumed to be

0.01 *dopant/atom* and the gate consists of HfO_2 having a dielectric constant $\kappa = 16$. The supply voltage is assumed to be $V_{DD} = 0.4\ V$. The transfer characteristics of a symmetrically doped GNR TFET exhibit ambipolar behavior. Therefore, developing techniques that can suppress the ambipolar conduction in GNR TFETs is important [46].

In a GNR TFET, to optimize the characteristics of the device, GNRs with different width and edge configurations can be chosen. As an illustration, the channel can be composed of 13-AGNR having a width of around 1.6 *nm* and a bandgap $E_g \approx 0.86\ eV$. It is observed that the edge bond relaxation leads to decrease in the minimal leakage current by about 10^6 and decrease in the subthreshold swing from 51 mV/dec to 14 mV/dec for 13-AGNR TFET [46]. These improvements in the characteristics of 13-AGNR TFET can be attributed to the increase in the bandgap from 0.71 eV to 0.86 eV due to edge bond relaxation. In general, the edge bond relaxation leads to an increase in the bandgap for the $(2p + 1)$AGNR family. Therefore, similar improvement due to edge bond relaxation in the characteristics of the TFET is expected when the channel is composed of any AGNR of the $(2p+1)$AGNR family. For a 12-AGNR based TFET, the edge bond relaxation leads to an increase in the minimal leakage current and a degradation of the subthreshold swing due to the decrease in the bandgap [46]. A similar degradation due to edge bond relaxation in the characteristics of the TFET is expected when the channel is composed of any AGNR of the $3p$-AGNR family.

It has also been demonstrated that the maximum achievable I_{ON}/I_{OFF} ratio increases considerably when the supply voltage V_{DD} is less than 0.4 V [46]. As an illustration, at $V_{DD} = 0.3\ V$, the maximum achievable I_{ON}/I_{OFF} ratio is found to be 10^{11}. At $V_{DD} = 0.5\ V$, the maximum achievable I_{ON}/I_{OFF} ratio decreases to 10^6. In the OFF-state, for $V_{DD} < 0.4\ V$, only the direct source-to-drain tunneling mechanism is effective resulting in an extremely small I_{OFF} and a high I_{ON}/I_{OFF} ratio. However, in the OFF-state, at higher supply voltages, BTBT is also effective resulting in a higher I_{OFF} and lower I_{ON}/I_{OFF} ratio. It is important to note that the studies in [46] assume GNR with smooth edges. As the edges of a GNR TFET become rougher, the I_{OFF} drastically increases while the I_{ON} remains almost constant [49]. The I_{OFF} increases due to the reduction in the bandgap and an increased source-to-drain tunneling leakage through the potential barrier of the gate. Therefore, the line edge roughness in GNR TFETs leads to a deterioration of the subthreshold swing and the I_{ON}/I_{OFF} ratio [49]. Currently, controlling the width of a GNR and its edge structure with atomic precision is an enormous challenge [31]. Therefore, realizing GNR TFETs with predictable behavior is very difficult.

6.7.2 Bilayer-Graphene TFET

Bilayer-graphene (BLG) can be used as a channel material in a TFET because a small and tunable bandgap can be opened up in BLG. A bandgap is obtained in bilayer graphene applying an electric field perpendicular to the graphene layer or by chemical doping [50–52]. The charge asymmetry between the graphene layers leads to a tunable bandgap in a BLG. The bandgap in a BLG can be modeled as [42, 50]

$$E_g = \frac{\Delta.\gamma}{\sqrt{\Delta^2 + \gamma^2}}, \tag{6.9}$$

where E_g is the bandgap in the BLG, Δ is the energy difference between the graphene layers due to charge asymmetry between the layers, and γ is the interlayer hopping energy. The perpendicular electric field in BLG TFET is obtained by employing a top and a bottom gate and maintaining a voltage difference between the top and the bottom gates [50–52].

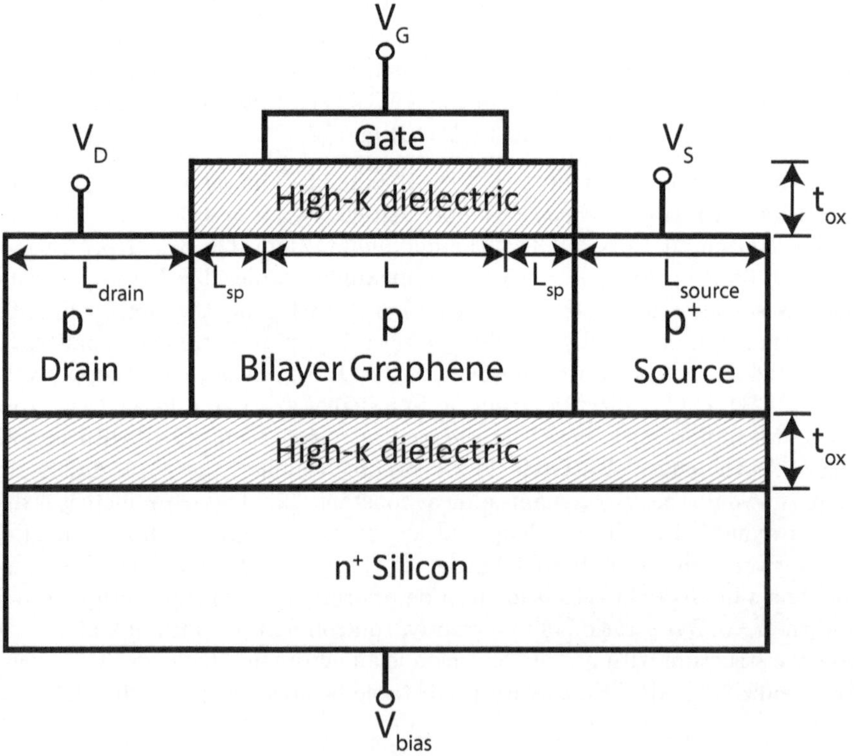

FIGURE 6.7
Schematic representation of a BLG TFET [50].

Figure 6.7 shows the schematic representation of a BLG TFET [50]. The bandgap is opened using both the vertical electric field due to back-gate voltage V_{bias} and the top-bottom asymmetric chemical doping. The source and the drain regions are formed by work function engineering of the metal-graphene contacts. The gate is composed of a high-κ dielectric with thickness $t_{ox} = 3.75\ nm$ and dielectric constant $\kappa = 25$. The performance limit of BLG TFET is investigated using quantum transport simulations [50]. Following parameters were assumed in the simulation [50]: channel length $L_G = 40\ nm$, source/drain extension lengths $L_{Source} = L_{Drain} = 20\ nm$, and top-gate underlap $L_{sp} = 5\ nm$. For the BLG TFET shown in Figure 6.7, at back-gate voltage $V_{bias} = 0.9\ V$ and supply voltage $V_{DD} = 0.2\ V$, the important electrical parameters computed using simulations are $I_{OFF} = 0.023\ \mu A/\mu m$, $I_{ON} = 67\ \mu A/\mu m$, $I_{ON}/I_{OFF}\ ratio = 2910$ and Subthreshold Swing= 35 $mV/decade$ [50]. As the supply voltage is increased, due to the ambipolar conduction, the I_{OFF} increases and the subthreshold swing degrades. The back-gate voltage V_{bias} serves the purpose of opening the bandgap in BLG, as well as, adjusting the threshold voltage of the TFET. However, it should be noted that increasing V_{bias} to adjust the threshold voltage, also results in an increase in the bandgap and the reduction in the ambipolar conduction.

An interesting aspect of BLG TFET is the electrostatic control of the bandgap. For a BLG TFET, when the transistor is in the OFF-state, the bandgap is large due to a greater difference in the potential of the top gate and the back gate. As a result, band-to-band is further suppressed due to an increase in the bandgap and a lower I_{OFF} is obtained. However, when the transistor is in the ON-state, the bandgap reduces due to a smaller difference in the potential of the top-gate and the back-gate bias [50]. As a result, the I_{ON} is further boosted in the ON-state as the bandgap is reduced. This is a key differentiating feature of a BLG TFET resulting from the dependence of the bandgap on the electrostatics of the device and can be exploited in the future TFETs.

6.7.3 Graphene-on-SiC-Based TFET

It is found that epitaxial graphene grown on the Si-terminated face of SiC exhibits a bandgap $E_g = 0.26\ eV$. Therefore, epitaxial graphene on SiC can be employed in TFETs [43, 53]. The graphene-SiC interaction leads to breaking of the sub-lattice symmetry resulting in an opening of the bandgap in graphene [53]. The bandgap decreases as the number of layers in the graphene sample increases [53].

A TFET employing graphene on SiC is proposed and is shown in Figure 6.8 [43]. Using an analytical model, it is shown that the proposed device can achieve an $I_{ON}/I_{OFF} = 10^4$ at $V_{DD} = 0.15\ V$ for channel length greater than 30 nm [43].

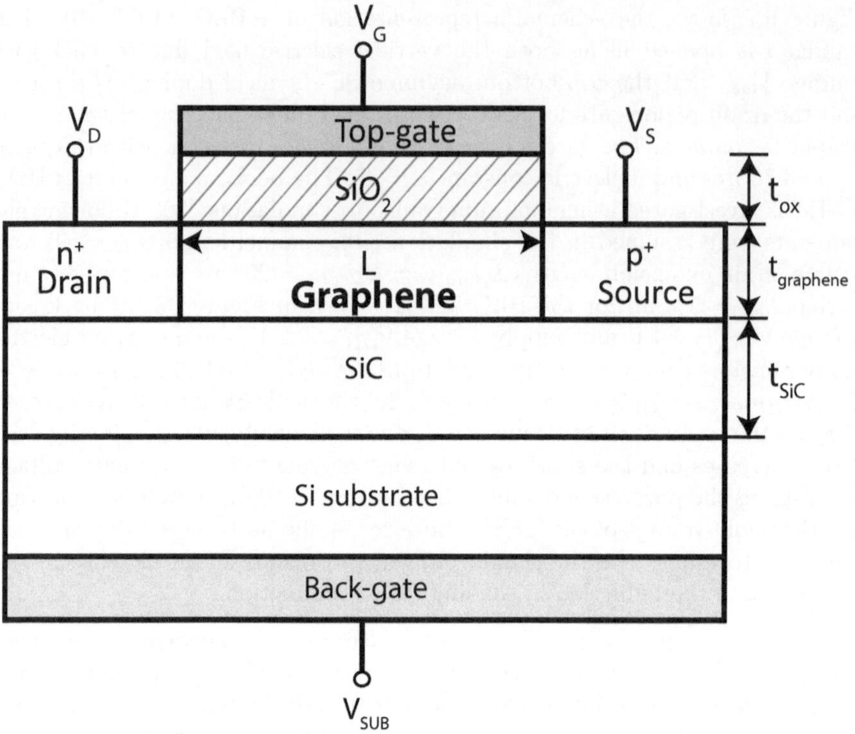

FIGURE 6.8
Schematic representation of a TFET based on epitaxial graphene on SiC. Device parameters are $t_{ox} = 1\ nm$, $t_{graphene} = 1.1\ nm$, $t_{SiC} = 100\ nm$ and $L = 30\ nm$ [43].

6.7.4 Zero-Bandgap Graphene TFET

The zero-bandgap in graphene is a serious handicap for digital applications since distinct ON-state and OFF-state behaviors are required. However, for operational amplifiers and RF applications, the zero-bandgap graphene can still be employed in FETs since the transistors are not required to be switched off in these applications. The tight gate control, improved scalability, and outstanding carrier mobility can be exploited in FETs for RF applications to obtain a high cutoff frequency. However, the absence of good saturation in the output characteristics of the FETs realized using zero-bandgap graphene is a serious problem [32, 54]. On the other hand, TFETs realized using zero-bandgap graphene can exhibit superior saturation in the output characteristics [54]. For a TFET realized using zero-bandgap graphene, with an appropriate gate-drain overlap and choosing an optimized channel/drain doping concentrations, the channel-to-drain tunneling current can be significantly suppressed resulting in improved current saturation and an increased intrinsic gain of the transistor [54].

6.8 Device Optimization of Graphene TFETs

To improve the characteristics of graphene TFETs, some of the following techniques can be employed:

1. **Heterojunctions:** Carbon based heterojunctions can be employed in TFETs to boost the I_{ON} and the I_{ON}/I_{OFF} ratio. By employing the small bandgap material of the heterojunction near the source and the large bandgap material in the channel, a high I_{ON} and a low I_{OFF} can be simultaneously obtained [55, 56]. A heterojunction can be obtained in a carbon-based material using several techniques. Since the bandgap in a GNR is dependent on its width, a GNR with spatially dependent width naturally yields a heterojunction [55]. Figure 6.9 shows a GNR TFET that employs a heterojunction obtained using GNRs of different widths [56]. The channel is composed of $(2p+1)$AGNR with a bandgap $E_g = 1.22\ eV$ and the gate is composed of SiO_2 with thickness $t_{ox} = 1\ nm$. The width of the channel and the length of the channel are $W_{CH} = 1.2\ nm$ and $L_{CH} = 14\ nm$, respectively. The heterojunction length L_{HJ}, the width of the GNR in the heterojunction W_{HJ}, and the gate underlap are varied to optimize the characteristics of the TFET. The variations in the length and the width of the heterojunction modulate the bands in the heterojunction region. It has been shown that a decrease in the width of the heterojunction leads to an increase in the bandgap, decreasing the overall current, with the I_{OFF} decreasing more sharply compared to the I_{ON} [56]. Further, the increase in heterojunction length decreases the I_{ON}/I_{OFF} ratio [56].

   A heterojunction can also be obtained by partially unzipping a CNT [57, 58]. Unzipping a CNT involves cutting open a given CNT using chemical or mechanical techniques to obtain a GNR. In a partially unzipped CNT TFET, the source and the drain consist of the GNRs, while the channel is either a semi-metallic or a semiconducting CNT. The carbon heterojunction TFET achieves an excellent I_{ON} because the materials can be subtly engineered to obtain a very small effective bandgap or barrier-free tunneling, thus obtaining a high I_{ON}/I_{OFF} ratio and improved subthreshold swing [57–59].

   GNR-hexagonal boron nitride (hBN) heterostructure-based TFETs

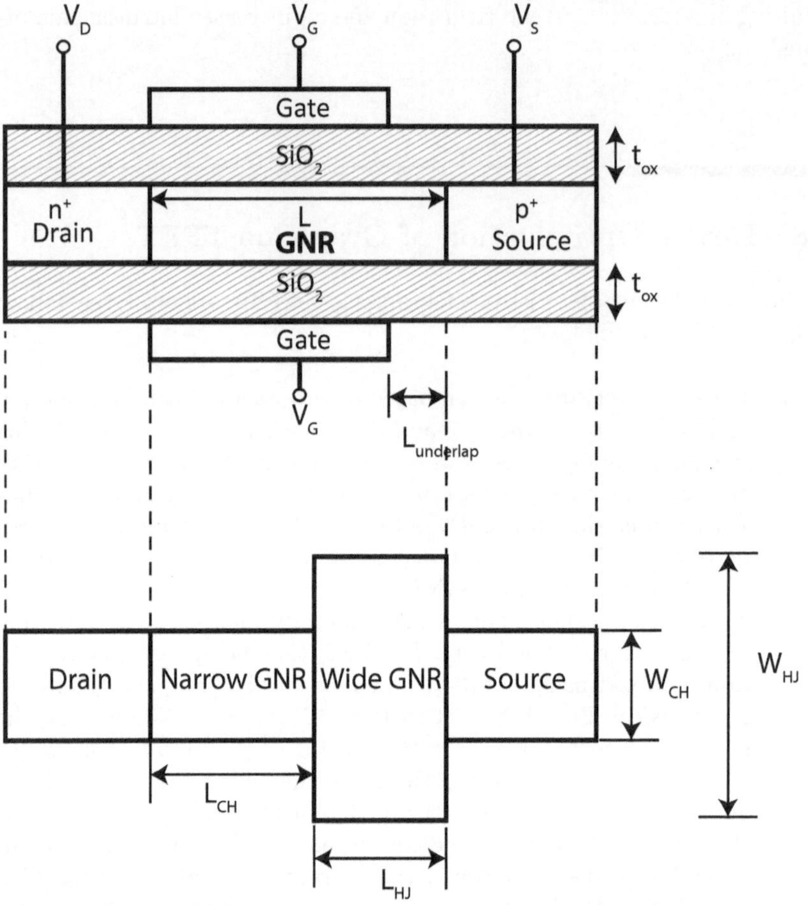

FIGURE 6.9
Schematic representation of a heterojunction GNR TFET [56].

have also been explored and have shown good I_{ON}/I_{OFF} ratio. It is found that in GNR-hBN TFET, with an increase in the width of the GNR, the I_{ON} of the TFET can be increased by trading off the I_{ON}/I_{OFF} ratio [60].

2. **Strain:** Strain can be applied in a GNR TFET to boost the I_{ON} since the bandgap is very sensitive to the strain in the GNR. As an illustration, for a 13-AGNR, 3% uniaxial strain can reduce the bandgap by 30% (from $E_g = 0.86\ eV$ to $E_g = 0.57\ eV$) [61]. It should also be noted that the impact of strain on the AGNR is strongly dependent on the width of the AGNR. For example, the trends of the bandgap variation with strain in 12-AGNR and 16-

AGNR are opposite [62]. Since the application of a uniform strain throughout a GNR TFET results in a high I_{OFF}, it is better to apply local strain in a GNR TFET [61].

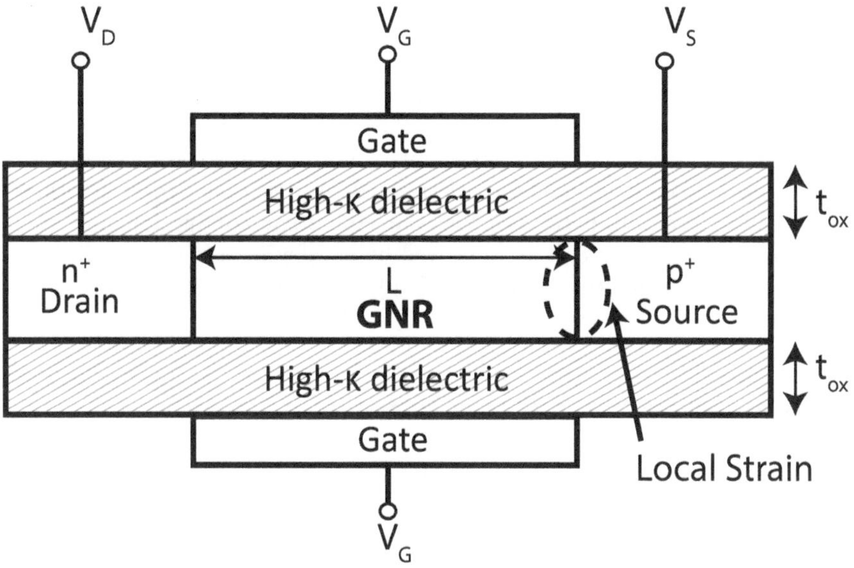

FIGURE 6.10
Schematic diagram of a locally strained GNR TFET [61].

Figure 6.10 shows a schematic diagram of a locally strained GNR TFET [61]. The local strain can be applied in a GNR TFET by placing graphene on a substrate or oxide fabricated with suitable patterns like trenches. It has been theoretically predicted that the application of strain in the GNR TFET can increase the I_{ON} by 10× while maintaining the same I_{OFF} [61].

3. **Source/drain doping:** It is difficult to dope carbon-based nanostructures using conventional doping techniques. Moreover, due to low density of states in CNTs and graphene, a high source doping leads to a high Fermi energy level in the source [12]. As a result, the carriers that are injected from the source into the channel correspond to the energy in the Boltzmann "tail" of the source Fermi–Dirac distribution function. This leads to a degradation in the subthreshold swing [12]. To obviate this problem, additional side gates in the source and the drain can be employed [12]. This allows optimization of the relative position of the Fermi energy level and the edges of the bands. The p-i-n profile required by the TFET

is realized electrostatically by the source/drain side gates and the main gate over the channel [12].

4. **Gate engineering:** Due to the atomic-layer-thick channel in a GNR TFET, the impact of lateral fringing fields at the source–channel interface is high [63]. As a result, employing a high-κ gate dielectric in a GNR-TFET results in a lower I_{ON} compared to a low-κ dielectric gate of the same thickness [63]. Therefore, low-κ spacers can be employed in GNR TFETs to reduce the fringe-field effect and exploit the full benefit of high-κ gate dielectrics in improving the device characteristics [63]. Additionally, gate underlap can be employed to suppress the ambipolar conduction and the I_{ON}/I_{OFF} ratio can be improved by employing a gate with two or three segments of different work functions [46, 64, 65].

6.9 Challenges and Future of Graphene TFETs

Due to the non-existence of a bandgap in graphene, the biggest challenge for graphene-based TFETs is to obtain a reasonable I_{OFF} while maintaining a high I_{ON}. Therefore, the primary goal for researchers working on graphene TFETs is to obtain a bandgap in a controlled way such that the transistors can be switched off. GNRs, having a reasonable bandgap, are being extensively explored for use in TFETs. However, fabricating GNRs with predictable properties such as width, edge disorder, and roughness is extremely challenging [32]. Additionally, several studies have shown that, due to the increase in the I_{OFF} at higher supply voltages, graphene TFETs can exhibit a reasonable I_{ON}/I_{OFF} ratio only at small supply voltages $V_{DD} \approx 0.2\ V$ and at a long channel length $L = 30\ nm$. Therefore, despite significant progress being made on graphene-based electronics, the ultimate goal of realizing graphene-based TFETs that can outperform conventional MOSFETs is extremely challenging.

6.10 Summary

In this chapter, we looked at some of the special properties of CNTs and graphene and how these properties can be tuned to meet the requirements of a TFET. We discussed techniques to exploit distinct properties of CNTs and graphene. We looked at some of the recently proposed implementations of TFETs based on CNTs and graphene. We also discussed some of the techniques to optimize the characteristics of carbon-based TFETs. Finally, we highlighted some of the technological and device challenges in realizing carbon-based TFETs. In the next chapter, we will look at TFETs that are realized using 1-D nanowire materials.

REFERENCES

[1] S. Iijima, "Helical microtubules of graphitic carbon," *Nature*, vol. 354, pp. 56–58, Nov. 1991.

[2] N. Hamada, S. Sawada, and A. Oshiyama, "New one-dimensional conductors: Graphitic microtubules," *Physical Review Letters*, vol. 68, pp. 1579–1581, Mar. 1992.

[3] S. Iijima and T. Ichihashi, "Single-shell carbon nanotubes of 1-nm diameter," *Nature*, vol. 363, pp. 603–605, June 1993.

[4] S. J. Tans, A. R. M. Verschueren, and C. Dekker, "Room-temperature transistor based on a single carbon nanotube," *Nature*, vol. 393, pp. 49–52, May 1998.

[5] S. J. Wind, J. Appenzeller, R. Martel, V. Derycke, and P. Avouris, "Vertical scaling of carbon nanotube field-effect transistors using top gate electrodes," *Applied Physics Letters*, vol. 80, May 2002.

[6] M. M. Shulaker, G. Hills, N. Patil, H. Wei, H.-Y. Chen, H.-S. P. Wong, and S. Mitra, "Carbon nanotube computer," *Nature*, vol. 501, pp. 526–530, Sept. 2013.

[7] S. O. Koswatta, D. Nikonov, and M. S. Lundstrom, "Computational study of carbon nanotube p-i-n tunnel FETs," in *IEDM Tech. Digest*, pp. 518–521, IEEE, 2005.

[8] M. Luisier and G. Klimeck, "Performance comparisons of tunneling field-effect transistors made of InSb, Carbon, and GaSb-InAs broken gap heterostructures," in *IEEE International Electron Devices Meeting (IEDM)*, pp. 1–4, IEEE, 2009.

[9] Y. Gao, T. Low, and M. Lundstrom, "Possibilities for VDD = 0.1 V logic using carbon-based tunneling field effect transistors," in *Symposium on VLSI Technology (VLSIT)*, pp. 180–181, IEEE, 2009.

[10] Y. Che, H. Chen, H. Gui, J. Liu, B. Liu, and C. Zhou, "Review of carbon nanotube nanoelectronics and macroelectronics," *Semiconductor Science and Technology*, vol. 29, pp. 1–17, May 2014.

[11] M. S. Arnold, A. A. Green, J. F. Hulvat, S. I. Stupp, and M. C. Hersam, "Sorting carbon nanotubes by electronic structure using density differentiation," *Nature Nanotechnology*, pp. 60–65, Oct. 2006.

[12] J. Knoch and M. R. Muller, "Electrostatic doping-controlling the properties of carbon-based FETs with gates," *IEEE Trans. on Nanotechnology*, vol. 13, pp. 1044–1052, Nov. 2014.

[13] J. Knoch and J. Appenzeller, "A novel concept for field-effect transistors: The tunneling carbon nanotube FET," in *Proc. 63rd DRC*, pp. 153–156, IEEE, June 2005.

[14] J. Appenzeller, Y.-M. Lin, J. Knoch, and P. Avouris, "Band-to-band tunneling in carbon nanotube field-effect transistors," *Physical Review Letters*, vol. 93, pp. 196805–1–196805–4, Nov. 2004.

[15] S. O. Koswatta, M. S. Lundstrom, and D. Nikonov, "Performance comparison between p-i-n tunneling transistors and conventional MOSFETs," *IEEE Trans. on Electron Devices*, vol. 56, pp. 456–465, Mar. 2009.

[16] S. F. S, P. Poncharal, Z. L. Wang, and W. A. Heer, "Carbon nanotube quantum resistors," *Science*, vol. 280, pp. 1744–46, June 1998.

[17] P. Poncharal, C. Berger, Y. Yi, Z. L. Wang, and W. A. de Heer, "Room temperature ballistic conduction in carbon nanotubes," *Journal of Phyical Chemistry B*, vol. 106, pp. 12104–12114, Nov. 2002.

[18] L. F. Register, M. M. Hasan, and S. K. Banerjee, "Stepped broken-gap heterobarrier tunneling field-effect transistor for ultralow power and high speed," *IEEE Electron Device Letters*, vol. 32, pp. 743–745, June 2011.

[19] S. Luryi, "Quantum capacitance in nanoscale device modeling," *Applied Physics Letters*, vol. 52, no. 501, 1988.

[20] D. L. John, L. C. Castro, and D. L. Pulfrey, "Quantum capacitance in nanoscale device modeling," *Journal of Applied Physics*, vol. 96, pp. 5180–5184, Nov. 2004.

[21] Z. L. Miskovic and N. Upadhyaya, "Modeling electrolytically top-gated graphene," *Nanoscale Research Letters*, vol. 5, pp. 505–511, Mar. 2010.

[22] J. Appenzeller, J. Knoch, M. T. Bjork, H. Riel, H. Schmid, and W. Riess, "Toward nanowire electronics," *IEEE Trans. on Electron Devices*, vol. 55, pp. 2827–2845, Nov. 2008.

[23] S. Poli, S. Reggiani, A. Gnudi, E. Gnani, and G. Baccarani, "Computational study of the ultimate scaling limits of CNT tunneling devices," *IEEE Trans. on Electron Devices*, vol. 55, pp. 313–321, Jan. 2008.

[24] J. Appenzeller, Y. Lin, J. Konch, Z. Chen, and P. Avouris, "Comparing carbon nanotube transistors-the ideal choice: A novel tunneling device design," *IEEE Trans. on Electron Devices*, vol. 52, pp. 2568–2576, Dec. 2005.

[25] T. Nakano, M. Ogawa, and S. Souma, "Strain induced modulation of switching behavior in carbon nanotube tunneling field effect transistors," in *IEEE International Meeting for Future Devices*, pp. 1–2, IEEE, 2012.

[26] T. Nakano, M. Ogawa, and S. Souma, "Strain induced modulation of switching behavior in carbon nanotube tunneling field effect transistors," in *IEEE International Meeting for Future Devices*, pp. 1–2, IEEE, 2012.

[27] A. D. Es-Sakhi and M. H. Chowdhury, "Multichannel tunneling carbon nanotube field effect transistor (MT-CNTFET)," in *IEEE International SoC Conference (SOCC)*, pp. 156–159, IEEE, Sept. 2014.

[28] J. Knoch and J. Appenzeller, "Tunneling phenomena in carbon nanotube field-effect transistors," *Physica Status Solidi (a)*, vol. 205, no. 4, pp. 679–694, 2008.

[29] C. Pan and A. Naeemi, "Interconnect design and benchmarking for charge-based beyond-CMOS device proposals," *IEEE Electron Device Letters*, vol. 37, pp. 508–511, Apr. 2016.

[30] K. S. Novoselov, A. K. Geim, S. V. Morozov, D. Jiang, Y. Zhang, S. V. Dubonos, I. V. Grigorieva, and A. A. Firsov, "Electric field effect in atomically thin carbon films," *Science*, vol. 306, pp. 666–669, Oct. 2004.

[31] A. K. Geim, "Graphene: Status and prospects," *Science*, vol. 324, pp. 1530–1534, June 2009.

[32] F. Schwierz, "Graphene transistors," *Nature Nanotechnology*, pp. 487–496, May 2010.

[33] Y.-W. Son, M. L. Cohen, and S. G. Louie, "Energy gaps in graphene nanoribbons," *Phys.Rev. Letters*, vol. 7, p. 216803, Nov. 2006.

[34] E. V. Castro, K. S. Novoselov, S. V. Morozov, N. M. R. Peres, J. M. B. L. dos Santos, J. Nilsson, F. Guinea, A. K. Geim, and A. H. C. Neto, "Biased bilayer graphene: Semiconductor with a gap tunable by the electric field effect," *Phys. Rev. Lett.*, vol. 99, p. 216802, Nov. 2007.

[35] P. Avouris, Z. Chen, and V. Perebeinos, "Carbon-based electronics," *Nature Nanotechnology 2*, pp. 605–615, Sept. 2007.

[36] L. Yang, C.-H. Park, T.-W. Son, M. L. Cohen, and S. G. Louie, "Quasiparticle energies and band gaps in graphene nanoribbons," *Physical Review Letters*, vol. 99, pp. 186801–1–186801–4, Nov. 2007.

[37] L. Tapaszto, G. Dobrik, P. Lambin, and L. P. Biro, "Tailoring the atomic structure of graphene nanoribbons by scanning tunnelling microscope lithography," *Nature Nanotechnology 3*, pp. 397–401, June 2008.

[38] H. Min, B. Sahu, S. K. Banerjee, and A. H. MacDonald, "Ab initio theory of gate induced gaps in graphene bilayers," *Phys.Rev. B*, vol. 75, p. 155115, Apr. 2007.

[39] J. B. Oostinga, H. B. Heersche, X. Liu, A. F. Morpurgo, and L. M. K. Vandersypen, "Gate-induced insulating state in bilayer graphene devices," *Nature Materials 7*, pp. 150–157, Dec. 2007.

[40] Y. Zhang, T. Tang, C. Girit, Z. Hao, M. C. Martin, A. Zettl, M. F. Crommie, Y. R. Shen, and F. Wang, "Direct observation of a widely tunable bandgap in bilayer graphene," *Nature*, vol. 459, pp. 820–823, June 2009.

[41] D. W. Boukhvalov and M. I. Katsnelson, "Tuning the gap in bilayer graphene using chemical functionalization: Density functional calculations," *Phys.Rev. B*, vol. 78, pp. 85413-1–85412-5, Aug. 2008.

[42] A. J. Samuels and J. D. Carey, "Molecular doping and band gap opening of bilayer graphene," *ACS Nano*, vol. 7, pp. 2790–2799, Feb. 2013.

[43] P. Michetti, M. Cheli, and G. Iannaccone, "Model of tunneling transistors based on graphene on SiC," *Applied Physics Letters*, vol. 96, no. 133508, 2010.

[44] X. Li, X. Wang, L. Zhang, S. Lee, and H. Dai, "Chemically derived, ultrasmooth graphene nanoribbon semiconductors," *Science*, vol. 319, pp. 1229–1232, Feb. 2008.

[45] Q. Zhang, T. Fang, H. Xing, A. Seabaugh, and D. Jena, "Graphene nanoribbon tunnel transistors," *IEEE Electron Device Letters*, vol. 29, pp. 1344–1346, Dec. 2008.

[46] P. Zhao, J. Chauhan, and J. Guo, "Computational study of tunneling transistor based on graphene nanoribbon," *Nano Letters*, pp. 684–688, 2009.

[47] M. S. Fahad, A. Srivastava, A. K. Sharma, and C. Mayberry, "Analytical current transport modeling of graphene nanoribbon tunnel field-effect transistors for digital circuit design," *IEEE Transactions on Nanotechnology*, vol. 15, pp. 39–50, Jan. 2016.

[48] R. Sako, H. Hosokawa, and H. Tsuchiya, "Computational study of edge configuration and quantum confinement effects on graphene nanoribbon transport," *IEEE Electron Device Letters*, vol. 32, pp. 6–8, Jan. 2011.

[49] M. Luisier and G. Klimeck, "Performance analysis of statistical samples of graphene nanoribbon tunneling transistors with line edge roughness," *Applied Physics Letters*, vol. 4, June 2009.

[50] T. K. Agarwal, A. Nourbakhsh, P. Raghavan, I. Radu, S. D. Gendt, M. Heyns, M. Verhelst, and A. Thean, "Bilayer graphene tunneling FET for sub-0.2 V digital CMOS logic applications," *IEEE Electron Device Letters*, vol. 35, pp. 1308–1310, Dec. 2014.

[51] G. Fiori and G. Iannaccone, "Ultralow-voltage bilayer graphene tunnel FET," *IEEE Electron Device Letters*, vol. 30, pp. 1096–1098, Oct. 2009.

[52] F. Chen, H. Ilatikhameneh, G. Klimeck, Z. Chen, and R. Rahman, "Configurable electrostatically doped high performance bilayer graphene tunnel FET," *IEEE Journal of the Electron Devices Society*, vol. 4, no. 3, pp. 124–128, 2016.

[53] S. Zhou, G.-H. Gweon, A. Fedorov, P. First, W. de Heer, D.-H. Lee, F. Guinea, A. C. Neto, and A. Lanzara, "Substrate-induced band gap opening in epitaxial graphene," *Nature Materials 6*, vol. 78, pp. 770–775, Sept. 2007.

[54] B. Rawat and R. Paily, "Analysis of graphene tunnel field-effect transistors for analog/RF applications," *IEEE Trans. on Electron Devices*, vol. 62, pp. 2663–2669, Aug. 2015.

[55] K.-T. Lam, D. W. Seah, S.-K. Chin, S. B. Kumar, G. Samudra, Y.-C. Yeo, and G. C. Liang, "A simulation study of graphene nanoribbon tunneling FET with heterojunction channel," *IEEE Electron Device Letters*, vol. 31, pp. 555–557, June 2010.

[56] H. Da, K.-T. Lam, G. Samudra, S.-K. Chin, and G. Liang, "Graphene nanoribbon tunneling field-effect transistors with a semiconducting and a semimetallic heterojunction channel," *IEEE Trans. on Electron Devices*, vol. 59, pp. 1454–1461, May 2012.

[57] Y. Yoon and S. Salahuddin, "Barrier-free tunneling in a carbon heterojunction transistor," *Applied Physics Letters*, vol. 97, no. 033102, 2010.

[58] Y. Yoon, S. H. Kim, and S. Salahuddin, "Performance analysis of carbon-based tunnel field-effect transistors for high frequency and ultralow power applications," *Applied Physics Letters*, vol. 97, no. 233504, 2010.

[59] L. Leem, A. Srivastava, S. Li, B. Magyari-Kope, G. Iannaccone, J. S. Harris, and G. Fiori, "Multi-scale simulations of partially unzipped CNT hetero-junction tunneling field effect transistor," in *IEEE International Electron Devices Meeting*, pp. 32.5.1–32.5.4, IEEE, 2010.

[60] N. Ghobadi and M. Pourfath, "A comparative study of tunneling FETs based on graphene and GNR heterostructures," *IEEE Trans. on Electron Devices*, vol. 61, pp. 186–192, Jan. 2014.

[61] Y. Lu and J. Guo, "Local strain in tunneling transistors based on graphene nanoribbons," *Applied Physics Letters*, vol. 97, no. 073105, 2010.

[62] J. Kang, Y. He, J. Zhang, X. Yu, X. Guan, and Z. Yu, "Modeling and simulation of uniaxial strain effects in armchair graphene nanoribbon tunneling field effect transistors," *Applied Physics Letters*, vol. 96, no. 252105, 2010.

[63] K. T. Lam, Y. Yang, G. S. Samudra, Y. C. Yeo, and G. Liang, "Electrostatics of ultimately thin-body tunneling FET using graphene nanoribbon," *Electron Device Letters*, vol. 32, pp. 431–433, Apr. 2011.

[64] S. Saurabh and M. J. Kumar, "Novel attributes of a nanoscale dual material gate tunnel field effect transistor," *IEEE Trans. on Electron Devices*, vol. 58, pp. 404–410, Feb. 2011.

[65] H. Wang, S. Chang, Y. Hu, H. He, J. He, Q. Huang, F. He, and G. Wang, "A novel barrier controlled tunnel FET," *IEEE Electron Device Letters*, vol. 35, pp. 798–800, July 2014.

7

Nanowire Tunnel Field-Effect Transistor

7.1 Introduction

Since the early 2000s, there has been tremendous progress in nanowire research. Various groups are exploring the possibility of applying novel properties of semiconductor nanowires for nanoelectronics. The intrinsic smallness of nanowires makes them a natural choice for continuously shrinking transistors. The tight gate control achievable in nanowire transistors is particularly beneficial in ameliorating short channel effects. The bottom-up synthesis methods possible for nanowires open doors for realizing structures that are extremely difficult to obtain using conventional top-down approaches. This has led device researchers into investigating nanowire FETs (NWFET) and significant advancements have been made in recent years.

The geometrical smallness of nanowires, excellent electrostatic control of the gate over the channel potential, and complicated structures realizable using nanowires can be exploited in TFETs also. Therefore, researchers are actively looking into realizing TFETs using different kinds of semiconductor nanowires. In this chapter, we will examine the various aspects of nanowire TFETs (NWTFETs). We will first study some of the basic concepts of nanowires and NWFETs. These concepts are important in developing an in-depth understanding of NWTFETs. Then we will discuss silicon-based NWTFETs, followed by NWTFETs realized using other semiconducting materials and nanowire heterostructures. We will also highlight some of the device optimization techniques that are particularly important for NWTFETs. Finally, the challenges and the future perspective of NWTFETs are presented.

7.2 Basics of Nanowires

A semiconductor nanowire (NW) is a single crystal material with diameters of a few nanometers and lengths up to a tens of microns. An NW is constrained in the lateral dimension and unconstrained along its length. Therefore, an NW can be considered as a one-dimensional (1-D) system and the unique properties inherent in a 1-D system are exhibited by NWs.

Nanowires can be synthesized using bottom-up approaches where the structure is assembled atom-by-atom from their basic components. This is in contrast to the conventional top-down approaches in which the nanostructure is fabricated from the bulk material by etching strategies. A big advantage of the bottom-up approach is the excellent control possible in at least one dimension of the structure [1].

The most common technique to produce nanowires is to employ a vapor-liquid-solid (VLS) growth process and use nanoclusters as a catalyst. Nanoclusters are used to define the site of nucleation and nanowire growth. The vapor-phase source material required for VLS growth can be provided by chemical vapor deposition (CVD), laser ablation, or chemical beam epitaxy (CBE). The synthesis process based on CVD is versatile and allows synthesis of NWs with a broad range of properties such as different morphology, composition, diameter, and electrical properties by controlling process parameters such as temperature, pressure, gas mixture, catalyst material, substrate orientation, etc. During NW synthesis, *in situ* doping profiles can be obtained by adjusting the doping concentration. The nanowire synthesis techniques are now capable of producing nanowires with high yield and with reproducible properties. The recent advancements in nanowire production techniques have been a big factor in the growing interest in nanowire-based electronics [1].

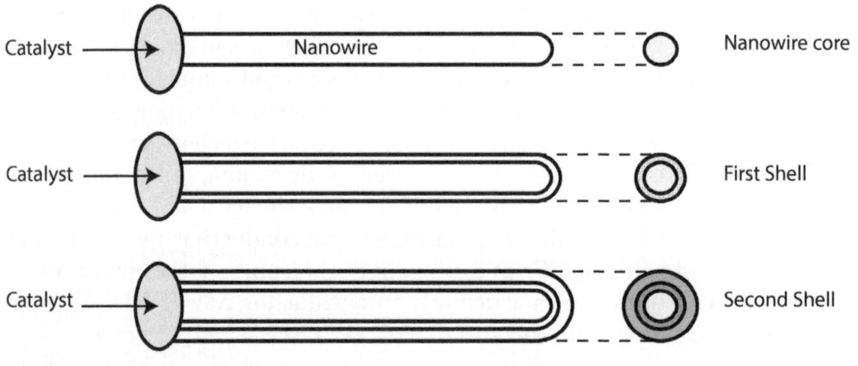

FIGURE 7.1
Radial nanowire heterostructure: reactant preferentially incorporated at the surface of the nanowire.

The nanowire growth process naturally allows relaxation of lattice strain and enables semiconductor growth on lattice-mismatched substrates without creating dislocations [1–4]. Therefore, both radial and axial heterostructures can be produced during the VLS growth process by switching on and switching off different source materials [1, 2, 4]. During the nanowire growth process, when the reactant vapor is changed, a shell will grow over the original nanowire if the reactant is preferentially incorporated on the surface of the semiconductor. By appropriately changing reactants during the radial growth process, the desired core-multishell radial heterostructure can be obtained as shown in Figure 7.1.

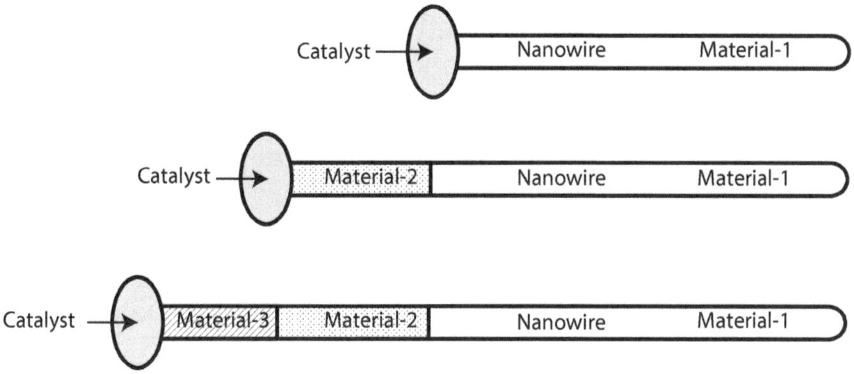

FIGURE 7.2
Axial nanowire heterostructure: reactant preferentially incorporated at the catalyst.

However, during the nanowire growth process, on changing the reactant vapor, nanowire of a different material will grow along the axial direction if the reactant adsorption/decomposition is exclusively favored at the surface of the catalyst nanocluster. Thus, by changing the reactants and the growth conditions appropriately, the required axial nanowire super-lattice can be formed as shown in Figure 7.2 [1, 2, 4]. Nanowire heterostructures have peculiar properties that can be exploited in NWFETs and NWTFETs, as we will discuss in the following sections.

Nanowires composed of silicon, germanium, and III-V materials are attractive for realizing transistors [5]. The effectiveness of nanowires in transistors can be appreciated by analyzing the electrostatics of a NWFET. The electrostatics of a NWFET can be explained by the 1-D modified Poisson's equation describing the potential at the channel–dielectric interface $\Phi_f(x)$ as follows [6]:

$$\frac{d^2\Phi_f(x)}{dx^2} - \frac{\Phi_f(x) - \Phi_g - \Phi_{bi}}{\lambda^2} = -\frac{q\,(\rho \pm N)}{\varepsilon_0 \varepsilon_{NW}}, \tag{7.1}$$

where Φ_g is the gate potential, Φ_{bi} is the built-in potential, ρ is the density of mobile carriers, N is the doping concentration in the NW (+ sign is used for donors and − sign for acceptors), ε_{NW} is the relative dielectric constant of the nanowire, ε_0 is the permittivity of free space, and x is the position along the channel length. When Equation (7.1) is solved analytically for constant charge density, the relationship between Φ_f and the position along the length of the channel can be given as [6]

$$\Phi_f(x) \propto \exp\left(-\frac{x}{\lambda}\right). \tag{7.2}$$

The screening length λ depends on the gate architecture and is a function of the diameter of the nanowire d_{NW}, the thickness of the gate oxide t_{ox}, the dielectric constant of the nanowire ε_{NW}, and the dielectric constant of the gate oxide ε_{ox}.

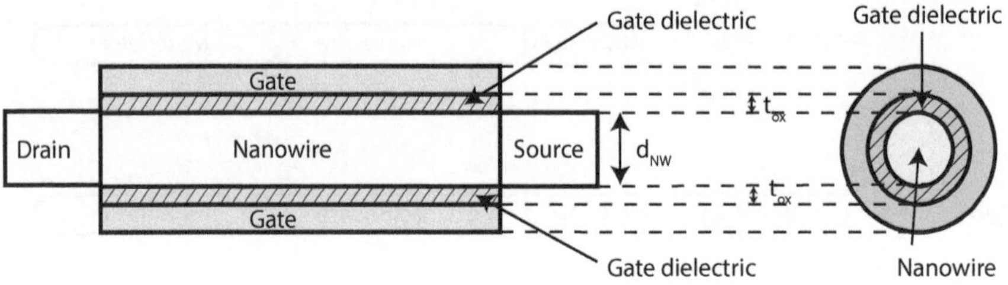

FIGURE 7.3
Schematic representation of a nanowire transistor with a gate-all-around gate.

For gate-all-around architecture as shown in Figure 7.3, the screening length λ is given as [6]

$$\lambda = \sqrt{\frac{\varepsilon_{NW} d_{NW}^2 \ln\left(1 + 2\frac{t_{ox}}{d_{NW}}\right)}{8\varepsilon_{ox}}}. \tag{7.3}$$

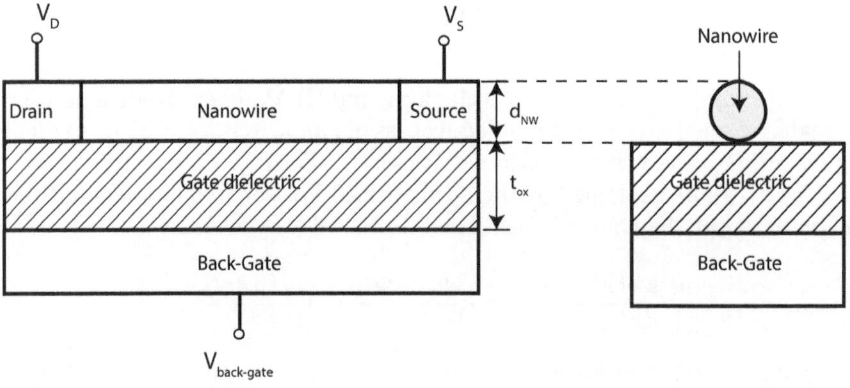

FIGURE 7.4
Schematic representation of a nanowire transistor with a large area back gate.

For the device with a large area back gate as shown in Figure 7.4, the screening length λ is found to be of the following form [6]

$$\lambda = \sqrt{\frac{\varepsilon_{NW} d_{NW} t_{ox}}{\varepsilon_{ox}}}. \tag{7.4}$$

The injection of carriers from the source into the channel is determined by the maximum value of the potential in the channel Φ_f^0. In the OFF-state, when the channel length L is larger than the screening length λ, the maximum value of the potential in the channel Φ_f^0 can be approximated as [6]

$$\Phi_f^0 \approx 2\sqrt{-\left(\Phi_g + \Phi_{bi}\right)\left(\Phi_d - \left(\Phi_g + \Phi_{bi}\right)\right)} \exp\left(-\frac{L}{2\lambda}\right) + \Phi_g + \Phi_{bi}, \tag{7.5}$$

where Φ_d is the drain potential given by $\Phi_d = -qV_{DS}$. For a MOSFET, in the OFF-state, it is desirable that the drain potential should not have any impact on the maximum potential in the channel. This can be ensured by keeping $L \gg \lambda$ such that the first term on the RHS of Equation (7.5) can be ignored.

From Equations (7.3) and (7.4), it can be inferred that when d_{NW} is decreased, the screening length λ also decreases. Therefore, by ensuring a small screening length λ, the condition $L \gg \lambda$ can be satisfied and the short-channel effects such as drain-induced barrier lowering (DIBL) can be suppressed in the NWFET. This is one of the biggest advantages of employing nanowires for highly scaled MOSFETs.

Additionally, as pointed out in Section 6.3, due to energy-decreasing 1-D density of states, the quantum capacitance limit (QCL) can easily be reached in NWFETs. When a transistor operates in QCL, the profile of the potential in the channel depends on the gate voltage rather than on the channel charge, which is a desirable property for a transistor [6].

The electrostatics of nanowire transistors allow realizing transistors with excellent characteristics with different mechanisms of injection of carriers into the channel. Different mechanisms of injection of carriers into the channel that are being investigated for NWFETs are [6, 7]:

1. A "conventional" type NWFET can be created by doping the source and drain with opposite dopant types as in the channel, similar to a conventional MOSFET as shown in Figure 7.5(a). The junction can also be created electrostatically using a back gate as shown in Figure 7.5(b). The potential of the back gate shifts the bands in the channel and sets the barrier for injection of carriers at the metal contacts. The electric field due to the back gate is screened by the top gate and the top gate modulates the carrier concentration below it.

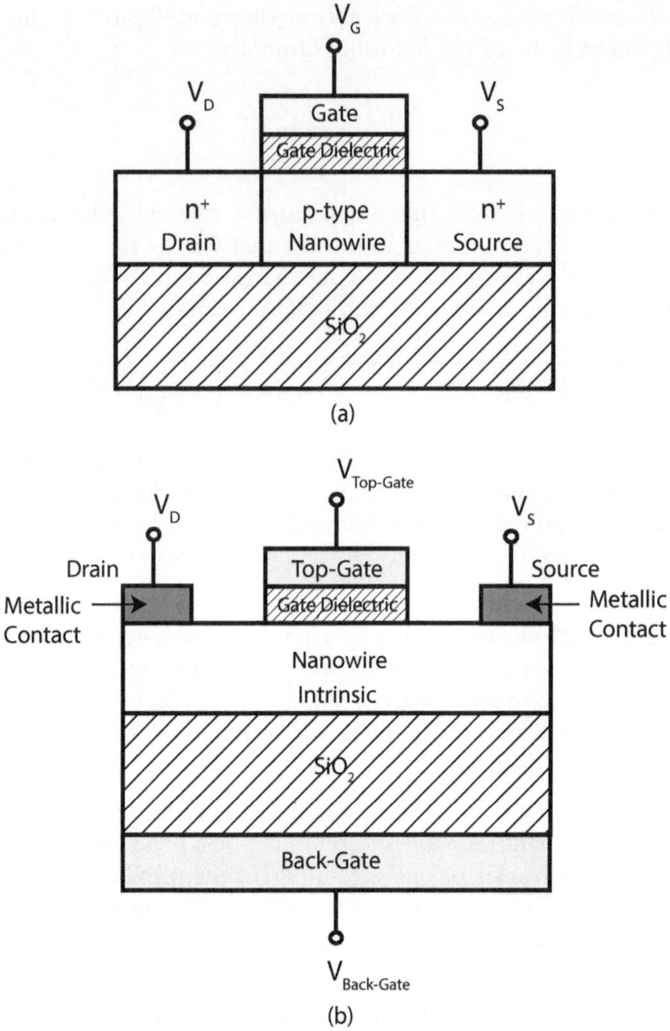

FIGURE 7.5
Schematic representation of nanowire transistor with conventional mechanisms of charge injection: (a) physical doping and (b) electrostatic doping.

2. A Schottky barrier NWFET can be created by utilizing a metal-semiconductor Schottky barrier at the source and the drain as shown in Figure 7.6. The height of the barrier is governed by the work function of the metal and the nature of the metal–semiconductor interface. The back gate controls the barrier for the injection of carriers. The mechanism of current injection can be through the thermal excitation over the Schottky barrier or tun-

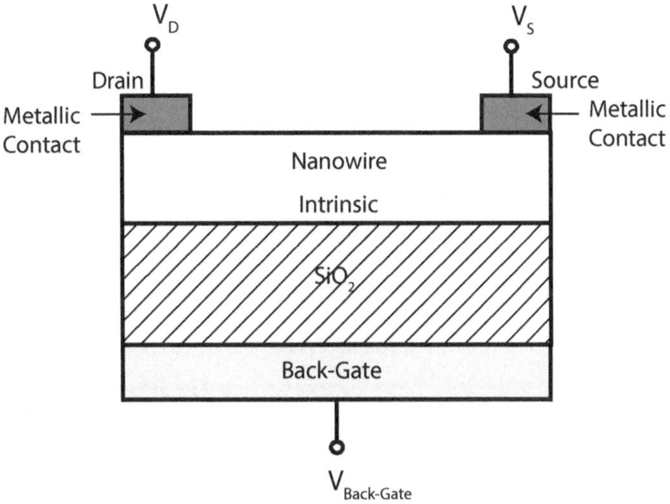

FIGURE 7.6
Schematic representation of a Schottky barrier nanowire transistor with only the back gate controlling the injection of carrier.

neling through the barrier depending on the gate voltage and the barrier height.

3. A junctionless NWFET, as shown in Figure 7.7, can be realized using tight gate control, thin body, metal/semiconductor ohmic contacts, and uniformly heavily doped channel [8, 9]. The device is turned off electrostatically by depleting the channel of carriers and the gate potential is used to modulate the resistance of the channel. The experimental results show that junctionless NWFETs can attain an ideal subthreshold swing of 60 $mV/decade$ at room temperature [8].

4. Other mechanisms of injection of carriers in nanowire transistors that are being investigated include impact ionization and band-to-band tunneling (BTBT). We shall discuss nanowire transistors based on BTBT in the rest of this chapter.

7.3 Silicon Nanowire TFETs

Nanowires are interesting for application in TFETs for these reasons:

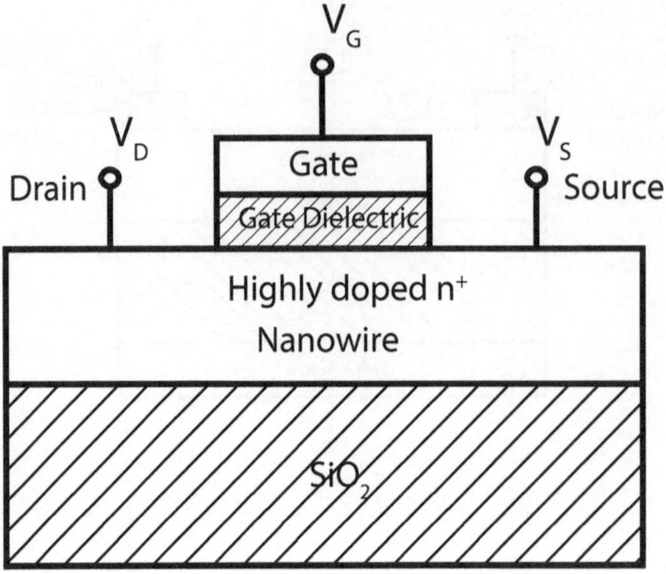

FIGURE 7.7
Schematic representation of a nanowire junctionless transistor.

1. There is excellent gate control over the channel electrostatics due to the cylindrical geometry and the small diameter of the nanowire. This suppresses short-channel effects and facilitates obtaining a steeper subthreshold swing.

2. In a nanowire the density of states (DOS) and the energy (E) are related as

$$DOS \propto \frac{1}{\sqrt{E}}. \tag{7.6}$$

This permits obtaining a high carrier density at the Fermi energy close to the band edges in the source of a NWTFET. As a result, a higher ON-state current (I_{ON}) and steeper subthreshold swing can simultaneously be realized in TFETs using 1-D systems rather than 2-D or 3-D systems [10, 11]. Additionally, for a 1-D system, the delay metrics ($\tau = (CV/I)$) decrease with the supply voltage (V_{DD}) and reach a minimum at ($V_{DD_{MIN}} \approx 6kT/q$), where k is the Boltzmann constant, T is the temperature in Kelvin, and q is the electronic charge [11].

3. In general, the tunneling probability is higher in a 1-D system and a higher I_{ON} is possible in NWTFETs. In 2-D and 3-D systems, the

tunneling probability is reduced due to the conservation of transversal momentum in the tunneling process [10].

Silicon nanowires (SiNWs) are particularly attractive for implementing TFETs because they can possibly be integrated in the existing CMOS processing flows. SiNW TFETs have been demonstrated using both bottom-up and top-down approaches [12–16].

A TFET employing SiNW prepared using the VLS method was recently demonstrated [13]. The p-i-n structure was obtained using *in situ* doping during NW growth employing diborane and phosphine for p-type and n-type dopants, respectively. The lengths of the p-type, intrinsic, and n-type regions in the p-i-n structure were 1.5 μm, 0.5 μm and 2.5 μm, respectively. The SiNWs were transferred to a pre-patterned insulating substrate and the gate dielectric was deposited. The results for the following two devices are presented below [13]:

1. Device-1: In this device, SiO_2 was deposited at $300°C$ using plasma-enhanced chemical vapor deposition (PECVD). Various other device parameters were thickness of gate oxide $t_{ox} = 20\ nm$, diameter of nanowire $d_{NW} = 40\ nm$, and gate material was aluminum.
2. Device-2: In this device, HfO_2 was deposited at $250°C$ using atomic layer deposition (ALD). Various other device parameters were thickness of gate oxide $t_{ox} = 5\ nm$, diameter of nanowire $d_{NW} = 50\ nm$, and Ti/Au was deposited as the gate material.

It was shown that for $V_{DS} = 0.5\ V$, the I_{ON} for device-1 and device-2 were 0.003 $\mu A/\mu m$ and 0.084 $\mu A/\mu m$, respectively [13]. The average subthreshold swing for four decades of current for device-1 and device-2 were found to be 200 $mV/decade$ and 120 $mV/decade$, respectively. These results show that a high-κ dielectric gate-stack results in an improvement in the I_{ON} and the subthreshold swing in a Si-NWTFET. However, the I_{ON} of the demonstrated SiNW TFETs is quite low, even when a high-κ gate dielectric is employed.

Although, the bottom-up approaches of realizing NW devices have their own advantages, the challenges in assembling the devices in the bottom-up approaches are significant. Top-down methods of realizing SiNW TFETs have also been demonstrated [14, 15, 17]. The biggest advantage of top-down approaches is that these processes are compatible with the existing CMOS technology. The fabrication of vertical gate-all-around structures using CMOS-compatible processes have been demonstrated for both p-type and n-type TFETs [14, 17]. The schematic representation of a TFET that was fabricated using the top-down approach is shown in Figure 7.8 [17]. It is reported that a steep doping gradient was obtained by keeping a low thermal budget and using a low-temperature dopant-segregated silicidation technique [17]. The

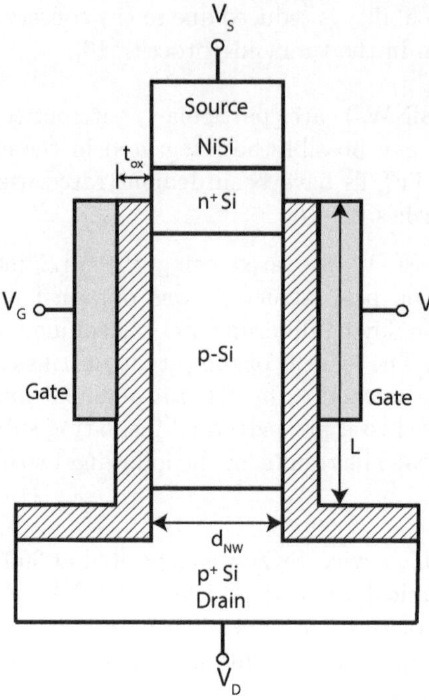

FIGURE 7.8
Schematic representation of a p-type SiNW TFET. Device parameters are $L = 140\ nm$, $d_{NW} = 18\ nm$, $t_{ox} = 4.5\ nm$ [17].

suppression of ambipolar conduction was achieved by asymmetric source and drain doping, which was facilitated by a vertical nanowire platform that allowed individual tuning of source and drain doping concentrations. The drain was implanted before the source and involved a high thermal budget. The source was implanted using a low-temperature dopant-segregated silicidation technique that resulted in piling up of dopants toward the source–channel junction. As a result, a subthreshold swing smaller than 50 $mV/decade$ was obtained for three decades of drain current for both n-type and p-type TFET operations.

The I_{ON} for the n-type TFET was reported to be 0.02 $\mu A/\mu m$ and for the p-type TFET it was reported to be 1.2 $\mu A/\mu m$ for $V_{DD} = -1\ V$ [17]. The extremely small I_{ON} for the n-type TFET was attributed to a large drain-gate underlap used in the n-type TFET. Additionally, different diffusivities of the p-type and n-type dopants are expected to result in different gradients of dopants in the n-type and the p-type TFET. This could result in a difference in the I_{ON} of the p-type and n-type TFETs. Therefore, the gate-drain overlap and doping profiles can be optimized to maximize the I_{ON} and suppress the ambipolar conduction in SiNW TFET.

The choice of annealing schemes and dopant profiles affect the characteristics of a SiNW TFET [15]. It has been demonstrated that by using excimer laser annealing in conjunction with a low-temperature rapid thermal annealing activation process to preserve the abruptness of the doping at the source–channel tunneling junction, the subthreshold swing and the I_{ON} can be improved significantly [15]. It may be noted that the smearing out of the source doping profile is expected to alleviate as the NW dimensions scale because a smaller implant energy will be required leading to a tighter lateral distribution of the dopants at smaller dimensions [15]. Therefore, high-κ dielectrics and reduced diameters of the nanowires are expected to improve the I_{ON} and the subthreshold swing of the SiNW TFET. Additionally, the characteristics of a SiNW TFET can be improved by applying uniaxial strain or using SiGe heterojunctions [18].

7.4 Nanowire Heterostructure-Based TFETs

The characteristics of a TFET can be greatly improved using heterojunctions as described in Section 4.4.3. Heterojunctions allow increasing the I_{ON} by employing low bandgap material on the source side and decreasing the OFF-state current (I_{OFF}) by using a high bandgap material in the channel and on the drain side. Therefore, TFETs based on nanowire heterostructures have been widely investigated. In this section, we will look at some of the nanowire heterostructures that have been investigated for realizing TFETs.

7.4.1 Ge–Si_xGe_{1-x} Core-Shell NWTFET

Germanium NW is a promising material for application in TFETs because it is compatible with CMOS technology, and has a small bandgap $E_g = 0.67\ eV$ and good carrier mobility. Ge–Si_xGe_{1-x} core-shell NW-based TFETs have been realized using bottom-up synthesis techniques [19, 20]. A Ge–Si_xGe_{1-x} core-shell NWTFET was grown by VLS technique using gold as a catalyst [19]. The Ge core was grown first in a CVD chamber and then the Si_xGe_{1-x} shell was epitaxially grown over the core in the same chamber. The Si mole-fraction was $x = 0.3$ and the thickness of the Si_xGe_{1-x} shell was 4 nm. The shell over the Ge core acts as a passivation layer for the Ge surface. For the activation of the dopants in the source and the drain, flash-assisted rapid thermal process (fRTP) can be employed [20]. This increases the abruptness of the doping profile by reducing the diffusion of the dopants in the lateral direction and increases the level of activation of the dopants [20]. As a result, a sharper band profile at the source–channel tunneling junction can be obtained and a thinner tunneling barrier can be realized [20]. The experimental results show that employing fRTP-assisted dopant activation in a Ge$-$$Si_xGe_{1-x}$ core-shell NWTFET at least one order of magnitude higher tunneling current can be obtained [20]. It is reported that using fRTP-assisted dopant activation, the drain current of around 28 $\mu A/\mu m$ was obtained for Ge$-$$Si_xGe_{1-x}$ core-shell NWTFET at $V_{DD} = -3\ V$ for an $HfAlO_x$ gate dielectric with thickness $t_{ox} = 5.5\ nm$ [20].

7.4.2 InAs-Si Heterostructure NWTFET

As described in Chapter 5, III-V semiconductors are especially suitable for TFETs due to their smaller bandgap and smaller carrier effective mass. However, obtaining high-quality III-V semiconductor–dielectric interface and the integration of III-V semiconductors into the existing CMOS technology are difficult challenges. By employing III-V/Si heterostructures in TFETs, the advantages of both III-V semiconductors and silicon can be exploited. Therefore, researchers are actively investigating InAs-Si heterostructure nanowire TFETs. The InAs-Si heterojunction provides a small effective bandgap, thus boosting the I_{ON} in a TFET. The larger bandgap of Si in the channel and drain regions results in a low I_{OFF} and lower interface trap density (D_{it}) [21]. Additionally, InAs-Si heterojunction NWTFETs can be realized on a Si platform, which allows exploiting the advantages of standard CMOS processes [21].

It should be noted that InAs and Si have an 11% lattice mismatch and cannot be grown by traditional techniques [22]. However, recent advancements in techniques to grow InAs nanowires (NWs) on Si using selective area growth (SAG) within patterned thin oxide films or using nanotube templates have made realizing InAs-Si heterostructure NWTFETs possible [21–26]. In order to obtain a low I_{OFF} and a steep subthreshold swing, it is crucial that the InAs-Si interface in the TFET not have defects and dislocations.

Figure 7.9 shows a schematic diagram of a p-type InAs-Si heterojunction TFET [21]. The TFET was fabricated using selective epitaxy in nanotube templates. The diameter of the nanowire was $d_{NW} = 100\ nm$. The doping of intrinsic-Si was $1 \times 10^{15}\ cm^{-3}$ and was grown over a heavily doped p-type substrate that worked as a drain in the TFET. The n-type InAs nanowire was Si-doped $N_D = 4 \times 10^{17}\ cm^{-3}$ and worked as the source in the TFET. The TFET had a gate-all-around architecture with the gate stack consisting of Al_2O_3 and HfO_2 with a thickness of 2.2 nm and 2.4 nm, respectively. The gate material was TiN/SiO_2 and was isolated from the drain and the source using tetraethyl orthosilicate (TEOS).

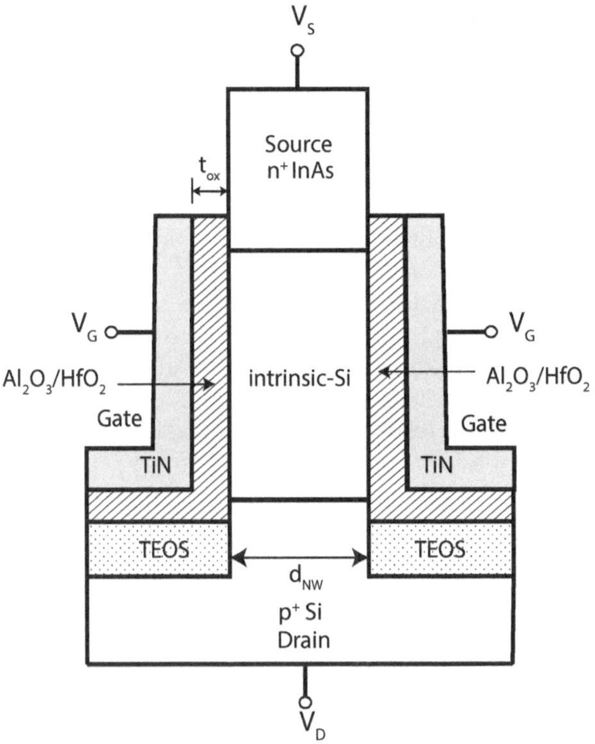

FIGURE 7.9
Schematic representation of an InAs-Si heterojunction TFET [21]. Device parameters are $d_{NW} = 100\ nm$, $t_{ox} = 4.6\ nm$.

The device in Figure 7.9 exhibited an I_{ON} of 0.2 $\mu A/\mu m$ at $V_{DD} = 0.5\ V$ and 6 $\mu A/\mu m$ at $V_{DD} = 1\ V$, an I_{ON}/I_{OFF} ratio of 10^6, and subthreshold swing of 160 $mV/decade$ for three decades of drain current [21]. The subthreshold swing improved with a decrease in the temperature and indicated the presence of traps at the InAs-Si heterojunction and also at the gate dielectric–semiconductor interface. Since the number of misfit dislocations at the InAs/Si heterojunction decreases with a decrease in the diameter of the NW, it is expected that, with the scaling of NW, the subthreshold swing of the NWTFET will improve [21, 24, 26]. However, the impact of reducing the diameter of NW on the increase in the effective bandgap needs to be carefully considered.

7.4.3 InGaAs-Si Heterostructure NWTFET

Recently, TFETs based on a InGaAs nanowire/Si heterojunction have been experimentally demonstrated [27]. There is a greater flexibility in tuning the bandgap in InGaAs, which facilitates extracting the benefits of narrow bandgap InAs, as well as large bandgap GaAs. These TFETs were grown using selective area growth and had diameter $d_{NW} = 30\ nm$. The source, channel and drain consisted of p-type silicon, intrinsic InGaAs, and n-type Si-doped InGaAs, respectively. The In composition in InGaAs was 70%. To ensure low carrier density in the channel, a compensation doping effect using diethylzinc (DEZn) was employed during the growth of the InGaAs NW channel. It should be noted that a low carrier density in the channel results in inducing a large electric field at the InGaAs/Si heterojunction making the subthreshold swing steeper. These experimental results show that InGaAs nanowire/Si heterostructure TFET can attain a subthreshold swing of 30 $mV/decade$ at $V_{DS} = 0.1\ V$ [27]. This is one of the best subthreshold swings reported for a TFET.

7.4.4 InAs-SiGe Heterostructure NWTFET

The InAs-Si heterojunction described above has an effective bandgap of around 0.4 eV for unconfined geometries. However, in confined geometries such as nanowires, due to quantum confinement effects, the effective bandgap is substantially higher [28]. However, a very low effective bandgap can be obtained using an InAs-SiGe heterostructure and a higher I_{ON} can be achieved in an InAs-SiGe heterostructure NWTFET [28]. The source, channel, and drain of the TFET consisted of InAs, $Si_{1\text{-}x}Ge_x$, and Si, respectively [28]. The simulation results show that the drive current improved fivefold for $I_{OFF} = 1\ nA/\mu m$ as the Ge composition in the heterojunction was increased from 0 to 50% and the subthreshold swing also improved. The I_{ON} also improved with the increase in the source doping concentration, though increasing the source doping also led to a degradation in the subthreshold swing due to the contribution of the Boltzmann "tail" in the tunneling process.

7.4.5 GaSb/InAs Heterostructure NWTFET

The band alignments in the GaSb/InAs nanowire system change with the diameter of the nanowire. A broken bandgap is expected to exist in the GaSb/InAs nanowire for diameters greater than 8 nm [29]. Theoretically, it has been shown that a broken-gap heterojunction can deliver I_{ON} comparable to a conventional MOSFET along with exhibiting a subthreshold swing smaller than 60 $mV/decade$, even in the presence of electron-phonon scattering [30]. The 1-D geometry of materials such as CNTs or nanowires is particularly beneficial for implementing broken-gap TFETs. The density of states (DOS) near the source–channel junction is suppressed due to the quantum confinement in 1-D materials. This results in a reduction of the thermally induced leakage current due to the deficiency of the available states and leads to a subthreshold swing smaller than 60 $mV/decade$ [30]. Additionally, for materials such as InAs NW where phonon energies are much smaller compared to CNT, phonon absorption-assisted leakage is suppressed and a sharper subthreshold swing can be obtained [30].

Recently, TFETs based on GaSb/InAs nanowires have been studied experimentally [29, 31, 32]. A TFET based on a GaSb/InAs(Sb) nanowire heterostructure that exploits the broken-gap band alignment has been recently demonstrated [32]. GaSb/InAs(Sb) nanowires were grown on a GaAs substrate using metalorganic vapor phase epitaxy (MOVPE). First, the GaSb portion of the nanowire was grown followed by InAs(Sb). While growing the segment of axial InAs(Sb), a thin shell of InAs is also formed around a GaSb segment, which reduces the resistance to the source electrode and could also provide a path for the leakage current. Therefore, nanowires were annealed in H_2 during the final stages of growth, which results in the formation of a constriction at the axial GaSb/InAs(Sb) heterointerface [32, 33]. The nanowires were dry-deposited onto pre-patterned Si chips and the gate dielectric of Al_2O_3/HfO_2 was deposited using ALD. Since the GaSb/InAs(Sb) heterostructure has a broken-gap band alignment that allows BTBT without a barrier, a high I_{ON} of 310 $\mu A/\mu m$ at $V_{DS} = 0.5\ V$ was observed. However, the minimum value of the I_{OFF} was limited by the narrow bandgap of InAs(Sb). It was observed that when the gate overlaps the heterointerface, the OFF-state characteristics of the device improved. The subthreshold swing was found to be 320 $mV/decade$ at $V_{DS} = 50\ mV$ and could have been limited by a high D_{it} at the high-κ/semiconductor interface [32].

A vertical InAs/GaSb NWTFET has been recently reported [29]. Metalorganic chemical vapor deposition (MOCVD)-based VLS-type of growth employing Au seeds as catalysts was used to grow an InAs drain/channel segment followed by GaSb source [29]. The I_{ON} observed in the vertical nanowire TFET was much lower compared to the similar lateral devices. This can be attributed to a different heterointerface at the GaSb/InAs junction due to the different dynamics of the growth processes [29]. However, vertical InAs/GaSb NWTFET allows precise placement of the individual devices and, therefore, could be an interesting device for future wide-scale applications [29].

7.5 Device Optimization of Nanowire TFETs

The most important parameter to optimize in a nanowire TFET is the diameter of the nanowire (d_{NW}). The diameter of the nanowire provides a tradeoff between the electrostatic control and the current injection [34, 35]. Decreasing d_{NW} improves the gate control of the transistor resulting in an increased I_{ON} and steeper subthreshold swing. However, decreasing d_{NW} increases the quantum confinement effects, leading to an increase in the bandgap and a decrease in the I_{ON} [35]. Additionally, the impact of the d_{NW} on the misfit dislocations in the heterostructure NWTFET and the smearing out of the source doping profile also need to be taken into account while optimizing the diameter.

The gate architecture has a profound impact on the subthreshold swing and the I_{ON} of an NWTFET [27, 29, 30, 36]. The gate structure in an NWTFET needs to be optimized to have an appropriate overlap/underlap with respect to the source and the tunneling junctions. Additionally, the quality of the gate dielectric–semiconductor interface is crucial for obtaining steep subthreshold swing, especially in III-V NWTFETs. The source doping and the contact properties also play an important role in determining the I_{ON} and the subthreshold swing of the TFET [34]. For TFETs in which the tunnel barrier dominates the overall conductance of the channel, the transport properties of the channel are less important. However, when the tunneling probability is high, such as in a broken-gap NWTFET, optimizing the transport properties of the channel also becomes important [30].

7.6 Challenges and Future of Nanowire TFETs

Various simulation and experimental results reported for nanowire TFETs have been quite encouraging, especially for III-V heterojunction NWTFETs. The possibility of decreasing the screening tunneling length using gate-all-around architecture, opportunity for simultaneously improving the I_{ON} and the subthreshold swing, geometric smallness, the possible compatibility with current CMOS technology, and massive research being focused on nanowire electronics make NWTFETs promising devices for future applications. The optimization of the nanowire growth process, improving contact properties and controlling interface trap density are some of the methods that can improve the characteristics of an NWTFET. However, several key challenges that are inhibiting deployment of nanowire FETs in the conventional CMOS technology, are also problematic for commercialization of NWTFETs. For example, in spite of great advancements in the production technology of nanowires, the control over the uniformity, yield, and the position registry of the nanowire devices still need to be more precise [37]. The discreteness of dopants combined with their random placement and variations in other device parameters in NWTFETs can cause large variation in the drain current and the subthreshold swing among devices [38, 39]. These issues are expected to be crucial in determining the future application of NWTFETs.

7.7 Summary

In this chapter, we have developed some basic understanding of nanowire and NWFETs. We looked at some of the implementations of NWTFETs that were reported recently. We discussed some of the properties of nanowires that can be exploited in TFETs. We also looked at some of the implementations of III-V NWTFETs and heterojunction NWTFETs. We highlighted the merits and demerits of different implementations of NWTFETs and the device optimization techniques that are particularly applicable for NWTFETs.

In the previous three chapters, we discussed TFETs that are implemented using different types of materials such as Si, III-V semiconductors, CNT, graphene, and nanowires. In the next chapter, we will look into how different kinds of TFETs can be modeled so as to develop a deeper understanding into the workings of TFETs and enable analysis of circuits in which TFETs are employed.

REFERENCES

[1] W. Lu, P. Xie, and C. M. Lieber, "Nanowire transistor performance limits and applications," *IEEE Trans. on Electron Devices*, vol. 55, pp. 2859–2876, Nov. 2008.

[2] L. J. Lauhon, M. S. Gudiksen, and C. M. Lieber, "Semiconductor nanowire heterostructures," *Philosophical Transactions of the Royal Society of London A: Mathematical, Physical and Engineering Sciences*, vol. 362, no. 1819, pp. 1247–1260, 2004.

[3] P. Yang, R. Yan, and M. Fardy, "Semiconductor nanowire: What's next?," *Nano Letters*, pp. 1529–1536, Apr. 2010.

[4] J. Johansson and K. A. Dick, "Recent advances in semiconductor nanowire heterostructures," *CrystEngComm*, vol. 13, pp. 7175–7184, 2011.

[5] M. J. Kumar, M. A. Reed, G. A. J. Amaratunga, G. M. Cohen, D. B. Janes, C. M. Lieber, M. Meyyappan, L.-E. Wernersson, K. L. Wang, R. S. Chau, T. I. Kamins, M. Lundstrom, B. Yu, and C. Zhou, "Special issue on nanowire transistors: Modeling, device design, and technology," *IEEE Transactions on Electron Devices*, vol. 55, pp. 2813–2819, Nov. 2008.

[6] J. Appenzeller, J. Knoch, M. T. Bjork, H. Riel, H. Schmid, and W. Riess, "Toward nanowire electronics," *IEEE Trans. on Electron Devices*, vol. 55, pp. 2827–2845, Nov. 2008.

[7] A. Fasoli and W. I. Milne, "Overview and status of bottom-up silicon nanowire electronics," *Materials Science in Semiconductor Processing*, vol. 15, pp. 601–614, Dec. 2012.

[8] J. P. Colinge, C. W. Lee, A. Afzalian, N. D. Akhavan, R. Yan, I. Ferain, P. Razavi, B. O'Neill, A. Blake, M. White, A. M. Kelleher, B. McCarthy, and R. Murphy, "Nanowire transistors without junctions," *Nature Nanotechnology 5*, pp. 225–229, Mar. 2010.

[9] S. Sahay and M. J. Kumar, "Realizing efficient volume depletion in SOI junctionless FETs," *IEEE Journal of the Electron Devices Society*, vol. 4, no. 3, pp. 110–115, 2016.

[10] J. Knoch, S. Mantl, and J. Appenzeller, "Impact of the dimensionality on the performance of tunneling fets: Bulk versus one-dimensional devices," *Solid-State Electronics*, vol. 51, pp. 572–578, 2007.

[11] Y. Taur, J. Wu, and J. Min, "Dimensionality dependence of TFET performance down to 0.1 V supply voltage," *IEEE Transactions on Electron Devices*, vol. 63, pp. 877–880, Feb. 2016.

[12] M. T. Bjork, J. Knoch, H. Schmid, H. Riel, and W. Riess, "Silicon nanowire tunneling field-effect transistors," *Applied Physics Letters*, vol. 92, no. 193504, 2008.

[13] M. T. Bjork, K. E. Moselund, H. Schmid, H. Ghoneim, S. Karg, E. Lortscher, J. Knoch, W. Riess, and H. Riel, "VLS-grown silicon nanowires: Dopant deactivation and tunnel FETs," in *2010 Silicon Nanoelectronics Workshop*, pp. 1–2, IEEE, June 2010.

[14] R. Gandhi, Z. Chen, N. Singh, K. Banerjee, and S. Lee, "Vertical Si-nanowire n-type tunneling FETs with low subthreshold swing ($\leq$ 50 $mV/decade$) at room temperature," *Electron Device Letters*, vol. 32, pp. 437–439, Apr. 2011.

[15] J. T. Smith, C. Sandow, S. Das, R. A. Minamisawa, S. Mantl, and J. Appenzeller, "Silicon nanowire tunneling field-effect transistor arrays: Improving subthreshold performance using excimer laser annealing," *IEEE Trans. on Electron Devices*, vol. 58, pp. 1822–1829, July 2011.

[16] V. Brouzet, B. Salem, P. Periwal, R. Alcotte, F. Chouchane, F. Bassani, T. Baron, and G. Ghibaudo, "Fabrication and electrical characterization of homo- and hetero-structure Si/SiGe nanowire tunnel field effect transistor grown by vapor-liquid-solid mechanism," *Solid-State Electronics*, vol. 118, pp. 26–29, 2016.

[17] R. Gandhi, Z. Chen, N. Singh, K. Banerjee, and S. Lee, "CMOS-compatible vertical-silicon-nanowire gate-all-around p-type tunneling FETs with ($\leq 50\ mV/decade$) subthreshold swing," *Electron Device Letters*, vol. 32, pp. 1504–1506, Nov. 2011.

[18] Q.-T. Zhao, S. Richter, C. Schulte-Braucks, L. Knoll, S. Blaeser, G. Luong, S. Trellenkamp, A. Schafer, A. Tiedemann, J. Hartmann, and K. Bourdelle, "Strained Si and SiGe nanowire tunnel FETs for logic and analog applications," *IEEE Journal of the Electron Devices Society*, vol. 3, pp. 103–114, May 2015.

[19] J. Nah, E.-S. Liu, K. M. Varahramyan, and E. Tutuc, "Ge–Si_xGe_{1-x} core-shell nanowire tunneling field-effect transistors," *IEEE Trans. on Electron Devices*, vol. 57, pp. 1883–1888, Aug. 2010.

[20] J. Nah, E.-S. Liu, K. Varahramyan, D. Dillen, S. MacCoy, J. Chan, and E. Tutuc, "Enhanced-performance germanium nanowire tunneling field-effect transistors using flash-assisted rapid thermal process," *IEEE Electron Device Letters*, vol. 31, pp. 1359–1361, Dec. 2010.

[21] D. Cutaia, K. E. Moselund, M. Borg, H. Schmid, L. Gignac, C. M. Breslin, S. Karg, E. Uccelli, and H. Riel, "Vertical InAs-Si gate-all-around tunnel FETs integrated on Si using selective epitaxy in nanotube templates," *IEEE Journal of the Electron Devices Society*, vol. 3, pp. 176–183, May 2015.

[22] H. Schmid, K. E. Moselund, M. T. Bjork, M. Richter, H. Ghoneim, and C. D. Bessire, "Fabrication of vertical InAs-Si heterojunction tunnel field effect transistors," in *DRC*, pp. 181–182, IEEE, 2011.

[23] K. Tomioka, J. Motohisa, S. Hara, and T. Fukui, "Control of InAs nanowire growth directions on Si," *Nano letters 8*, no. 10, pp. 3475–3480, 2008.

[24] K. Tomioka and T. Fukui, "Tunnel field-effect transistor using InAs nanowire/Si heterojunction," *Applied Physics Letters*, vol. 98, no. 083114, 2011.

[25] H. Riel, K. E. Moselund, C. Bessire, M. T. Bjrk, A. Schenk, H. Ghoneim, and H. Schmid, "InAs-Si heterojunction nanowire tunnel diodes and tunnel FETs," in *IEDM*, pp. 16.6.1–16.6.4, IEEE, Dec. 2012.

[26] K. Tomioka, M. Yoshimura, and T. Fukui, "Steep-slope tunnel field-effect transistors using III-V nanowire/Si heterojunction," in *Symposium on VLSI Technology (VLSIT)*, pp. 47–48, IEEE, 2012.

[27] K. Tomioka and T. Fukui, "Current increment of tunnel field-effect transistor using InGaAs nanowire/Si heterojunction by scaling of channel length," *Applied Physics Letters*, vol. 104, pp. 073507–1–073507–4, Feb. 2014.

[28] C. Kshirsagar and S. J. Koester, "InAs/SiGe on Si nanowire tunneling field effect transistors," in *69th Annual Device Research Conference*, pp. 151–152, IEEE, 2011.

[29] E. Lind, E. Memisevic, A. Dey, and L.-E. Wernersson, "III-V heterostructure nanowire tunnel FETs," *Journal of the Electron Devices Society*, vol. 3, pp. 96–102, May 2015.

[30] S. O. Koswatta, S. J. Koester, and W. Haensch, "On the possibility of obtaining MOSFET-like performance and sub-60-mv/dec swing in 1-D broken-gap tunnel transistors," *IEEE Trans. on Electron Devices*, vol. 57, pp. 3222–3230, Dec. 2010.

[31] B. M. Borg, K. A. Dick, B. Ganjipour, M.-E. Pistol, L.-E. Wernersson, and C. Thelander, "InAs/GaSb heterostructure nanowires for tunnel field-effect transistors," *Nano Letters*, vol. 10, no. 10, pp. 4080–4085, 2010.

[32] A. W. Dey, B. M. Borg, B. Ganjipour, M. Ek, K. A. Dick, E. Lind, C. Thelander, and L.-E. Wernersson, "High current GaSb/InAs(Sb) nanowire tunnel field-effect transistors," *Electron Device Letters*, vol. 34, pp. 211–213, Feb. 2013.

[33] B. M. Borg, M. Ek, K. A. Dick, B. Ganjipour, A. W. Dey, C. Thelander, and L.-E. Wernersson, "Diameter reduction of nanowire tunnel heterojunctions using in situ annealing," *Applied Physics Letters*, vol. 99, no. 20, p. 203101, 2011.

[34] M. Luisier and G. Klimeck, "Investigation of $In_xGa_{1-x}As$ ultra-thin-body tunneling FETs using a full-band and atomistic approach," in *SISPAD*, vol. 30, pp. 1–4, IEEE, Sept. 2009.

[35] Y. Lu, A. Seabaugh, P. Fay, S. J. Koester, S. E. Laux, W. Haensch, and S. O. Koswatta, "Geometry dependent tunnel FET performance: Dilemma of electrostatics vs. quantum confinement," in *DRC*, pp. 17–18, IEEE, June 2010.

[36] A. S. Verhulst, W. G. Vandenberghe, K. Maex, and G. Groeseneken, "Tunnel field-effect transistor without gate-drain overlap," *Applied Physics Letters*, vol. 91, no. 053102, 2007.

[37] International Technology Roadmap for Semiconductors (ITRS), "Emerging research devices, 2013 edition." http://www.itrs.net, 2013.

[38] S. S. Sylvia, K. M. M. Habib, M. A. Khayer, K. Alam, M. Neupane, and R. K. Lake, "Effect of random, discrete source dopant distributions on nanowire tunnel FETs," *IEEE Trans. on Electron Devices*, vol. 61, pp. 2208–2214, June 2014.

[39] S. Datta, H. Liu, and V. Narayanan, "Tunnel FET technology: A reliability perspective," *Microelectronics Reliability*, vol. 54, pp. 861–874, Mar. 2014.

8

Models for Tunnel Field-Effect Transistor

8.1 Introduction

TFETs are being investigated intensely using experimental setups and numerical simulations. Our understanding of TFETs has been enriched immensely during the last ten years. As we have seen in previous chapters, several novel TFET device architectures employing different materials have been proposed and are being actively investigated. As a result, the electrical characteristics of TFETs have been improving consistently over the years. However, in order to apply TFETs for various applications, it is essential that appropriate models of TFETs are developed. The models not only give a physical insight into the working of the device, but are also essential for carrying out device-level optimization and circuit-level analysis. The models for TFETs can be hooked into the circuit simulators to explore various applications of TFETs and evaluate the competitive advantages of TFETs over conventional MOSFETs and other exploratory devices. The models also facilitate abstracting out the details of the device and exploring various trade-offs by performing analysis at the circuit and system levels.

In this chapter, we discuss various aspects of modeling TFETs. Firstly, we will examine some of the tunneling models, especially the models that are widely used in Technology Computer Aided Design (TCAD) tools. Then we discuss some of the techniques that are employed to simulate TFETs. We also review various types of models of TFETs that have been proposed and evaluate their merits and demerits. We illustrate the technique to develop an analytical model of a TFET using a model for a nanowire TFET that has been recently proposed.

8.2 Tunneling Models

The physics of tunneling was described in Chapter 2 and the techniques to solve the tunneling problems were described in Section 2.4. In this section, we will review some of the band-to-band tunneling models that are commonly employed in simulating TFETs. Band-to-band tunneling (BTBT) involves tunneling of electrons in the valence band to the conduction band, which can be viewed as creation of electron–hole pairs: electrons are created in the conduction band and holes are created in the valence band. In semi-classical simulators, BTBT is modeled by including an additional generation term in the drift-diffusion equation. There are two types of tunneling models: local tunneling and non-local tunneling.

8.2.1 Local Tunneling Models

In the local tunneling model, the carrier generation rate at a particular position in the device is calculated using a local electric field at that location. Some of the local tunneling models are as follows:

1. **Kane's model**: One of the earliest models for BTBT was developed by E.O. Kane assuming a uniform electric field in the semiconductor [1]. The Kane's tunneling model is the most popular local tunneling model used in the simulation of TFETs. The generation rate due to BTBT G_{BTBT} for a direct bandgap semiconductor in Kane's model is given as

$$G_{BTBT} = \frac{E_{ext}^2 m_r^{\frac{1}{2}}}{18\pi\hbar^2 E_g^{\frac{1}{2}}} \exp\left(\frac{-\pi m_r^{\frac{1}{2}} E_g^{\frac{3}{2}}}{2\hbar|E_{ext}|}\right), \tag{8.1}$$

where E_{ext} is the uniform electric field, E_g is the bandgap of the semiconductor and m_r is the reduced mass of the charge carrier. For an indirect bandgap semiconductor, where the phonons are involved in the conservation of momentum, the generation rate due to BTBT G_{BTBT} is given as

$$G_{BTBT} = \frac{E_{ext}^{\frac{5}{2}} m_r^{\frac{1}{2}}}{18\pi\hbar^2 E_g^{\frac{1}{2}}} \exp\left(\frac{-\pi m_r^{\frac{1}{2}} E_g^{\frac{3}{2}}}{2\hbar|E_{ext}|}\right). \tag{8.2}$$

In TCAD tools, a general form of Kane's model is provided as follows:

$$G_{BTBT} = A E_{ext}^{\alpha} \exp\left(-\frac{B}{|E_{ext}|}\right), \tag{8.3}$$

where A, B, and α are the fitting parameters that the user can specify. These fitting parameters are calibrated such that the tunnel model yields a close agreement with some experimental data or already calibrated simulation data. Kane's model is a simple and numerically robust model. However, when Kane's model is employed in TFETs, it may sometimes yield non-physical results due to non-uniform electric fields or due to ignoring the actual tunneling paths.

2. **Hurkx's model**: This model multiplies a prefactor to the generation rate computed by the local tunneling model. The pre-factor is given as [2, 3]

$$D = \frac{np - n_i^2}{(n + n_i)(p + p_i)}, \tag{8.4}$$

where n and p are electron and the hole concentrations, respectively, and n_i and p_i are intrinsic electron and hole concentrations, respectively, for the given semiconductor. The Hurkx's model accounts for recombination and it can be employed to eliminate the problem of non-zero current observed at $V_{DS} = 0\ V$ when simulating TFETs using Kane's model [4].

8.2.2 Non-Local Tunneling Models

BTBT is physically a non-local process since the rate of BTBT depends on quantities that are spatially separated. Therefore, local tunneling models, though simple to implement in the simulator, cannot capture the physics of BTBT precisely. Local tunneling models assume that the electric field is constant over the tunneling path. For a non-uniform electric field in TFETs, this assumption can become invalid and yield non-physical results.

In a non-local tunneling model, the spatial variation of the energy bands in the semiconductor is taken into account in computing the BTBT rate. In general, non-local tunneling models can be implemented using two types of techniques: mesh node-pair based non-local BTBT model and tunnel path-based non-local BTBT model [3].

1. **Node-pair-based BTBT model**: In node-pair-based non-local BTBT implementation, for all given mesh nodes in a device, the simulator identifies all pairs of mesh nodes A and B that satisfy [3, 5]

$$E_C^A < E_V^B, \tag{8.5}$$

where E_C^A is the conduction band energy at node A and E_V^B is the valence band energy at node B. For these pairs of mesh nodes, the tunneling probability is computed and the carrier-pairs are generated/recombined at mesh nodes A and B, assuming that the BTBT rate is constant within the control volume enclosed by mesh nodes A and B [3, 5].

2. **Tunnel path-based BTBT model**: In the tunnel path-based non-local BTBT model, a tunnel path is identified to compute the BTBT rate. A tunnel path is a physical path between two points in the semiconductor where the electron energy corresponding to the conduction band and the electron energy corresponding to the valence band are equal [6]. As a result, in non-local BTBT models, electrons and holes are generated at the extremities of a tunnel path in contrast to a local BTBT model where both electron–hole pairs are generated at the same physical location [7]. The tunnel path in a non-local tunneling model can be defined in several ways [7]:

   (a) In one of the simplest implementations, the user defines a special-purpose mesh which is used by the simulator to identify locations where non-local BTBT takes place. In a TFET, the special-purpose mesh is usually defined in the source–channel tunneling junction. Therefore, the direction of the tunnel path is locally close to the gradient of the doping. For simple devices, the user can have a fair idea of the tunneling regions and the tunnel paths. Therefore, the user can specify special-purpose mesh in simple tunneling devices. However, manual specification of special-purpose mesh can be problematic in TFETs with complicated architectures. Additionally, since the definition of special-purpose mesh specified by the user is static during simulation, if the tunneling region and the tunneling direction change with the applied bias, then the simulation results can be inaccurate. It should be noted that in order to capture the ambipolar conduction in a TFET, the special-purpose mesh should be defined in the drain-channel junction region also.

   (b) The tunnel path can be identified in the direction of the gradient of the valence band at the start point of the tunnel path and can be dynamically updated during simulation on the basis of computed potential.

   (c) The tunnel path can be identified such that an electron at a location with certain energy in the valence band jumps to the nearest spatial point with equal energy in the conduction band. This corresponds to the tunnel path that minimizes the tunnel length.

   The results obtained using a non-local tunneling model with the above three definitions of tunnel path can be appreciably different

[7]. Therefore, it is suggested that the choice of the tunnel path be compared with a full-quantum simulation [7]. Recently, a robust algorithm that computes current due to BTBT independent of the mesh grid has been proposed and identifies the most probable tunneling paths automatically [3].

8.3 Simulation of TFETs

Numerical simulation plays an important role in the research of exploratory devices such as TFETs. Using numerical simulation, various device architectures can be explored quickly and in a cost-effective way, in contrast to performing experiments. Additionally, various physical quantities for the device such as electric field, tunnel length, carrier concentrations etc. can easily be extracted using simulators. Some of the physical quantities are very difficult or impossible to extract using actual experiments and are important in developing insights into the working of the device.

The quality of results obtained using a simulator is as good as the models used in the simulation. The simulation of TFETs should be done carefully such that all the relevant physics of the device is captured by the simulator. It should be noted that, in simulations, there is typically a trade-off between the computer resource used by the simulator and the accuracy of the results obtained. Therefore, a user needs to intelligently choose models and other parameters of the simulator to get sufficiently accurate results within a reasonable runtime and memory consumed by the simulator. In general, simulation of TFETs are done using one of two techniques: semi-classical simulation or full quantum simulation.

8.3.1 Semi-Classical Simulation

In semi-classical simulation of TFETs, the devices are simulated using drift-diffusion theory. The band-to-band tunneling is incorporated as an additional generation/recombination mechanism either as a local BTBT model or a non-local BTBT model. The semi-classical simulation of TFETs is widely done to investigate novel TFET architectures and perform device optimizations. In semi-classical simulators, the structure of the device is specified using meshes or grids. Fine meshes are defined over the region where greater accuracy is required and where the electrical parameters such as potential and carrier concentrations change rapidly. For TFETs, the meshes should be defined more finely in the source–channel tunneling junction. The simulator formulates the problem in terms of differential equations for each mesh node. These equations are solved self-consistently to compute potential, carrier concentrations, and other quantities of interest.

The choice of the tunneling model plays the most important role in determining the accuracy of results obtained using semi-classical simulation of TFETs. In general, local tunneling models are inaccurate for simulating TFETs. For complicated TFET architectures, tunneling paths are not obvious and the trajectories of tunneling paths can vary with the applied bias. Therefore, automatic identification of tunneling paths by the simulators is important for simulating TFETs [3].

An important step in the simulation of a TFET using the semi-classical method is the calibration of the tunneling models. Since tunneling plays a significant role in the operation of a TFET, this calibration is done on the basis of experimental results for a Zener diode or a TFET. However, as the trap-assisted tunneling, direct tunneling, and phonon-assisted tunneling can occur simultaneously in a TFET, the calibration of the tunneling model can be difficult [8].

In general, the source is degenerately doped in TFETs. Therefore, bandgap narrowing effects are normally incorporated in simulating TFETs.

A couple of examples of input files that can be used for simulating TFETs are provided in Appendix A and Appendix B for illustration.

8.3.2 Full Quantum Simulation

Since BTBT is a quantum-mechanical phenomenon, rigorous simulation of TFETs can be done only using quantum transport formalism. Moreover, with shrinking device dimensions, quantum confinement effects become prominent and quantum transport simulation becomes necessary to obtain accurate results. However, full-band quantum transport simulators take excessively long runtime compared to semi-classical simulators [3, 9–11]. For example, simulations based on WKB approximation take a few seconds on a single CPU computer, while a full-band quantum transport simulator for the same device takes several hours on a super-computer [9]. Therefore, full-band quantum simulations are not always desirable but become necessary under certain conditions. For example, when multiple, tightly coupled, imaginary paths exist between the valence band and the conduction band, as in Si or Ge nanowire TFETs, quantum effects and phonon-assisted tunneling need to be taken into account [9]. For TFETs where quantum confinement effects are dominant such as CNT TFET, GNR TFET or thin nanowire TFET, full-band quantum transport simulations become indispensable.

In an atomistic full-band quantum transport simulation of TFETs, each atom in the device is incorporated explicitly using tight-binding formalism [3, 12–25]. The atomistic simulator calculates carrier density self-consistently with the Poisson equation. Non-equilibrium Green's function (NEGF) formalism is a popular means to obtain a self-consistent solution of the Poisson and the Schrödinger equations in the simulation of nanoscale devices [26, 27]. The NEGF formalism is widely used for the transport description within ballistic approximation in TFETs [9, 12, 19, 23, 28, 29].

8.4 Types of TFET Models

During the last five years, there has been a great deal of research on developing models for TFETs. As the understanding of TFETs is gradually developing, the models for TFETs are becoming sophisticated and are able to capture the physics of TFETs more accurately. Depending on the TFET architectures considered in the simulation, these models differ in the level of accuracy and the computational resources required to compute the results. In broad terms, models for TFETs can be classified into the following categories: lookup table-based models, empirical/semi-empirical models, and analytical models.

8.4.1 Lookup Table (LUT)-Based Models

To perform simulation of TFET circuits, LUT-based TFET models are often created. Figure 8.1 depicts a methodology to simulate TFET circuits using LUT-based models. Firstly, the transfer characteristics and various capacitive components of a TFET at different bias points are computed or experimentally measured. Using these characteristics, a Verilog-A model for the TFET is derived [11, 30–36]. The Verilog-A model for a TFET typically contains the following two-dimensional LUTs:

$$I_{DS} = f_1 (V_{GS}, V_{DS}) \tag{8.6}$$

$$C_{GS} = f_2 (V_{GS}, V_{DS}) \tag{8.7}$$

$$C_{GD} = f_3 (V_{GS}, V_{DS}) \tag{8.8}$$

where f_1, f_2 and f_3 are two-dimensional tables of V_{GS} and V_{DS}. The Verilog-A models of a TFET, combined with a TFET circuit description, are given as inputs to a circuit simulator such as SPICE to perform DC analysis and transient analysis. The characteristics of TFETs for bias points that have not been characterized in the LUTs are obtained by interpolation. The accuracy of the LUT-based TFET models depends on the number of bias points for

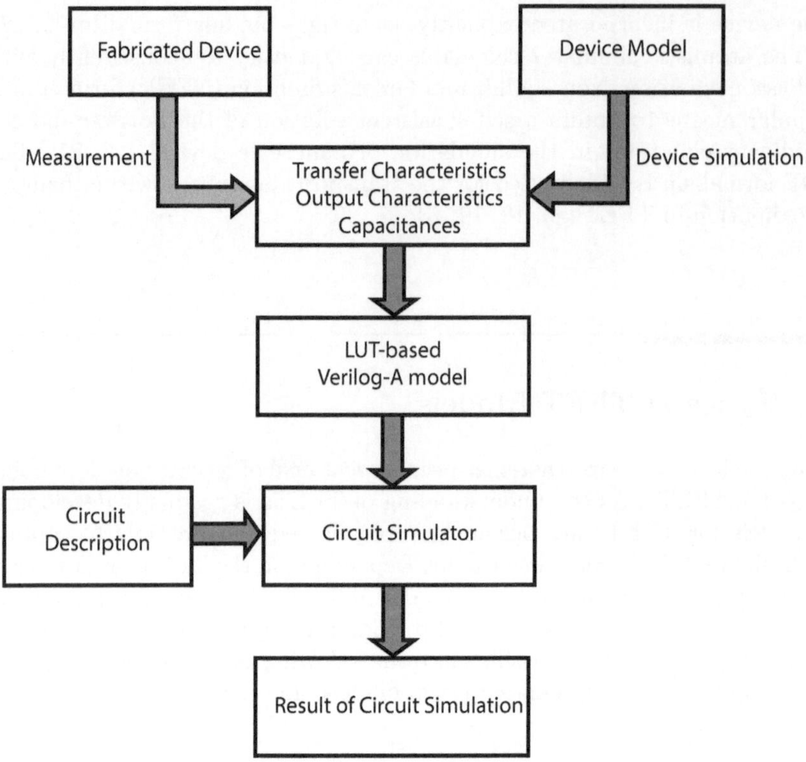

FIGURE 8.1
Methodology to simulate circuits using LUT-based TFET models.

which the 2-D tables have been characterized. For exploratory devices such as TFETs that do not have an established analytical model, simulation using LUT-based models is a fast and effective technique to explore their circuits. However, LUT-based models are not scalable and need to be derived for each device separately. Additionally, device simulations may be required at several bias points to derive a LUT with acceptable accuracy. This can be a severely time-consuming process.

8.4.2 Empirical and Semi-Empirical Models

Empirical and semi-empirical models do not capture the physics of TFET rigorously but are able to show reasonably good agreement with the observed characteristics of the device. These models are in terms of equations, not necessarily derived from the physics of the TFETs, but are based on some observation about the characteristics of the TFETs. In contrast to a LUT-based model, these models are not defined at discrete bias points and typically do not require interpolation. In general, empirical and semi-empirical models are simple, computationally efficient, and of moderate accuracy. The empirical and semi-empirical models of a TFET can be developed in several ways [37–39]. For example, a TFET can be modeled using a high-level SPICE behavioral model in which the tunneling behavior is modeled using a Zener diode and the dependent sources are incorporated to model the dependence of the drain current on the terminal voltages [37]. The model can also include voltage-dependent parasitic capacitances to enable transient circuit simulation. A TFET can also be modeled as a MOSFET connected in series with a gate-controllable tunneling diode [38]. In this model, the tunneling diode dominates at low inter-band tunneling current and the MOSFET component dominates at high tunneling current. Recently, a semi-empirical model based on the Kane–Sze formula is also proposed [39]. This model is able to capture the dependence of the subthreshold swing on the drain current and the super-linear onset of drain current in the output characteristics of the TFET.

8.4.3 Analytical Models

Analytical models are derived from the physics of a device. Analytical models are easy to use in the simulators, are scalable, and give an insight into the working of the device. Therefore, during the last five years, a great deal of effort has been put in by researchers in developing analytical models for TFETs [4, 6, 29, 39–59]. There are some distinct characteristics of a TFET compared to a MOSFET that must be carefully considered when developing an analytical model for a TFET. These characteristics are as follows:

1. Complicated dependence of the subthreshold swing on the gate voltage
2. Dual control of the gate voltage and the drain voltage on the tunnel barrier
3. Non-linear or exponential onset of drain current in the output characteristics
4. Different mechanism of saturation in the output characteristics
5. Different contribution of the various capacitive components to the total gate capacitance in a TFET

A widely used approach to develop an analytical model for a TFET is to extend an existing analytical model of a corresponding MOSFET. The special characteristics of a TFET, as described above, are accounted for by incorporating the relevant physics of the TFETs into the model of a MOSFET. In the rest of this section, we will review the analytical models of TFETs that have been recently proposed. However, we will not discuss these analytical models in detail in this book. Interested readers are directed to the references given below to further explore the analytical models for a TFET.

One of the earliest analytical models for a TFET was presented in [6]. This model was proposed for the vertical tunneling in a TFET where the gate is directly over the source. The tunneling in the region where the gate overlaps the source is referred to as "line tunneling." The drain current was computed by first finding the average electric field on the tunnel path and then using Kane's formula to compute the drain current. Subsequently, an analytical model for tunneling in the localized region at the source–channel junction, which is referred to as "point tunneling" was also proposed [42]. It should be noted that the generalized case of tunneling can be considered as a combination of line tunneling and point tunneling. Further improvements in line tunneling– and point tunneling– based TFET models were made by incorporating the impact of the drain bias and using Wentzel–Kramer–Brillouin (WKB) approximation to compute the BTBT rate rather than using Kane's model [49, 53]. It should be noted that an analytical model proposed for a particular TFET architecture is often not valid for a different TFET architecture. In general, developing a model for a TFET starts with modeling the potential and the electric field profile of the device. Since, different TFET architectures can have different potential and electric field profiles, the models for different TFET architectures are typically distinct. For example, distinct tunneling models for three different gate architectures (single-gate, double-gate, and gate-all-around configurations) have been developed in [45].

Several models for TFET that have been recently proposed are developed in the following two steps:

1. Compute the surface potential and the electric field profile in the TFET.
2. Find the tunneling generation rate for the computed electric field and integrate it over the tunneling volume to compute the drain current.

Therefore, surface potential models for different TFET architectures have been intensely investigated [41, 43, 57]. Recently, analytical models for SOI TFETs have been proposed in [4, 46, 50, 51, 60], models for dual material gate (DMG) TFETs [61, 62] have been proposed in [54, 55, 63], and models for nanowire TFETs have been proposed in [55, 56, 59, 64].

In addition to the analytical models for drain current in a TFET, there have been several pioneering works that have investigated or modeled specific phenomena of a TFET. These works have provided a foundation for developing analytical models for TFETs. Some of the noteworthy works that have examined and modeled the physics of TFETs are the dependence of tunneling current on various device parameters has been described in [65], expressions for the subthreshold slope have been derived in [6, 40, 66], the dependence of the drain current on the drain voltage is modeled in [41, 48], the exponential onset in the output characteristics is investigated in detail in [52], modeling of capacitances in a TFET is developed in [44], and modeling of small-signal parameters in a TFET is described in [47]. Readers are encouraged to refer to these works to develop further insight into the physics and modeling of TFETs.

It should be noted that to obtain a reasonable agreement of results obtained using the analytical models with experimental data, several fitting parameters of the models might be required to be computed. This can be a cumbersome and computationally intensive process.

8.5 Drain Current Model

As mentioned above, a general approach for developing a compact TFET model is to find the surface potential for the device and then compute drain current using an appropriate BTBT model. In this section, we illustrate this technique using the analytical drain current model for a gate-all-around (GAA) nanowire TFET that is recently proposed [56]. Nanowire TFETs with GAA architecture are being extensively investigated by various research groups because of enhanced electrostatic control, excellent subthreshold swing, and high ON-state current (I_{ON}).

Figure 8.2 shows a cross-sectional view of a p-type GAA nanowire TFET. The co-ordinate in the radial direction is denoted as r and along the axial direction it is denoted as z. The diameter of the nanowire is d_{NW}, the channel length is L, the gate oxide thickness is t_{ox}, the doping in the channel is N_S, and the gate work function is ϕ.

The nature of surface potential in an NW TFET is shown in Figure 8.3. The surface potential in the TFET can be divided into two regions: R1 and R2. In region R1, the channel is depleted and in region R2, the potential drop in the channel is negligible due to small drain current. The surface potential in region R2 can be taken as a constant ψ_C. The behavior of a TFET is similar to a MOSFET regarding the formation of the channel and the current transport in the channel, except in the region around the source where band-to-band tunneling occurs. Therefore, the surface potential ψ_C in region R2 can be given as [56]

$$\psi_C = V_{DS} + \psi_B \text{ when } |V_{DS}| < |V_{GS} - V_{Th}| \tag{8.9}$$

$$\psi_C = V_{GS} - V_{Th} + \psi_B \text{ when } |V_{DS}| \geq |V_{GS} - V_{Th}|, \tag{8.10}$$

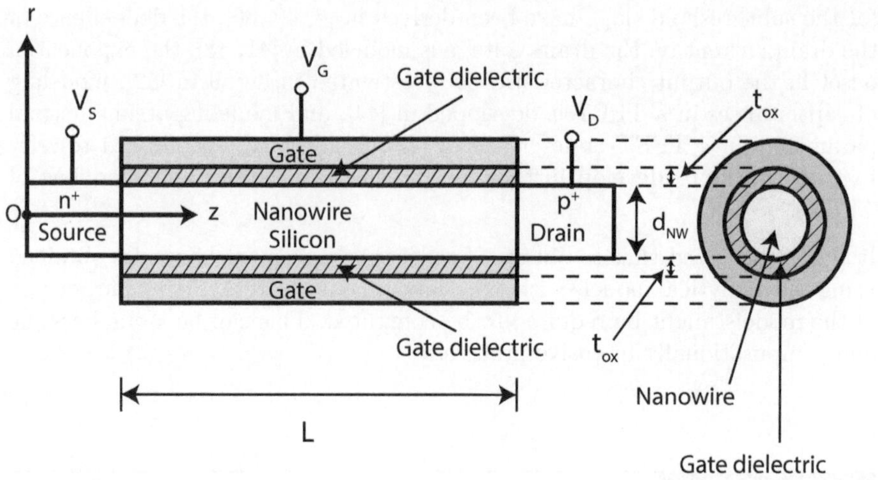

FIGURE 8.2
Cross-sectional view of a p-type GAA nanowire TFET [56].

where ψ_B is the amount of band-bending caused by V_{GS} and V_{Th} is the threshold voltage of the MOSFET with the gate work function the same as that of the TFET. The potential at the source and the drain can be given as

$$\psi_{Source} = V_S + V_{bi} \tag{8.11}$$

$$\psi_{Drain} = V_{DS}, \tag{8.12}$$

where V_{bi} is the built-in potential at the source–channel junction and V_S is the ground potential.

In region R1, the 2-D Poisson equation in the cylindrical co-ordinates is given as follows

$$\frac{1}{r}\frac{\partial}{\partial r}\left(r\frac{\partial\psi\left(r,z\right)}{\partial r}\right)+\frac{\partial^2\psi\left(r,z\right)}{\partial z^2}=\frac{qN_S}{\varepsilon_{Si}}, \tag{8.13}$$

where $\psi\left(r,z\right)$ is the potential, q is the charge of an electron, and ε_{Si} is the permittivity of silicon. The nature of potential along the radial direction is shown in Figure 8.4. It can be approximated as a second-order polynomial. Let us represent the potential profile of the circular disc of the NW at an axial distance z as:

$$\psi\left(r,z\right)=a_0\left(z\right)+a_1\left(z\right)r+a_2\left(z\right)r^2, \tag{8.14}$$

where $a_0\left(z\right)$, $a_1\left(z\right)$ and $a_2\left(z\right)$ are the coefficients that need to be determined. The potential profile given by Equation (8.14) must satisfy the following boundary conditions:

1. The potential at $r = d_{NW}/2$ is the same as the surface potential

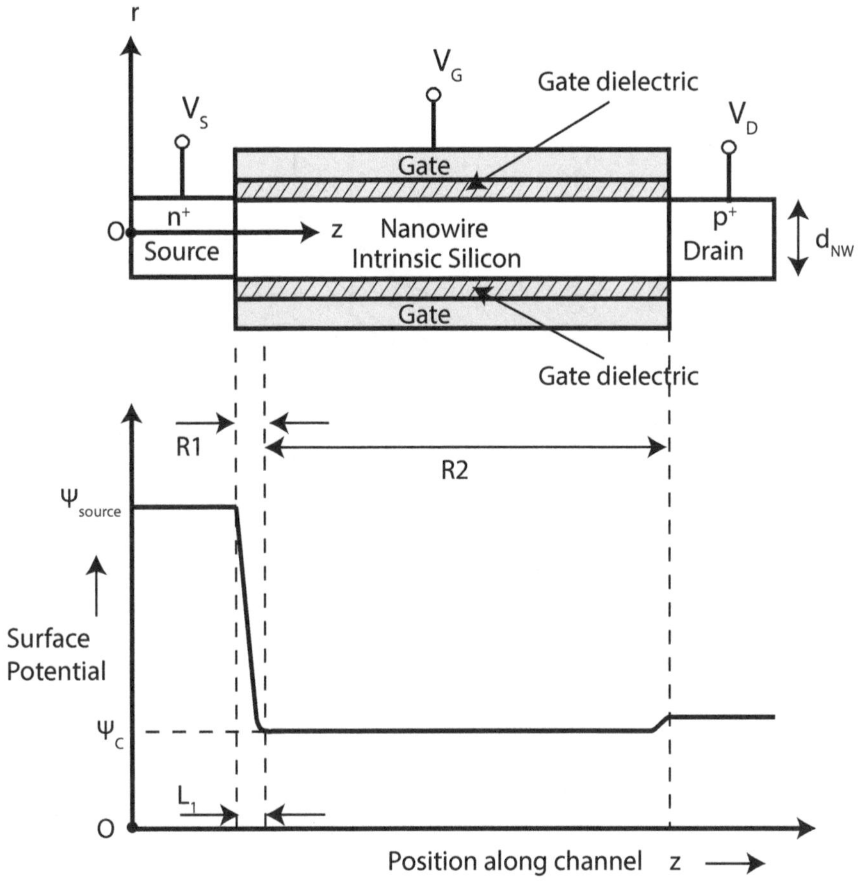

FIGURE 8.3
Nature of surface potential along the axial direction [56].

$\psi_S(z)$:

$$\psi(d_{NW}/2, z) = \psi_S(z). \tag{8.15}$$

2. The electric field at $r = 0$ is equal to zero:

$$E_r(0, z) = 0. \tag{8.16}$$

3. The electric field displacements across the oxide–silicon interface are the same:

$$E_r(d_{NW}/2, z) = -C_{ox}(\psi_G - \psi_S(z))/\varepsilon_{Si}, \tag{8.17}$$

where C_{ox} is the capacitance per unit area at $r = d_{NW}/2$ and can

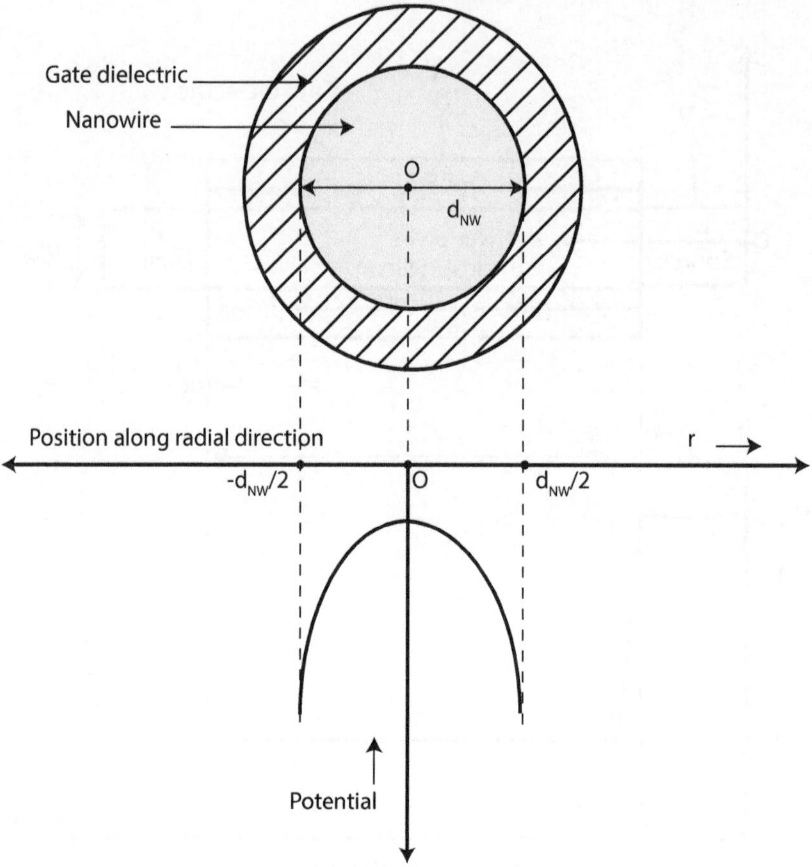

FIGURE 8.4
Nature of potential along the radial direction [56].

be given as

$$C_{ox} = \frac{2\varepsilon_{Si}}{d_{NW} \ln\left(1 + \frac{2t_{ox}}{d_{NW}}\right)}. \tag{8.18}$$

Using the above boundary conditions, the coefficients $a_0(z)$, $a_1(z)$ and $a_2(z)$ are computed as [56]

$$a_0(z) = \psi_S(z)\left(1 + \frac{d_{NW}C_{ox}}{\varepsilon_{Si}}\right) - \frac{d_{NW}C_{ox}}{4\varepsilon_{Si}}\psi_G \tag{8.19}$$

$$a_1(z) = 0 \tag{8.20}$$

$$a_2(z) = \frac{C_{ox}}{d_{NW}\varepsilon_{Si}}\left(\psi_G - \psi_S(z)\right). \tag{8.21}$$

Substituting the values of the coefficients from Equations (8.19)–(8.21) into Equation (8.14) and using (8.13) yields [56]:

$$\frac{\partial^2 \psi_S}{\partial z^2} - \frac{4C_{ox}}{d_{NW}\varepsilon_{Si}}\psi_S = \frac{qN_S}{\varepsilon_{Si}} - \frac{4C_{ox}}{d_{NW}\varepsilon_{Si}}\psi_G, \tag{8.22}$$

The solution for Equation (8.22) can be written as

$$\psi_S(z) = C\exp\left(\frac{z-L_1}{L_d}\right) + D\exp\left(\frac{-(z-L_1)}{L_d}\right) + \psi_G - \frac{qN_S L_d^2}{\varepsilon_{Si}}, \tag{8.23}$$

where C and D are coefficients that need to be determined, L_1 is the length of region R1, and L_d is the characteristics length that can be given as

$$L_d = \sqrt{\frac{d_{NW}^2}{4}\ln\left(1 + \frac{2t_{ox}}{d_{NW}}\right)\left(\frac{\varepsilon_{Si}}{2\varepsilon_{ox}}\right)}. \tag{8.24}$$

The surface potential $\psi_S(z)$ given by Equation (8.23) must satisfy the following boundary conditions:

1. Surface potential at the source is equal to the built-in potential barrier:
$$\psi_S(z) = V_{bi}. \tag{8.25}$$

2. Surface potential at the boundary of regions R1 and R2 should be equal:
$$\psi_S(L_1) = \psi_C. \tag{8.26}$$

3. The electric field at the boundary of regions R1 and R2 is equal to zero:
$$\frac{\partial \psi_S(L_1)}{\partial z} = 0. \tag{8.27}$$

With these boundary conditions, the coefficients C and D can be computed as follows:

$$C = D = \left(\psi_C - \psi_G + \frac{qN_S L_d^2}{\varepsilon_{Si}}\right)/2. \tag{8.28}$$

Using C and D in Equation (8.23), the length of region R1, i.e., L_1, can be computed. Considering that the doping in the channel, i.e., N_S is low, the term $(qN_S L_d^2)/\varepsilon_{Si}$ is negligible compared to other terms in Equation (8.23) and (8.28).

Since the slope of the surface potential decreases along the channel, the shortest tunneling length L_T must be from the source to the location where the surface potential decreases by (E_g/q) where E_g is the energy bandgap of silicon. Therefore, the shortest tunneling length L_T can be written as [56]

$$L_T = x(\psi_{Source}) - x(\psi_{Source} - E_g/q) \tag{8.29}$$

$$x(\psi_S) = L_d \cosh^{-1}\left(\frac{\psi_S - \psi_G}{\psi_C - \psi_G}\right). \tag{8.30}$$

The average electric field can now be computed as follows

$$E_{avg} = \frac{E_g}{qL_T}. \tag{8.31}$$

By substituting the average electric field (E_{avg}) in Kane's model, the drain current I_D can be computed as

$$I_D = A_K \times L_T \times E_{avg}^{\alpha} \times \exp\left(-\frac{B_K}{E_{avg}}\right), \tag{8.32}$$

where A_K, B_K and α are the parameters of Kane's model that depend on the material properties. The term ($A_K \times L_T$) can be treated as the tunneling volume and is modulated by the gate voltage [56].

To estimate the accuracy of the TFET model developed above, the drain current computed using the model is compared with the drain current obtained using three-dimensional numerical simulations [56]. For gate voltages above threshold, the drain current obtained using the model shows a good agreement with the drain current obtained using simulations. However, in the subthreshold region, the drain current obtained using the model is higher than that obtained using simulations. This is attributed to the model assuming that the shortest tunneling length is valid over the entire tunneling volume [56]. However, in the subthreshold region, this assumption becomes invalid due to longer tunneling lengths [56]. To avoid this problem, an analytical model for the drain current is developed using a tangent line approximation method to integrate the tunneling generation rate rather than taking the tunnel length constant over the tunneling volume [59]. The model proposed in [59] is able to compute the drain current accurately in the subthreshold region also.

8.6 Summary

In this chapter, we discussed various local and non-local tunneling models. We also discussed semi-classical and full-band quantum transport simulation techniques. We studied various types of models for TFETs: LUT-based models, semi-empirical models, and analytical models. We also developed an analytical drain current model for a nanowire TFET. In the next chapter, we will look at the circuits that can be realized using TFETs. Most of the circuits that are presented in the next chapter have been analyzed using simulations and LUT-based models for TFETs.

REFERENCES

[1] E. O. Kane, "Theory of tunneling," *Journal of Applied Physics*, vol. 32, pp. 83–91, Jan. 1961.

[2] G. A. M. Hurkx, D. B. M. Klassens, and M. P. G. Knuvers, "A new recombination model for device simulation including tunneling," *IEEE Trans. Electron Devices*, vol. 39, pp. 331–338, Feb. 1992.

[3] C. Shen, L. T. Yang, G. Samudra, and Y. C. Yeo, "A new robust non-local algorithm for band-to-band tunneling simulation and its application to Tunnel-FET," *Solid-State Electronics*, pp. 23–30, Mar. 2011.

[4] R. Vishnoi and M. J. Kumar, "An accurate compact analytical model for the drain current of a TFET from subthreshold to strong inversion," *IEEE Trans. on Electron Devices*, vol. 62, pp. 478–484, Feb. 2015.

[5] J. Peng, S. Haddad, J. Hsu, J. Chen, S. Longcor, and C. Chang, "Accurate simulation on band-to-band tunneling induced leakage current using a global non-local model," in *International Conference on Solid-State and Integrated Circuit Technology*, pp. 141–143, IEEE, Oct. 1995.

[6] W. G. Vandenberghe, A. S. Verhulst, G. Groeseneken, B. Soree, and W. Magnus, "Analytical model for a tunnel field-effect transistor," in *Proc. 14th IEEE MELECON*, pp. 923–928, IEEE, 2008.

[7] L. D. Michielis, M. Iellina, P. Palestri, A. M. Ionescu, and L. Selmi, "Effect of the choice of the tunnelling path on semi-classical numerical simulations of TFET devices," *Solid-State Electronics*, vol. 71, pp. 7–12, May 2012.

[8] C. Kampen, A. Burenkov, and J. Lorenz, "Challenges in TCAD simulations of tunneling field effect transistors," in *Proceedings of the European Solid-State Device Research Conference (ESSDERC)*, pp. 139–142, IEEE, Sept. 2011.

[9] M. Luisier and G. Klimeck, "Simulation of nanowire tunneling transistors: From the Wentzel-Kramers-Brillouin approximation to full-band phonon-assisted tunneling," *Journal of Applied Physics*, vol. 107, no. 084507, 2007.

[10] J. Z. Huang, L. Zhang, W. C. Chew, C.-Y. Yam, L. J. Jiang, G.-H. Chen, and M. Chan, "Model order reduction for quantum transport simulation of band-to-band tunneling devices," *IEEE Trans. on Electron Devices*, vol. 61, pp. 561–568, Feb. 2014.

[11] D. H. Morris, U. E. Avci, R. Rios, and I. A. Young, "Design of low voltage tunneling-fet logic circuits considering asymmetric conduction characteristics," *IEEE Journal on Emerging and Selected Topics in Circuits and Systems*, vol. 4, pp. 380–388, Dec. 2014.

[12] M. Luisier and G. Klimeck, "Performance analysis of statistical samples of graphene nanoribbon tunneling transistors with line edge roughness," *Applied Physics Letters*, vol. 4, pp. 223505–1–223505–3, June 2009.

[13] M. Luisier and G. Klimeck, "Investigation of $In_xGa_{1-x}As$ ultra-thin-body tunneling FETs using a full-band and atomistic approach," in *SISPAD*, vol. 30, pp. 1–4, IEEE, Sept. 2009.

[14] D. Basu, L. F. Register, M. J. Gilbert, and S. K. Banerjee, "Atomistic simulation of band-to-band tunneling in III-V nanowire field-effect transistors," in *SISPAD*, pp. 1–4, IEEE, 2009.

[15] M. Luisier and G. Klimeck, "Performance comparisons of tunneling field-effect transistors made of InSb, Carbon, and GaSb-InAs broken gap heterostructures," in *IEEE International Electron Devices Meeting (IEDM)*, pp. 1–4, IEEE, 2009.

[16] S. Agarwal, G. Klimeck, and M. Luisier, "Leakage-reduction design concepts for low-power vertical tunneling field-effect transistors," *IEEE Electron Device Letters*, vol. 27, pp. 621–623, June 2010.

[17] L. Liu, D. Mohata, and S. Datta, "Scaling length theory of double-gate interband tunnel field-effect transistors," *IEEE Trans. on Electron Devices*, vol. 59, pp. 902–908, Apr. 2012.

[18] U. E. Avci, S. Hasan, D. E. Nikonov, R. Rios, K. Kuhn, and I. A. Young, "Understanding the feasibility of scaled III-V TFET for logic by bridging atomistic simulations and experimental results," in *Symposium on VLSI technology*, pp. 183–184, IEEE, June 2012.

[19] M. Fujimatsu, H. Saito, and Y. Miyamoto, "71 mV/dec of sub-threshold slope in vertical tunnel field-effect transistors with GaAsSb/InGaAs heterostructure," in *Int. Conf. on Indium Phosphide and Related Materials (IPRM)*, pp. 27–30, IEEE, Aug. 2012.

[20] N. Ghobadi and M. Pourfath, "A comparative study of tunneling FETs based on graphene and GNR heterostructures," *IEEE Trans. on Electron Devices*, vol. 61, pp. 186–192, Jan. 2014.

[21] H. Wang, S. Chang, Y. Hu, H. He, J. He, Q. Huang, F. He, and G. Wang, "A novel barrier controlled tunnel FET," *IEEE Electron Device Letters*, vol. 35, pp. 798–800, July 2014.

[22] T. K. Agarwal, A. Nourbakhsh, P. Raghavan, I. Radu, S. D. Gendt, M. Heyns, M. Verhelst, and A. Thean, "Bilayer graphene tunneling FET for sub-0.2 *V* digital CMOS logic applications," *IEEE Electron Device Letters*, vol. 35, pp. 1308–1310, Dec. 2014.

[23] A. Sharma, A. A. Goud, and K. Roy, "Sub-10 nm FinFETs and tunnel-FETs: From devices to systems," in *Design, Automation and Test in Europe Conference and Exhibition (DATE) 2015*, pp. 1443–1448, IEEE, Mar. 2015.

[24] U. E. Avci, D. H. Morris, and I. A. Young, "Tunnel field-effect transistors: Prospects and challenges," *IEEE Journal of the Electron Devices Society*, vol. 3, pp. 88–95, May 2015.

[25] F. Liu, J. Wang, and H. Guo, "Atomistic simulations of device physics in monolayer transition metal dichalcogenide tunneling transistors," *IEEE Transactions on Electron Devices*, vol. 63, pp. 311–317, Jan. 2016.

[26] S. Datta, "Nanoscale device modeling: The Green's function method," *Superlattices and Microstructures*, vol. 28, no. 4, pp. 253–278, 2000.

[27] E. Polizzi and S. Datta, "Multidimensional nanoscale device modeling: the finite element method applied to the non-equilibrium Green's function formalism," in *IEEE Conference on IEEE-NANO*, vol. 2, pp. 40–43, IEEE, Aug. 2003.

[28] A. Sharma, A. A. Goud, and K. Roy, "GaSb-InAs n-TFET with doped source underlap exhibiting low subthreshold swing at sub-10-nm gate-lengths," *IEEE Electron Device Letters*, vol. 35, pp. 1221–1223, Dec. 2014.

[29] M. S. Fahad, A. Srivastava, A. K. Sharma, and C. Mayberry, "Analytical current transport modeling of graphene nanoribbon tunnel field-effect transistors for digital circuit design," *IEEE Transactions on Nanotechnology*, vol. 15, pp. 39–50, Jan. 2016.

[30] D. Kim, Y. Lee, J. Cai, I. Lauer, L. Chang, S. J. Koester, D. Sylvester, and D. Blaauw, "Low power circuit design based on heterojunction tunneling transistors (HETTs)," in *IEEE International Symposium on Low Power Electronics and Design (ISLPED)*, pp. 219–224, IEEE, Aug. 2009.

[31] V. Saripalli, S. Datta, V. Narayanan, and J. P. Kulkarni, "Variation-tolerant ultra low-power heterojunction tunnel FET SRAM design," in *IEEE/ACM International Symposium on Nanoscale Architectures (NANOARCH)*, pp. 45–52, IEEE, June 2011.

[32] R. Mukundrajan, M. Cotter, V. Saripalli, M. J. Irwin, S. Datta, and V. Narayanan, "Ultra low power circuit design using tunnel FETs," in *IEEE Computer Society Annual Symposium on VLSI (ISVLSI)*, pp. 153–158, IEEE, Aug. 2012.

[33] M. Cotter, H. Liu, S. Datta, and V. Narayanan, "Evaluation of tunnel FET-based flip-flop designs for low power, high performance applications," in *Proc. International Symposium on Quality Electronic Design*, pp. 430–437, IEEE, Mar. 2013.

[34] S. Datta, H. Liu, and V. Narayanan, "Tunnel FET technology: A reliability perspective," *Microelectronics Reliability*, vol. 54, pp. 861–874, Mar. 2014.

[35] K. Swaminathan, H. Liu, X. Li, M. S. Kim, J. Sampson, and V. Narayanan, "Steep slope devices: Enabling new architectural paradigms," in *Proceedings of the 51st Annual Design Automation Conference*, pp. 1–6, IEEE, June 2014.

[36] H. Liu, V. Saripalli, V. Narayanan, and S. Datta, "The III-V tunnel FET model." http://www.ndcl.ee.psu.edu/downloads.asp.

[37] Y. Hong, Y. Yang, L.Yang, G. Samudra, C.-H. Heng, and Y.-C. Yeo, "SPICE behavioral model of the tunneling field-effect transistor for circuit simulation," *IEEE Trans. on Circuits and Systems-II*, vol. 56, pp. 946–950, Dec. 2009.

[38] J. Wan, C. L. Royer, A. Zaslavsky, and S. Cristoloveanu, "A tunneling field effect transistor model combining interband tunneling with channel transport," *Journal of Applied Physics*, vol. 110, no. 104503, 2011.

[39] H. Lu, J. W. Kim, D. Esseni, and A. Seabaugh, "Continuous semiempirical model for the current-voltage characteristics of tunnel FETs," in *International Conference on Ultimate Integration on Silicon (ULIS)*, pp. 25–28, IEEE, Apr. 2014.

[40] Q. Zhang, W. Zhao, and A. Seabaugh, "Low-subthreshold-swing tunnel transistors," *IEEE Electron Device Letters*, vol. 27, pp. 297–300, Apr. 2006.

[41] C. Shen, S.-L. Ong, C.-H. Heng, G. Samudra, and Y.-C. Yeo, "A variational approach to the two-dimensional nonlinear Poisson's equation for the modeling of tunneling transistors," *IEEE Electron Device Letters*, vol. 29, pp. 1252–1255, Nov. 2008.

[42] W. G. Vandenberghe, A. S. Verhulst, G. Groeseneken, B. Soree, and W. Magnus, "Analytical model for point and line tunneling in a tunnel field-effect transistor," in *SISPAD*, pp. 137–140, IEEE, 2008.

[43] M. G. Bardon, H. P. Neves, R. Puers, and C. V. Hoof, "Pseudo-two-dimensional model for double-gate tunnel FETs considering the junctions depletion regions," *IEEE Trans. on Electron Devices*, vol. 57, pp. 827–834, Apr. 2010.

[44] Y. Yang, X. Tong, L. T. Yang, P. F. Guo, L. Fan, and Y. C. Yeo, "Tunneling field-effect transistor: Capacitance components and modeling," *Electron Device Letters*, vol. 31, pp. 752–754, July 2010.

[45] A. S. Verhulst, B. Soree, D. Leonelli, W. G. Vandenberghe, and G. Groeseneken, "Modeling the single-gate, double-gate, and gate-all-around tunnel field effect transistor," *Journal of Applied Physics*, vol. 107, no. 024518, 2010.

[46] M. J. Lee and W. Y. Choi, "Analytical model of single-gate silicon-on-insulator (SOI) tunneling field-effect transistors (TFETs)," *State Electronics*, pp. 110–114, Sept. 2011.

[47] S. Cho, J. S. Lee, K. R. Kim, B.-G. Park, J. S. Harris, and I. M. Kang, "Analyses on small-signal parameters and radio-frequency modeling of gate-all-around tunneling field-effect transistors," *IEEE Trans. on Electron Devices*, vol. 58, pp. 4164–4171, Dec. 2011.

[48] A. Mallik and A. Chattopadhyay, "Drain-dependence of tunnel field-effect transistor characteristics: The role of the channel," *IEEE Trans. on Electron Devices*, vol. 58, pp. 4250–4257, Dec. 2011.

[49] A. S. Verhulst, D. Leonelli, R. Rooyackers, and G. Groeseneken, "Drain voltage dependent analytical model of tunnel field-effect transistors," *Journal of Applied Physics*, vol. 110, no. 024510, 2011.

[50] V. Dobrovolsky and F. Sizov, "Analytical model of the thin-film silicon-on-insulator tunneling field effect transistor," *Journal of Applied Physics*, vol. 110, no. 114513, 2011.

[51] B. Bhushan, K. Nayak, and V. R. Rao, "DC compact model for SOI tunnel field-effect transistors," *IEEE Trans. on Electron Devices*, vol. 59, pp. 2635–2642, Oct. 2012.

[52] L. D. Michielis, L. Lattanzio, and A. M. Ionescu, "Understanding the superlinear onset of tunnel-FET output characteristic," *Electron Device Letters*, vol. 33, pp. 1523–1525, Nov. 2012.

[53] W. G. Vandenberghe, A. S. Verhulst, K.-H. Kao, K. D. Meyer, B. Soree, W. Magnus, and G. Groeseneken, "A model determining optimal doping concentration and material's band gap of tunnel field-effect transistors," *Applied Physics Letters*, vol. 100, no. 193509, 2012.

[54] R. Vishnoi and M. J. Kumar, "Compact analytical model of dual material gate tunneling field-effect transistor using interband tunneling and channel transport," *IEEE Trans. on Electron Devices*, vol. 61, pp. 1936–1942, June 2014.

[55] R. Vishnoi and M. J. Kumar, "A pseudo-2-D-analytical model of dual material gate all-around nanowire tunneling FET," *IEEE Trans. on Electron Devices*, vol. 61, pp. 2264–2270, July 2014.

[56] R. Vishnoi and M. J. Kumar, "Compact analytical drain current model of gate-all-around nanowire tunneling FET," *IEEE Trans. on Electron Devices*, vol. 61, pp. 2599–2603, July 2014.

[57] C. Wu, R. Huang, Q. Huang, C. Wang, J. Wang, and Y. Wang, "An analytical surface potential model accounting for the dual-modulation effects in tunnel FETs," *IEEE Trans. on Electron Devices*, vol. 61, pp. 2690–2696, Aug. 2014.

[58] P. Pandey, R. Vishnoi, and M. J. Kumar, "Drain current model for SOI TFET considering source and drain side tunneling," in *IEEE International Conf. on Emerging Electronics (ICEE)*, pp. 1–5, IEEE, Dec. 2014.

[59] R. Vishnoi and M. J. Kumar, "A compact analytical model for the drain current of gate-all-around nanowire tunnel FET accurate from sub-threshold to ON-state," *IEEE Trans. on Nanotechnology*, vol. 14, pp. 358–362, Mar. 2015.

[60] R. Vishnoi and M. J. Kumar, "2-D analytical model for the threshold voltage of a tunneling FET with localized charges," *IEEE Transactions on Electron Devices*, vol. 61, pp. 3054–3059, Sept. 2014.

[61] S. Saurabh and M. J. Kumar, "Novel attributes of a nanoscale dual material gate tunnel field effect transistor," *IEEE Trans. on Electron Devices*, vol. 58, pp. 404–410, Feb. 2011.

[62] C. Y. Hsu, C. Y. Chang, E. Y. Chang, and C. Hu, "Suppressing non-uniform tunneling in InAs/GaSb TFET with dual-metal gate," *IEEE Journal of the Electron Devices Society*, vol. 4, pp. 60–65, Mar. 2016.

[63] P. Pandey, R. Vishnoi, and M. J. Kumar, "A full-range dual material gate tunnel field effect transistor drain current model considering both source and drain depletion region band-to-band tunneling," *Journal of Computational Electronics*, vol. 14, no. 1, pp. 280–287, 2014.

[64] J. Wu and Y. Taur, "A continuous semianalytic current model for DG and NW TFETs," *IEEE Transactions on Electron Devices*, vol. 63, pp. 841–847, Feb. 2016.

[65] J. Knoch and J. Appenzeller, "A novel concept for field-effect transistors: The tunneling carbon nanotube FET," in *Proc. 63rd DRC*, pp. 153–156, IEEE, June 2005.

[66] J. Knoch, S. Mantl, and J. Appenzeller, "Impact of the dimensionality on the performance of tunneling FETs: Bulk versus one-dimensional devices," *Solid-State Electronics*, vol. 51, pp. 572–578, 2007.

9

Applications of Tunnel Field-Effect Transistors

9.1 Introduction

TFETs are being investigated for use in different types of applications such as conventional logic gates, memories, sensors, and various analog circuits. Several theoretical studies have demonstrated that circuits employing TFETs can outperform conventional CMOS circuits at low supply voltages, especially in terms of energy efficiency. These results have encouraged researchers to explore circuits where the special properties of TFETs can be exploited to obtain energy and power savings. Since TFETs exhibit electrical characteristics that are quite distinct from a conventional MOSFET, novel circuit applications based on TFETs are possible. Moreover, in contrast to a conventional MOSFET, TFETs exhibit electrical characteristics such as unidirectional current flow and exponential onset of drain current in the output characteristics. These electrical characteristics pose new challenges in circuit applications of TFETs. A given CMOS circuit cannot be made functional just by replacing MOSFETs with similar TFETs. As a result, researchers are exploring innovative circuits and TFET architectures to tackle the challenges in the circuit applications of TFETs.

In this chapter, first of all, we review some of the important electrical characteristics of TFETs that are relevant for circuit applications. Then we discuss different applications of TFETs in digital circuits, memories, and analog circuits. We highlight the strengths and the weaknesses of TFETs in these applications. We also describe a few techniques that can overcome the challenges in the circuit applications of TFETs.

9.2 Electrical Characteristics of TFETs

There are certain electrical characteristics of a TFET that distinguish it from a conventional MOSFET. In exploiting TFETs in circuit applications, the following properties play an important role.

9.2.1 Steep Subthreshold Swing

In contrast to a conventional MOSFET, TFETs can exhibit subthreshold swings lower than 60 $mV/decade$ at room temperature. The low subthreshold swing of the TFETs allows voltage scaling without degradation in performance.

9.2.2 Unidirectional Conduction

In contrast to a conventional MOSFET, the current flows in a single direction in a TFET due to its asymmetric p-i-n structure. For a given gate voltage, in an n-type TFET, an appreciable current flows only when the p-i-n structure is reverse-biased or V_{DS} is positive [1–6]. When V_{DS} is negative, the p-i-n structure is forward biased and behaves as a forward-biased diode. When V_{DS} is negative and below the turn-on voltage $V_{Turn-On}$ of the forward-biased diode ($V_{DS} < 0$ and $|V_{DS}| < V_{Turn-On}$), a very small drain current flows in the TFET. When the drain bias is negative and is greater than the turn-on voltage $V_{Turn-On}$ of the forward-biased diode, an appreciable drain current can flow independent of the gate voltage. However, TFETs are operated in circuits with a supply voltage smaller than 0.5 V, which is normally lower than the turn-on voltage of the forward-biased diode. Therefore, typically, the drain current is negligible when V_{DS} is negative. It is worthwhile to note that when V_{DS} is negative and very small, the forward-biased diode can exhibit negative differential resistance as in a tunnel diode or Esaki diode described in Section 2.6. In the Esaki diode region of the TFET, some gate control over the drain current can be observed [2]. Figure 9.1 qualitatively shows the difference in the output characteristics of a TFET and a MOSFET. A conventional MOSFET allows current to flow when V_{DS} is positive and when V_{DS} is negative, since the drain and the source terminals are interchangeable. However, a TFET allows current to flow only when V_{DS} is positive.

A p-type TFET also exhibits unidirectional current flow. An appreciable current flows in a p-type TFET only when the p-i-n structure is reverse-biased or V_{DS} is negative. The unidirectional flow of current in a TFET has a profound implication for the application of TFETs in circuits. It poses problems in

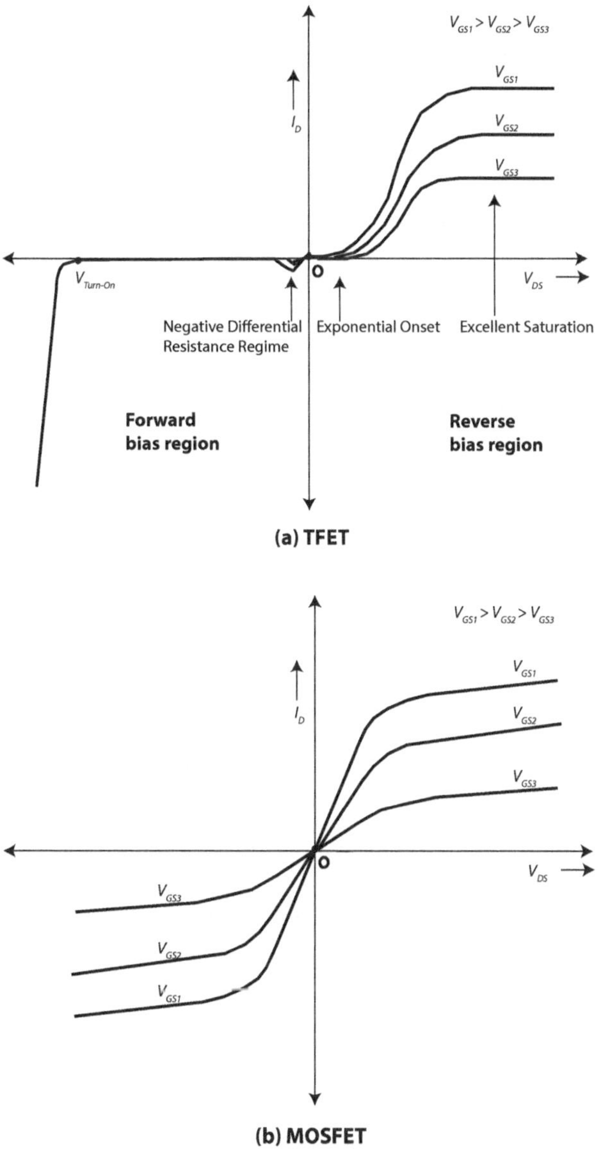

FIGURE 9.1
Output characteristics of (a) TFET and (b) MOSFET.

implementing circuits such as pass transistor logic and SRAM. The unidirectional property of the TFET can be exploited in improving the efficiency of RF rectifier circuits by preventing reverse leakage paths. It should be noted that due to the asymmetric current conduction, the drain and the source terminals

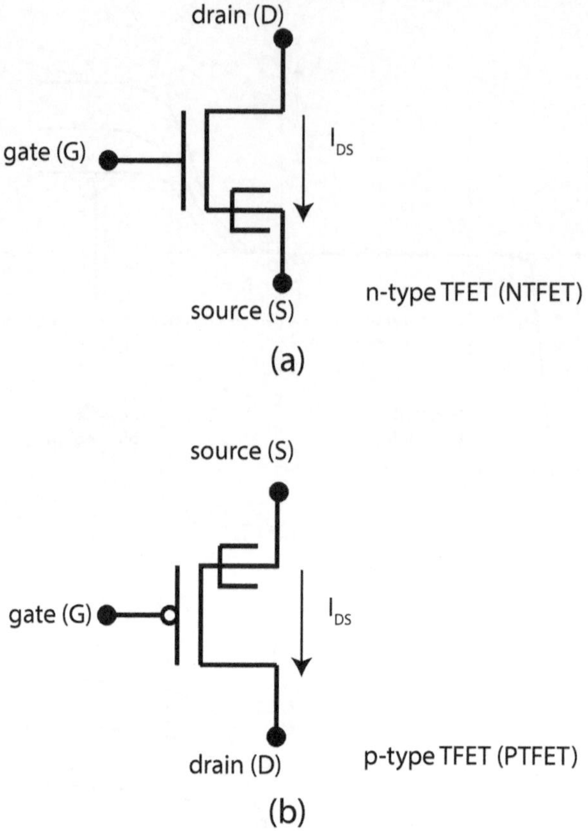

FIGURE 9.2
Schematic representation of (a) an n-type TFET (NTFET) and (b) a p-type TFET (PTFET). The direction of current flow is also shown in this figure.

in a TFET are not interchangeable. Therefore, in the schematic representation of a TFET, the drain and the source terminals need to be distinguished. Figure 9.2 shows the schematic representation of an n-type TFET (NTFET) and a p-type TFET (PTFET). A bracket-like symbol is added to the source terminal to distinguish it from the drain terminal.

9.2.3 Onset of Current in the Output Characteristics

The output characteristics of a TFET show an exponential onset of drain current in contrast to a linear onset in a conventional MOSFET as shown in Figure 9.1 [6–11]. As a result, for low V_{DS}, the drain current in a TFET is extremely small. In fact, a drain threshold voltage can be identified in a

TFET as the V_{DS} at which the drain current changes its form in the output characteristics from exponential behavior to linear behavior [7].

In addition to the tunneling probability, the drain current in a TFET is strongly dependent on the availability of carriers at the source of the tunneling path and the availability of empty states at the destination of the tunneling path [11]. The exponential onset in the output characteristics has been shown to be related to the source-degeneracy level and is mainly determined by the modulation of the availability of carriers and the empty states at the source and the destination of the tunneling path, respectively [10, 11].

Due to the exponential onset of current in the output characteristics, the output resistance offered by a TFET at low V_{DS} is much higher compared to the output resistance in a conventional MOSFET [6]. The high output resistance at low V_{DS} is undesirable since it has a detrimental effect on the delay of the circuits, as we will discuss in the following sections. By optimizing the source and the drain doping concentrations and by choosing an appropriate material for TFET such as InAs nanowire, the output resistance at low V_{DS} can be appreciably decreased [10].

9.2.4 Gate-to-Drain Miller Capacitance

In a TFET, under all bias conditions, the total gate capacitance (C_GG) is dominated by the gate-to-drain capacitance (C_{GD}) [2, 6, 9, 12, 13]. When the TFET is in the OFF-state, C_{GD} is much greater than C_{GS} because the channel-to-drain resistance is much lower compared to the channel-to-source resistance. When the TFET is in the ON-state, there is little potential drop between the channel and the drain, resulting in a large C_{GD} [12]. This is in contrast to a conventional MOSFET where the gate-to-source capacitance (C_{GS}) and gate-to-drain capacitance (C_{GD}) contribute almost equally in the linear region and C_{GS} dominates in the saturation region [12]. Due to the large C_{GD} in a TFET, a stronger coupling exists between the input pin (gate of the TFET) and the output pin (drain of the TFET) in the circuits employing TFETs and has serious implications for the behavior of the circuit due to the Miller effect.

9.2.5 Matching Characteristics of NMOS-Type TFETs and PMOS-Type TFETs

To apply TFETs in CMOS-type circuits, it is desirable that the characteristics of an NTFET and a PTFET match. For a conventional MOSFET, the drive strengths of PMOS and NMOS transistors are matched by appropriately scaling the ratios of the width (W) and length (L) of the transistors. However, for an NTFET and a PTFET, in addition to the mismatch in the drive strengths, there are other differences in electrical characteristics such as the subthreshold swing [14]. To increase the I_{ON} in a TFET, the source doping must be high. In a PTFET the source is an n^+ material, while in an NTFET the source is a p^+ material. Due to low density of states in the conduction band, for the same doping in the n-type material and the p-type material, the Fermi energy level moves deeper into the conduction band in an n-type material compared to the Fermi energy level moving into the valence band in a p-type material. As a result, the contribution of the Boltzmann "tail" in the tunneling process increases in a PTFET and the subthreshold swing degrades [3, 15]. In general, a PTFET has a lower I_{ON} and a gentler subthreshold swing compared to an NTFET. This mismatch in the characteristics of an NTFET and a PTFET has a significant impact on the energy consumption and delay characteristics of the CMOS-type circuits employing TFETs. Therefore, this mismatch needs to be reduced by device optimization or the impact of the mismatch on the circuit needs to be mitigated using clever circuit designs, as we shall discuss in the following sections.

9.2.6 Other Electrical Characteristics

In addition to the above distinguishing electrical characteristics of a TFET, the other properties of a TFET that have serious implications for the behavior of the circuit are:

1. **ON-current (I_{ON}):** In general, TFETs suffer from a low I_{ON} and this has been the most intensively studied problem for TFETs. The low I_{ON} in a TFET has several serious implications for the circuit such as increased delay. It is desirable that the low I_{ON} of the TFET is boosted using appropriate techniques, which are described in Chapter 4.
2. **Saturation in the output characteristics:** In general, the saturation of the drain current in the output characteristics occurs at a higher drain voltage in a TFET compared to a MOSFET [16, 17]. Since the drain saturation voltage plays a vital role in determining the noise margin of the digital circuits, the delayed saturation in the output characteristics of a TFET is an important issue, especially for circuits operating at lower supply voltages.
3. **Ambipolar conduction:** Ambipolar conduction occurs when appreciable band-to-band tunneling (BTBT) takes place on the drain side of the TFET at low V_{GS}. The ambipolar conduction is undesirable since it increases the I_{OFF} and may lead to a circuit failure, as described in Section 3.11.

The characteristics of a TFET described above suggest that there are considerable challenges in employing them. A circuit designed using conventional MOSFETs cannot be directly implemented in TFETs, since the peculiar electrical characteristics of TFETs may not allow achieving the same functionality that was obtained using conventional MOSFETs. This aspect of circuit design has been investigated by several researchers. These studies have demonstrated that TFETs can be employed to realize conventional CMOS circuits with appropriate device optimizations and innovative circuit designs. Some of the CMOS circuits have been experimentally realized using TFETs and have shown competitive performance compared with the conventional MOSFET circuits [18, 19]. However, it may be noted that most of the studies of TFET circuits are based on simulations. Many of these simulations are performed using either device-circuit mixed-mode simulation offered by technology computer aided design (TCAD) tools or are performed using lookup table (LUT)-based Verilog-A models of a TFET. In the following sections, we will discuss some of the applications of TFETs in digital circuits, memories, and analog circuits.

9.3 Digital Circuits

The key matrices that are used to benchmark digital circuits are the delay characteristics such as propagation delay and rise/fall times, noise resilience, power dissipation and the area of the circuit [20]. These characteristics are derived for the circuits based on TFETs and compared with conventional CMOS circuits to assess the competitive advantage of TFETs over MOSFETs for digital applications. Since an inverter is the simplest digital circuit, the characteristics of inverters realized using TFETs have been investigated in greater detail by researchers. In this section, we will first examine a TFET-based inverter and then look at some other digital circuit applications of TFETs.

9.3.1 TFET Inverters

The circuit diagram of an inverter based on a TFET is shown in Figure 9.3. It is composed of an NTFET forming the pull-down network (PDN) and a PTFET forming the pull-up network (PUN). In general, the drive strength of a PTFET is lower than an NTFET. As a result, the voltage transfer characteristics (VTC) of an inverter that employs a PTFET with a lower drive strength exhibits an asymmetrical behavior. To make the transfer characteristics symmetrical and switch at $V_{DD}/2$, device-level techniques are employed to boost the current in the PTFET or multiple PTFETs are employed in the pull-up network [21]. Experimentally, inverters and simple circuits employ-

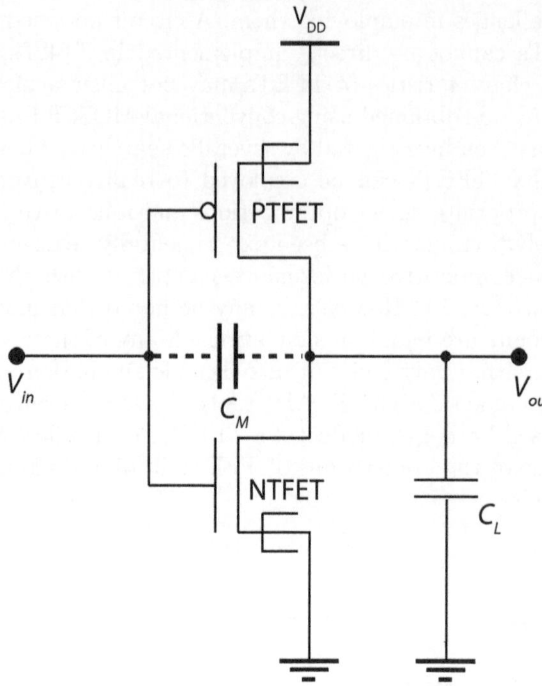

FIGURE 9.3
Circuit diagram of a TFET inverter. The Miller capacitance is shown in the figure as a broken line.

ing TFETs have been fabricated and characterized [18, 19, 22]. It has been shown that an inverter realized using strained-Si NW TFETs exhibits sharp transition in the voltage transfer characteristics even when the V_{DD} is as low as 0.2 V [19]. However, for a low input voltage (corresponding to logic state "0"), the output voltage of the inverter is less than V_{DD} and for high input voltage (corresponding to logic state "1"), the output voltage is greater than 0 V. This degradation in the characteristics of the inverter can be attributed to the ambipolar behavior of the TFETs [22]. If the ambipolar current of the NTFET and the PTFET are suppressed using appropriate device optimization techniques, then the voltage transfer characteristics of the inverter exhibit rail-to-rail transitions. Another important point to note is that the delayed onset of saturation in the TFET can result in a considerable degradation in the voltage transfer characteristics of the TFET inverter compared to that of a CMOS inverter [17]. This is manifested as a reduced gain of the TFET inverter. The degradation in the voltage transfer characteristics of the inverter due to delayed saturation in the TFETs causes serious problems in utilizing TFETs in SRAM circuits in which two cross-coupled inverters are typically employed.

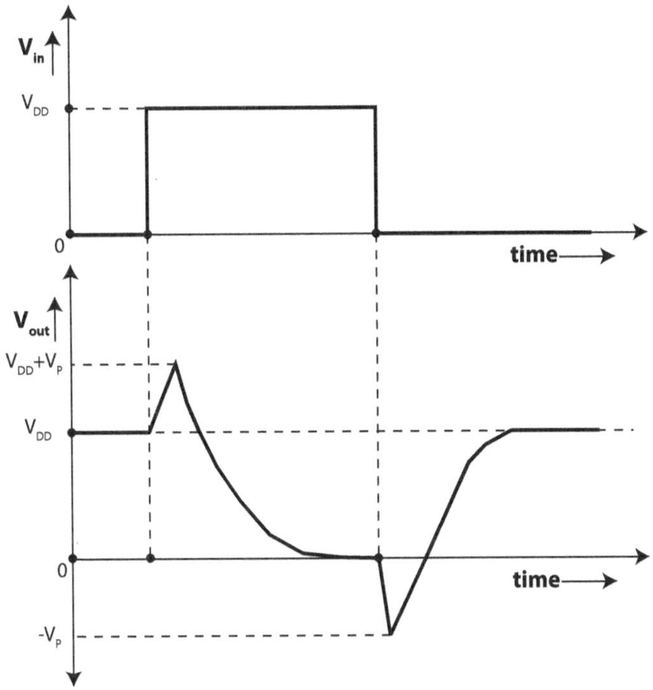

FIGURE 9.4
Transient behavior of a TFET inverter showing the overshoot and the undershoot [12].

The transient response of a typical inverter realized using TFETs is shown in Figure 9.4. The transient response of a TFET inverter generally exhibits undershoot and overshoot transients. The overshoot and undershoot in the transient response of a TFET inverter have been observed in simulation and in experiments [6, 12, 19, 23]. The inverter exhibits an overshoot when the input makes a low-to-high transition and it exhibits an undershoot when the input makes a high-to-low transition [12]. This behavior of the inverter can be attributed to the large coupling between the input of the inverter and the output of the inverter due to the high gate-to-drain capacitance (C_{GD}) of the TFETs. Although, the C_{GD} of the NTFET and the PTFET changes as the output of the inverter undergoes the transition, the total Miller capacitance due to the sum of C_{GD} of the NTFET and the PTFET remains large throughout the transition and results in a high overshoot and undershoot. The peak voltage V_P in the overshoot/undershoot can be computed by taking into account charge conservation and can be given as [12]

$$V_P = \frac{C_M}{C_M + C_L} V_{DD}, \tag{9.1}$$

where C_M is the Miller capacitance and C_L is the capacitive load as shown in Figure 9.3. The overshoot/undershoot in the TFET inverter can be suppressed by reducing the Miller capacitance, for example, by using a material with low density of states such as InAs [12]. Another noteworthy feature of

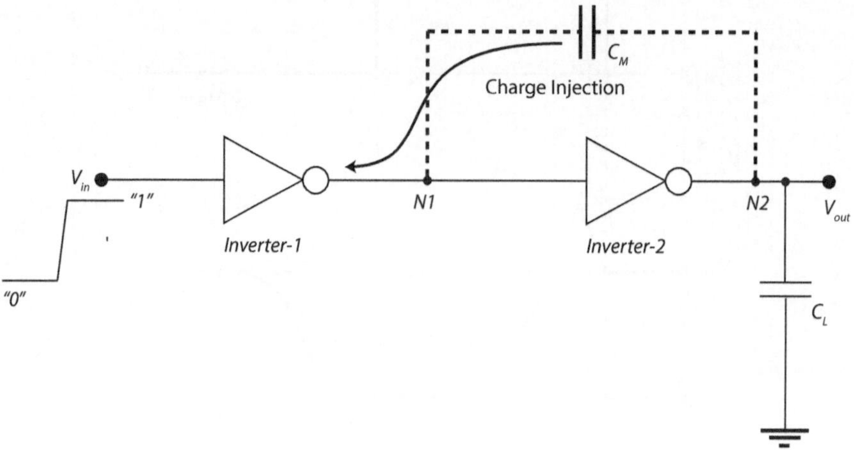

FIGURE 9.5
Circuit showing the impact of Miller feedback on the cascaded inverters.

the transient response of a TFET inverter is the relatively longer settling time compared to a conventional CMOS inverter [6]. The settling time is longer because the TFET enters the exponential region of the output characteristics as the V_{DS} becomes smaller, and the drain current reduces drastically. Additionally, when two inverters are in cascade, the charge injection from the load inverter further increases the settling time [6]. To understand this, let us assume that the input V_{in} is initially in the low state and then it transitions to a high state as shown in Figure 9.5. When the input is in the low state, the Miller capacitance C_M between nodes $N1$ and $N2$ is initially charged. With the change in V_{in}, when the output of the second inverter starts to transition from low to high, the Miller capacitance C_M starts discharging. The discharge path is through the NTFET of the first inverter, which is forced to sink more current due to charge injection from C_M. As a result, the output node $N1$ is forced to remain high for a longer time and the settling time of the first inverter is further increased [6]. The undershoot/overshoot and the long settling time of the TFET inverters not only degrades the speed of the circuit, but also increases the energy consumption of the circuit and the distortion of the output waveform [6].

An important issue that needs to be considered when employing TFETs for CMOS circuits is the susceptibility of the circuits to crosstalk and glitches. Figure 9.6 shows a TFET inverter with a coupling capacitor of an aggressor at the output of the inverter Z and the PTFET and the NTFET modeled

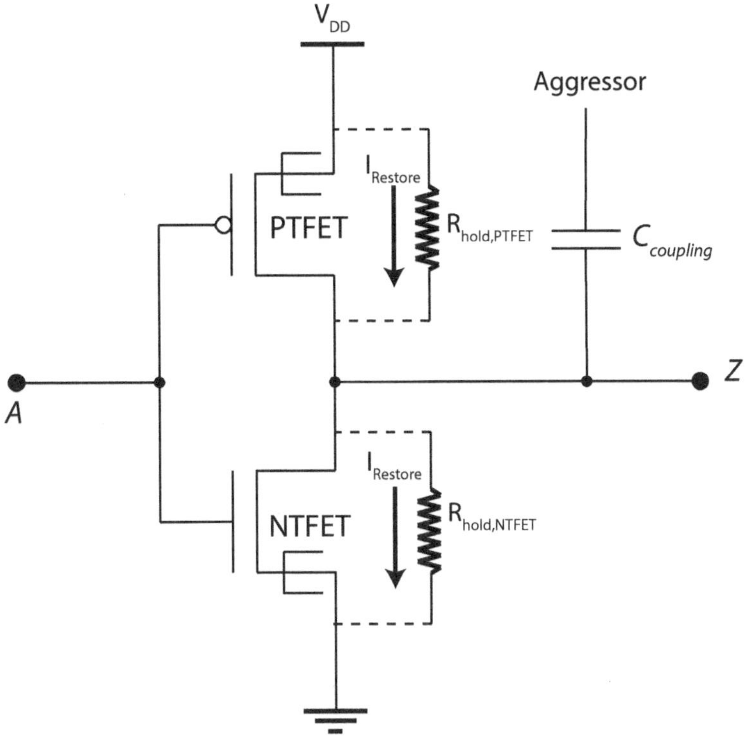

FIGURE 9.6
Circuit showing how the currents flow from/to the V_{DD}/ground due to the aggressor coupling [6].

as holding resistances $R_{hold,PTFET}$ and $R_{hold,NTFET}$, respectively. When the output of the inverter is low, the output node Z is connected to the ground through a small resistance $R_{hold,NTFET}$ and the PTFET is in the OFF-state. When the output of the inverter is high, the output node Z is connected to the power supply voltage V_{DD} through a small resistance $R_{hold,PTFET}$ and the NTFET is in the OFF-state. Normally, the voltage of the output node will be held close to the ground potential or the supply voltage V_{DD} due to the low-resistance path of either the NTFET or the PTFET. However, due to the noise or coupling from the neighboring wires, the voltage of the output node Z can change and can even go beyond the rail voltage. Whenever such an event occurs, the NTFET or the PTFET should be able to provide a low-resistance path to restore the voltage of the output node Z to the ground potential or the supply voltage V_{DD}. The NTFET is expected to provide the low-resistance path when the output node voltage is at ground potential and the PTFET is expected to provide the low-resistance path when the output node voltage is at supply voltage V_{DD}. However, due to the unidirectional nature of current flow in a TFET and the exponential onset of drain current at small drain bias, a low-resistance path in both directions cannot be provided by the TFETs and the voltage of the output node Z cannot be quickly restored at the occurrence of a glitch [6]. Figure 9.6 shows the direction in which the TFETs in the ON-state allow restoring currents to flow through the low holding resistances. On the other hand, a conventional MOSFET, being symmetrical in structure, can provide a low-resistance path in both the directions, when it is in the ON-state. Therefore, circuits realized using TFETs are more susceptible to noise and glitches compared to the CMOS circuits realized using conventional MOSFETs.

Several researchers have assessed the competitive advantages of TFET inverters over CMOS inverters realized using MOSFETs [2, 15, 17, 21, 23–27]. Most of the studies indicate that TFET inverters are faster and are more energy efficient at lower supply voltages, while the advanced MOSFET-based CMOS inverters are faster at higher supply voltages. For instance, at low supply voltage $V_{DD} = 0.25\ V$, for the same static power dissipation and ITRS-specified Low Operating Power (LOP) and Low Standby Power (LSTP) environment, TFET inverters perform 10× and 100× faster, respectively, compared to the inverter realized using FinFET [21]. However, at higher supply voltages $V_{DD} > 0.4\ V$, TFET inverters do not offer any competitive advantage over inverters realized using FinFETs [21]. Additionally, the competitive advantages of TFET inverters strongly depend on the choice of TFET material and TFET architecture, circuit conditions such as the switching activity factor and loading of interconnects, and the impact of process-induced variations.

9.3.2 Applications in Digital Circuits

TFETs are being explored for application in digital circuits other than inverters [3, 22, 25, 28]. TFETs can be used to implement pass transistor logic (PTL). However, there are some problems associated with the conventional implementation of PTL using TFETs due to the unidirectional current flow in TFETs [3, 4, 25]. These problems can be illustrated using the circuit shown in Figure 9.7 implemented using an NTFET and a PTFET with the source of the transistors connected to the input and the drain of the transistors connected to the output. The gate of the NTFET is connected to the signal *select* and the gate of the PTFET is connected to $\overline{select}$ (complement of *select*). When the signal *select* is high, only one of the TFETs (either the NTFET or the PTFET) conducts, depending on the polarity of the drain-to-source voltage ($V_{DS} = (V_{out} - V_{in})$). If V_{DS} is positive, then the NTFET conducts and if V_{DS} is negative then the PTFET conducts. This is in contrast to the implementation of PTL using conventional MOSFETs, where both the PMOS and the NMOS conduct when the signal *select* is high, thus providing a strong drive current and a strong "0" and "1" logic. Therefore, the implementation

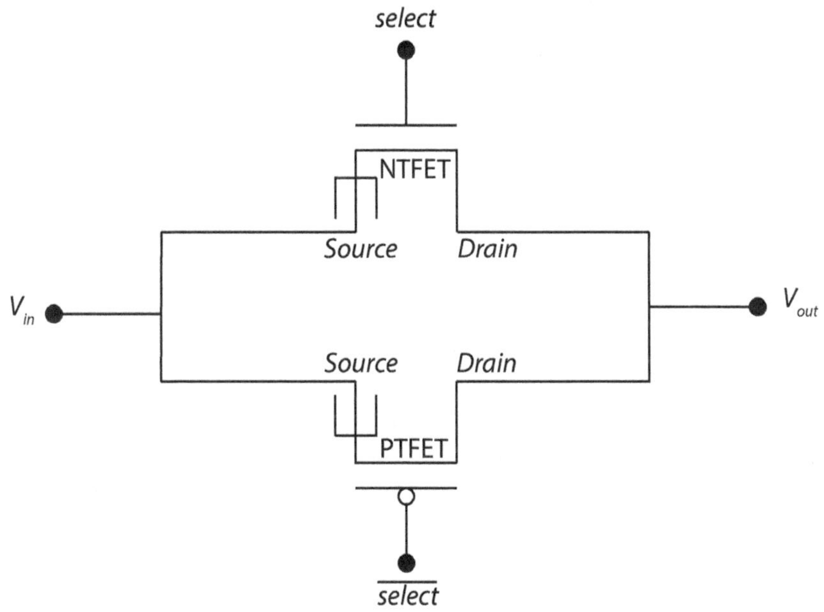

FIGURE 9.7
Circuit diagram of pass transistor logic implemented using TFET.

of PTL using TFETs is expected to be slower compared to the implementation of PTL using MOSFETs.

The problem of implementation of PTL using TFETs described above can be circumvented by connecting two NTFETs in parallel with the drains oriented in opposite directions as shown in Figure 9.8 [3]. If the signal *select* is high, NTFET-TOP conducts when its drain-to-source voltage is greater than zero ($V_{DS,NTFET-TOP} = (V_{out} - V_{in})) > 0$ and NTFET-BOTTOM conducts when its drain-to-source voltage is greater than zero ($V_{DS,NTFET-BOTTOM} = (V_{in} - V_{out})) > 0$. Thus, a bidirectional switch is realized in Figure 9.8. However, adding an extra NTFET leads to an increase in the area of the circuit and an increase in the input capacitance.

The pass transistor logic can be used to implement multiplexors (MUX) and the unidirectional current flow in a TFET can be exploited to obtain a compact implementation [3, 25]. The unidirectional current flow in a TFET allows combining the pull-up network and the pull-down network in a topology that would normally create a short circuit when implemented using conventional MOSFETs [25]. As a result, the number of transistors in the compact-MUX implementation can be reduced from ten to eight.

The distinctive electrical property of a TFET can be exploited to obtain a dual clocked dynamic logic that is implemented using only NTFETs [3]. For

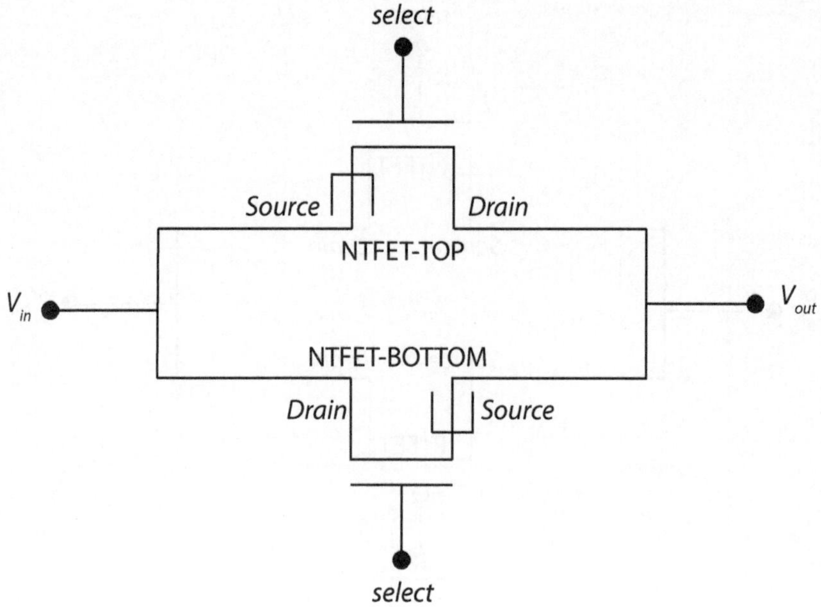

FIGURE 9.8
Circuit implementing pass transistor logic using a bidirectional switch composed of two NTFETs.

a conventional MOSFET-based CMOS circuit, PMOS is normally preferred over NMOS in the pull-up network, since an NMOS can pull up the voltage only to $(V_{DD} - V_T)$ where V_T is the threshold voltage of the NMOS and a PMOS can pull up the voltage to V_{DD}. However, in TFET-based circuits, an NTFET can also be used in the pull-up network, since there is negligible voltage drop between the output node and the supply rails. This property of the NTFET is utilized in the dual clocked dynamic logic yielding a simpler level-restorer circuit and an improvement in the energy consumption and the delay of the circuit [3]. The technique of employing only NTFETs in implementing a Manchester carry-chain adder dynamic logic circuit is shown to outperform FinFET-based implementation in terms of energy consumption and delay of the circuit when $V_{DD} < 0.6\ V$ [3].

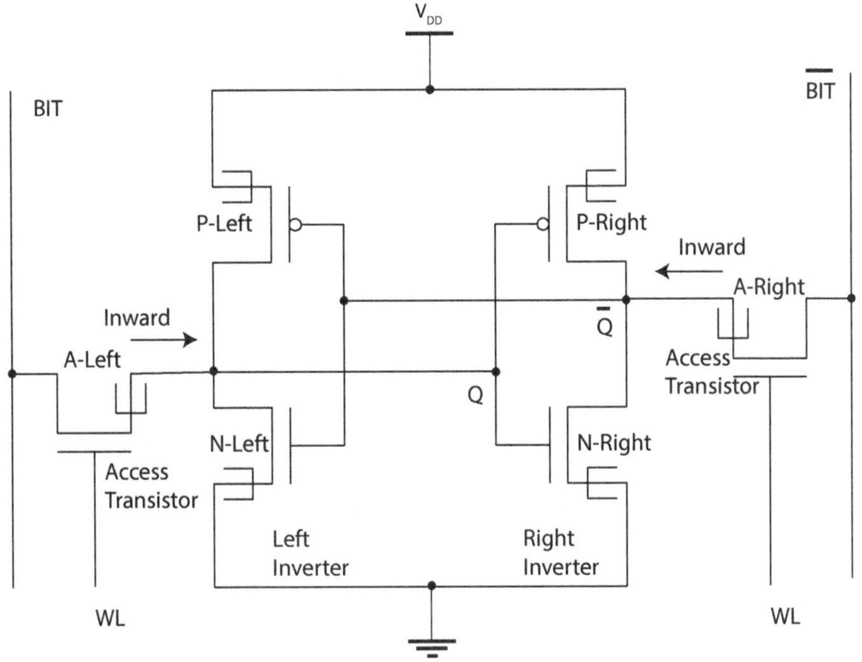

FIGURE 9.9
Circuit diagram showing conventional 6T SRAM implemented using TFETs.

9.4 Applications in Memories

TFETs have been widely investigated for realizing SRAMs due to their extremely low leakage current and excellent subthreshold swing [1, 4, 17, 29–32]. However, the unidirectional current conduction in a TFET and the delayed saturation in the output characteristics create problems in the realization of SRAM circuits using TFETs. The problem in implementing SRAMs using TFETs can be elucidated using the circuit shown in Figure 9.9, which has the topology of a conventional 6-transistor (6T) SRAM widely implemented using CMOS technology.

The read operation for the SRAM is carried out by first pre-charging the bit lines BIT and $(\overline{BIT})$ to supply voltage V_{DD} and then turning on the write line WL high. As a result, the access transistors A-Left and A-Right are expected to switch on and depending on the value of the stored bits Q and $\overline{Q}$, the bit lines BIT or $(\overline{BIT})$ will be pulled down. For example, if $Q = 0$ and $\overline{Q} = 1$, bit line BIT discharges through the transistors A-Left and N-Left, and the bit line $(\overline{BIT})$ will remain high. The pull-down current or voltage can be detected by a sense amplifier and the stored value can be determined. To ensure that the read operation does not flip the stored value, the drive strength of the pull-down transistor N-Left should be greater than the access transistor A-Left. Therefore, the sizing ratio of the transistors N-Left to A-Left, which is commonly referred to as cell β *ratio*, should be kept high making the voltage at Q stable at ground potential during the read operation.

The write operation for the SRAM is carried out by driving the bit lines BIT and ($\overline{BIT}$) with the value that needs to be written and then turning on the write line WL high. The bit lines are expected to overpower the inverters and overwrite the stored values Q and $\overline{Q}$. For example, when the stored bit $Q = 0$, $\overline{Q} = 1$ and the bit to be written is BIT=1, ($\overline{BIT} = 0$), the node $\overline{Q}$ is discharged to the ground potential by the transistor A-Right overpowering the pull-up transistor P-Right and the node $\overline{Q}$ is charged to the power supply voltage by the transistor A-Left. To ensure proper writing of the values, the drive strength of the access transistor A-Right should be greater than the pull-up transistor P-Right. Therefore, the sizing ratios of the transistors A-Right to P-Right should be kept high to ensure a reliable write operation.

The implementation of a 6T SRAM using conventional MOSFETs is straightforward and proper sizing of all the transistors ensures a reliable operation of the circuit. However, due to the unidirectional flow of current in TFETs, it is difficult to ensure simultaneous stability of the read and the write operations in 6T TFET SRAM. The access transistors A-Left and A-Right shown in Figure 9.9 can conduct only in one direction, i.e., from the bit lines to the inputs of the inverters. This orientation of the access transistors is called inward access transistors. The TFET access transistors A-Left and A-Right cannot conduct from the inverter toward the bit line. Therefore, when writing BIT=1 with the stored data $Q = 0$, $\overline{Q} = 1$, the node $\overline{Q}$ cannot be effectively discharged to the ground potential by the transistor A-Right. As a result, the success of writing $BIT = 1$ solely depends on the ability of the access transistor A-Left to pull up the node Q by overpowering the transistor N-Left. Therefore, to ensure a proper write operation, the drive strength of the access transistor A-Left should be much greater than the drive strength of the pull-down transistor N-Left. Thus, for ensuring a proper write operation, the cell β *ratio* needs to be kept low. However, as discussed above, to ensure a proper read operation, a high cell β *ratio* is required. Therefore, the TFET implementation of a conventional 6T SRAM circuit has conflicting read and write requirements and it is not possible to simultaneously ensure acceptable read and write static noise margins in a 6T TFET SRAM [4, 29]. A similar problem exists when the access transistors are outward, i.e., access transistors conduct current from the inverter to the bit lines [4].

To tackle the problem of stable read and write operation in a 6T TFET SRAM various solutions have been suggested [1, 4]. Among different read-assist and write-assist techniques, a practical design strategy is to size the 6T TFET

SRAM to obtain a stable write operation with the cell β *ratio* = 0.6 and to use ground-lowering read-assist techniques to improve the dynamic stability of the read operation [4]. By lowering the ground potential, the relative strength of the cross-coupled inverters compared to the access transistors increases, resulting in improved stability of the read operation. In SRAM design, ground-lowering is usually implemented by employing another power supply or on-chip regulator [1].

The problem of ensuring both read and write stability in 6T TFET SRAM can be addressed by adding an additional transistor to decouple the constraints of read and write stability. A 7T TFET SRAM is proposed in which a separate NTFET, called "NRD," is used for reading out the stored value from the SRAM [4]. For writing, outward access transistors are used similar to the 6T TFET SRAM. The read and write operations are performed by distinct current paths and the device sizes for all the transistors other than the NRD can be chosen to facilitate a stable write operation. The NRD is optimized for ensuring read stability. A 7T TFET SRAM achieves $7 - 37\times$ improvement in the leakage power reduction compared to a CMOS 6T SRAM [4].

While writing, another technique to ensure both read and write stability in a 6T TFET SRAM is by providing a virtual ground to the internal transistor N-Left (the NTFET of the left cross-coupled inverters in Figure 9.9) [29]. The virtual ground is established using an additional write-enable line WRA. During the write operation, WRA is also raised and this weakens the regenerative action of the cross-coupled inverters and improves the write noise margin. The 6T TFET SRAM with virtual ground has been shown to be better than 7T TFET SRAM [29]. The 6T TFET SRAM is also suitable for ultralow voltage such as 0.3 V and exhibits significant improvement in leakage power, noise margin, and performance compared to a 6T CMOS SRAM [29]. However, it exhibits greater delay than the 6T CMOS SRAM in the read operation due to smaller drive current in the TFETs [29].

The problem arising due to unidirectional current flow in the access transistors in a 6T TFET SRAM can also be addressed by adding two additional access transistors in opposite directions. The 8T TFET SRAM, with two additional TFET access transistors, can remove some of the limitations of the 6T TFET SRAM [17]. However, it is shown that the read static noise margin of the 8T TFET SRAM can be considerably reduced because of the delayed saturation in the output characteristics [17]. Another TFET-based implementation of the 8T SRAM cell utilizing dedicated read buffers, as proposed earlier for the $32 - nm$ CMOS technology node, has been investigated [33]. The dedicated read buffer decouples the read current from the cell storage node, thus avoiding read upset and enhancing the read stability and is also viable with the unidirectional current transfer property of TFETs [32]. In addition to the 6T SRAM topologies, there are other promising TFET-based SRAM topologies that are being actively investigated [17, 30–32].

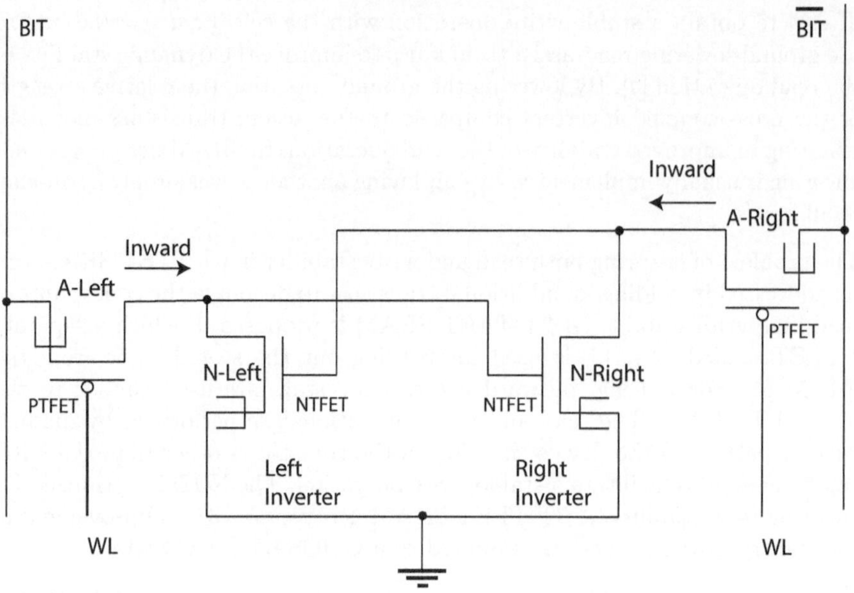

FIGURE 9.10
Circuit diagram showing implementation of 4T SRAM using NTFET [30].

An $In_{0.53}Ga_{0.47}As$ TFET-based implementation of 4T SRAM that was originally proposed for $0.18-\mu m$ CMOS technology has been recently investigated [30, 34]. In the 4T TFET SRAM, two inward-oriented PTFETs are utilized as access transistors and load transistors as shown in Figure 9.10. The cross-coupled NTFETs are used to store the data for the SRAM. To retain the state of the SRAM, the leakage current of the PTFET must be higher than the leakage current of the NTFET. This is ensured by intelligently exploiting the properties of the $In_{0.53}Ga_{0.47}As$. Since the effective density of states (DOS) in the conduction band for $In_{0.53}Ga_{0.47}As$ is much lower than the effective DOS in the valence band, the Fermi level in the n^+ doped source in the PTFET lies deep into the conduction band, while for the same doping in the p^+ source of the NTFET, the Fermi level lies close to the edge of the valence band. As a result, the subthreshold swing of the PTFET is limited to 60 $mV/decade$ at room temperature and the subthreshold swing of the NTFET is lower than 60 $mV/decade$ at room temperature. The difference in the subthreshold swing of the PTFET and the NTFET ensures that the leakage current of the PTFET is higher than the leakage current of the NTFET. It has been demonstrated that the 4T TFET SRAM cell is 100× faster than a low static power (LSP) 6T CMOS SRAM cell and consumes 3× less standby power than a high-performance (HP) 6T CMOS SRAM cell [30].

A Schmitt–Trigger (ST)-based TFET SRAM cell composed of 10 transistors has been recently proposed and it is shown that the ST feedback mechanism improves the read and write noise margins, thus facilitating ultralow voltage operation for the TFET SRAMs [17]. The ST TFET SRAM cell has sufficiently high variation tolerance to operate at a supply voltage as low as 124 mV [17].

A mixed TFET-MOSFET 8T SRAM cell can also be realized [32]. The cross-coupled inverters comprise the MOSFET and dedicated TFET read stack and TFET write access transistors are employed. The mixed TFET-MOSFET cell can exploit the merits of TFETs and MOSFETs and results in an improvement in stability and performance. It has been demonstrated that the mixed TFET-MOSFET cell is feasible for ultralow voltage operation and the pure MOSFET-based cell provides improved stability and performance for high-voltage operation [32].

In addition to SRAMs, interesting applications of TFETs in other memory devices and flip-flops are also being investigated [28, 35, 36].

9.5 Analog Circuits

TFETs have been investigated predominantly for digital applications and as a replacement for conventional MOSFETs for logic applications. However, recently, researchers have evaluated the merits of applying TFETs in different kinds of analog circuits [2, 37–51]. These studies have shown that TFETs are promising candidates for low-power analog circuits because of the small subthreshold swing and excellent saturation in the output characteristics. However, exponential onset in the output characteristics, the unidirectional current flow, and delayed saturation in the output characteristics observed in TFETs can have detrimental effects on some analog circuits. On the other hand, special electrical properties of TFETs can be exploited in analog circuits to realize new topologies [45]. In this section, we will review some of the intrinsic parameters of TFETs that are crucial for analog circuits such as transconductance (g_m), transconductance generation efficiency (g_m/I_D), output resistance (r_O), and intrinsic voltage gain ($g_m r_O$). We will also review the merits and demerits of implementing some of the analog circuits using TFETs instead of MOSFETs.

9.5.1 Intrinsic Analog Parameters of TFETs

The intrinsic parameters of a TFET that are relevant for analog circuits such as transconductance and intrinsic voltage gain strongly depend on the particular implementation of the TFET. However, a few general observations can be made regarding the intrinsic analog parameters of the TFETs, which we shall discuss in this section. Additionally, some of the analog parameters can be adjusted by intelligent device optimization [37, 39, 40, 42, 46–49, 52]. Therefore, it is important to appreciate the impact of various device and material properties on the analog parameters of the TFETs.

1. **Transconductance and transconductance generation efficiency:** The transconductance (g_m) in the subthreshold region of a transistor can be defined as [41]

$$g_m = \frac{\partial I_D}{\partial V_{GS}} = \frac{\ln(10)\, I_D}{SS_{POINT}}, \tag{9.2}$$

where SS_{POINT} is the point subthreshold swing of the transistor at a given drain current I_D. As we discussed in Chapter 3, the SS_{POINT} of a TFET varies with the drain current: it is small at low drain current and gradually increases as the drain current increases. Therefore, the g_m of a TFET in the subthreshold region, as defined by Equation (9.2), shows a variation with the drain current [45]. In the subthreshold region, for a given low drain current, the transconductance of an InAs TFET is found to be more than the transconductance of a similar MOSFET because of the lower subthreshold swing of the TFET compared to a MOSFET at low drain current [45]. However, at higher drain current, the transconductance of the MOSFET is higher compared to the InAs TFET [45].

The transconductance generation efficiency (g_m/I_D) is defined as

$$\frac{g_m}{I_D} = \frac{\ln(10)}{SS_{POINT}}. \tag{9.3}$$

Since the SS_{POINT} in a TFET is extremely small when the drain current starts to take off, (g_m/I_D) is very high at low drain current and it gradually decreases with the increase in the drain current [40, 45, 48]. This is in contrast to a conventional MOSFET where SS_{POINT} is fairly constant and is limited to a minimum value of 60 $mV/decade$ at room temperature. Therefore, the maximum transconductance generation efficiency (g_m/I_D) in a conventional MOSFET is 38.4 V^{-1} at room temperature. The high transconductance generation efficiency in a TFET can be exploited in low-power analog circuits as described later in this section.

In the saturation region, the transconductance g_m in a Si TFET is found to be approximately an order of magnitude lower than that of a similar MOSFET [40]. However, the transconductance generation efficiency g_m/I_D is found to be higher for the Si TFET compared to the transconductance generation efficiency g_m/I_D of a similar MOSFET [40]. The low g_m of a Si TFET can be attributed to the low drain current in Si-based TFETs. Therefore, increasing the drain current by employing low bandgap materials in TFETs is expected to improve the transconductance of the TFETs [40]. It has been demonstrated that the NW TFETs with gate-all-around architecture exhibit a higher g_m/I_D and intrinsic gain compared to the state-of-the-art 20 nm FinFETs at low voltages [22].

2. **Output resistance:** The current saturation mechanism in the output characteristics of a TFET is quite different from the saturation mechanism in a conventional MOSFET. As the channel becomes devoid of carriers, the current in a TFET saturates and the impact of the drain potential on the source–channel tunnel junction becomes negligible. Therefore, TFETs exhibit excellent current saturation and do not show channel length modulation that is typically exhibited by conventional MOSFETs. As a result, the output resistance of the TFETs in the saturation region is typically very high [40, 41, 45]. For a Si TFET, the output resistance is found to be more than two orders of magnitude higher than a similar conventional MOSFET [40]. Due to the extremely high output resistance, the intrinsic small-signal voltage gain ($g_m R_O$) of a TFET is found to be higher than a similar MOSFET, even though the g_m of the TFET might be low [40, 45].

3. **Cutoff frequency:** As described in Section 9.2, in general, the gate-to-drain capacitance dominates total capacitance in a TFET. However, the total capacitance in a TFET ($C_{GS} + C_{GD}$) strongly depends on the properties of the material employed in the TFET. For example, the total capacitance in a Si-based TFET is found to be greater than the total capacitance in a conventional MOSFET [40]. However, the total capacitance in an InAs TFET is found to be smaller than the total capacitance in a comparable MOSFET because of the lower density of states in InAs [45]. The combined effect of the transconductance and the capacitance can be accounted for in the unity gain cutoff frequency (f_T) of the transistor, which is defined as

$$f_T = \frac{g_m}{2\pi\left(C_{GS} + C_{GD}\right)}. \tag{9.4}$$

Since the transconductance g_m and the total capacitance (C_{GS}+C_{GD}) change with the drain current and the bias point, the cutoff frequency f_T also changes with the drain current and the bias point. In general, the maximum attainable f_T in a TFET is lower compared to a similar MOSFET. For a Si TFET, the maximum attainable f_T is found to be an order of magnitude lower than a similar MOSFET and can largely be attributed to the smaller transconductance in a Si TFET [40]. For a GaSb-InAs heterojunction TFET, the maximum attainable f_T is found to be approximately 300 GHz, which is of a similar order as the maximum attainable f_T in a comparable MOSFET [45]. However, the maximum cutoff frequency f_T is obtained at a smaller current density in a GaSb-InAs heterojunction TFET compared to a similar MOSFET [45]. This suggests that TFETs can be employed for low-power analog applications [45]. In an InN TFET, at a higher drain bias ($V_{DS} = 1\ V$), the gate-drain capacitance C_{GD} is found to be lower, which results in a higher maximum attainable cutoff frequency f_T [43].

9.5.2 Applications in Analog Circuits

The application of TFETs in various kinds of analog circuits are being actively investigated and the competitive advantages of TFETs over conventional MOSFETs for analog applications are being examined [5, 37–40, 45, 48, 52, 53]. In this section, we will review some of the applications of TFETs in analog circuits that have been recently proposed, highlighting the merits and demerits of TFETs for analog applications.

TFETs can be employed in amplifier circuits similar to a conventional MOSFET. The application of Si TFETs in common source amplifiers has been recently compared with a CMOS implementation of the same circuit [40]. It is demonstrated that the high output resistance of the Si TFET results in a larger voltage gain for the Si TFET common source amplifier compared to a similar CMOS amplifier. However, the smaller g_m of a Si TFET combined with its larger total capacitance results in a lower value of gain bandwidth product of the Si TFET common source amplifier compared to a similar CMOS amplifier. The high output resistance of the TFET can also be exploited in a cascode current mirror to obtain a simpler TFET implementation by removing the cascode transistors [45]. Removing the cascode transistors allows reducing power supply and relaxing the minimum voltage constraint on the output voltage.

The unidirectional current flow and bias-dependent subthreshold swing can be problematic for analog applications. For example, the asymmetric property of current conduction in a TFET creates problems in track-and-hold (T/H) circuits similar to the problems described for the pass transistor logic in Section 9.3. The unidirectional current flow in a TFET results in a large difference in the charge time and the discharge time of the hold capacitor in the T/H circuit [45].

The negative differential resistance observed in TFETs for a small positive drain bias can be exploited to realize a Schmitt-trigger inverter [45]. When an NTFET configured as a forward-biased diode is employed as a load in a common source amplifier, the total resistance of the circuit can be negative. In this configuration, the circuit has two stable operating points and can be used as a Schmitt-trigger inverter [45].

The small subthreshold swing, enhanced drive capability at lower supply voltages, reduced turn-on voltage, and the unidirectional current flow observed in high-performing TFETs, can be exploited in RF rectifier designs for RF-powered systems [5]. It is demonstrated that the GaSb-InAs heterojunction TFET-based RF rectifier exhibits higher power conversion efficiency and improved sensitivity compared to a CMOS implementation in ambient RF-power-scavenging applications [5].

The smaller subthreshold swing and larger transconductance generation efficiency g_m/I_D in the subthreshold region compared to a conventional MOSFET can be exploited to realize low-power analog circuits drawing a smaller current from the power supply [39, 45, 48]. However, due to the large impact of Miller capacitances in TFET-based circuits, the bandwidth is typically lower compared to the MOSFET-based circuits. Therefore, TFETs are interesting devices for low-power applications in a moderate frequency range [40]. For example, a $14-nm$ III-V TFET-based operational transconductance amplifier (OTA) exhibits a fivefold reduction in the power dissipation and more than an order of magnitude greater voltage gain compared with Si-MOSFET-based designs [45]. It is also demonstrated that graphene nanoribbon-based TFETs can offer a large bandwidth at a low voltage drive due to their capability to deliver a high current density in the subthreshold region [39].

An interesting application of a TFET is in realizing temperature-independent current biasing [38]. It exploits the distinct dependence of the drain current temperature coefficient on the drain and gate voltages in a TFET [38]. A voltage reference circuit employing TFETs has been proposed that exhibits a lower dependence on the supply voltage and the temperature [37]. Additionally, TFETs can be employed in circuits operating at high voltages due to the weak dependence of the subthreshold swing on the temperature [54].

Other applications where TFETs are proposed for use include digital-to-analog convertors, OTA-based neural amplifiers, and as optical sensors and biosensors [41, 50, 52, 53, 55].

9.6 Future Perspective of TFETs in Circuits

As discussed in the previous sections, most of the studies have demonstrated that TFETs offer distinct advantages over conventional MOSFETs in terms of power savings at low to medium frequencies. When TFETs operate under the quantum capacitance limit, TFETs offer smaller switching energies [56]. TFETs are also promising devices for SRAMs and for ultralow analog circuit applications. However, many of the studies have demonstrated that the superiority of TFETs compared to CMOS in terms of energy efficiency is limited to low supply voltage. Due to the mechanism of current transport in a TFET, there is an inherent limitation of TFETs in delivering a higher I_{ON} compared to a CMOS, which results in restricting the maximum operating frequency that can be obtained in circuits realized using TFETs. However, TFETs can be employed in heterogeneous CMOS-TFET multicore circuits [57, 58]. In the heterogeneous CMOS-TFET multicore circuits, the cores composed of TFETs are optimized for operating at low voltages and the cores composed of CMOS are optimized to deliver higher performance at comparatively higher supply voltages [58]. Therefore, TFETs can be used to address the problem of "dark silicon" or under-utilization of processing power due to power constraints [57, 58]. TFETs also exhibit the potential to be employed in Near-Threshold Voltage circuits to obtain lower power and higher throughput [26].

It should be noted that TFETs are promising devices for supply voltages lower than 0.5 V and technology nodes below 10 nm, where the reliability issues such as soft errors, electrical noise, and the impact of process-induced variations can have a stronger impact on the circuit design [59, 60]. These issues must be taken into account in future applications of TFETs. Another important point to note is that most of the studies that have been carried out to investigate the circuit applications of TFETs are based on simulations. There have been only a few experimental demonstrations of TFETs in circuits. Therefore, practical problems such as obtaining matched transistors for analog/mixed signal circuits, impact of interconnects in TFET-based circuits, signal integrity issues, and impact of process-induced variations are still largely unexplored [61].

9.7 Summary

In this chapter, we studied the distinct electrical characteristics of TFETs that have a great impact on the application of TFETs in various kinds of circuits. We discussed some of the applications of TFETs in digital circuits, memories, and analog circuits. We highlighted the problems arising due to unidirectional current flow, exponential onset of current in the output characteristics, greater gate-to-drain capacitance C_{GD}, and lower ON-state current I_{ON} on the application of TFETs in circuits. We also discussed some of the techniques to solve these problems. We further looked into some of the applications of TFETs where their distinct properties such as unidirectional current flow, excellent subthreshold swing, and perfect saturation in the output characteristics can be exploited. We also highlighted the key benefits and limitations of TFETs over MOSFETs in different circuits. In the next chapter, we will look at the future prospects for TFETs, assess their competitive advantages over CMOS, and highlight the problems that need to be solved before the commercialization of TFETs becomes possible.

REFERENCES

[1] X. Yang and K. Mohanram, "Robust 6T Si tunneling transistor SRAM design," in *2011 Design, Automation and Test in Europe*, pp. 1–6, IEEE, 2011.

[2] P. M. Solomon, I. Lauer, A. Majumdar, J. T. Teherani, M. Luisier, J. Cai, and S. J. Koester, "Effect of uniaxial strain on the drain current of a heterojunction tunneling field-effect transistor," *Electron Device Letters*, vol. 32, pp. 464–466, Apr. 2011.

[3] R. Mukundrajan, M. Cotter, V. Saripalli, M. J. Irwin, S. Datta, and V. Narayanan, "Ultra low power circuit design using tunnel FETs," in *IEEE Computer Society Annual Symposium on VLSI (ISVLSI)*, pp. 153–158, IEEE, Aug. 2012.

[4] D. Kim, Y. Lee, J. Cai, I. Lauer, L. Chang, S. J. Koester, D. Blaauw, and D. Sylvester, "Low-power circuit analysis and design based on heterojunction tunneling transistors (HETTs)," *IEEE Transactions Very Large Scale (VLSI) Systems*, vol. 21, pp. 1632–1643, Sept. 2013.

[5] H. Liu, X. Li, R. Vaddi, K. Ma, S. Datta, and V. Narayanan, "Tunnel FET RF rectifier design for energy harvesting applications," *IEEE Journal on Emerging and Selected Topics in Circuits and Systems*, vol. 4, pp. 400–411, Dec. 2014.

[6] N. Dagtekin and A. M. Ionescu, "Impact of super-linear onset, off-region due to uni-directional conductance and dominant CGD on performance of TFET-based circuits," *IEEE Journal of the Electron Devices Society*, vol. 3, pp. 233–239, May 2015.

[7] K. Boucart and A. M. Ionescu, "A new definition of threshold voltage in tunnel FETs," *Solid-State Electronics*, vol. 52, pp. 1318–1323, 2008.

[8] C. Sandow, J. Knoch, C. Urban, Q. T. Zhao, and S. Mantl, "Impact of electrostatics and doping concentration on the performance of silicon tunnel field-effect transistors," *Solid-State Electronics*, vol. 53, pp. 1126–1129, Oct. 2009.

[9] A. Pal, A. B. Sachid, H. Gossner, and V. R. Rao, "Insights into the design and optimization of tunnel-FET devices and circuits," *IEEE Trans. on Electron Devices*, vol. 58, pp. 1045–1053, Apr. 2011.

[10] E. Gnani, S. Reggiani, A. Gnudi, and G. Baccarani, "Drain conductance optimization in nanowire TFETs," in *Proceedings of the European Solid-State Device Res. Conf. (ESSDERC)*, pp. 105–108, IEEE, Sept. 2012.

[11] L. D. Michielis, L. Lattanzio, and A. M. Ionescu, "Understanding the superlinear onset of tunnel-FET output characteristic," *Electron Device Letters*, vol. 33, pp. 1523–1525, Nov. 2012.

[12] S. Mookerjea, R. Krishnan, S. Datta, and V. Narayanan, "On enhanced Miller capacitance effect in interband tunnel transistors," *IEEE Electron Device Letters*, vol. 30, pp. 1102–1104, Oct. 2009.

[13] Y. Yang, X. Tong, L. T. Yang, P. F. Guo, L. Fan, and Y. C. Yeo, "Tunneling field-effect transistor: Capacitance components and modeling," *Electron Device Letters*, vol. 31, pp. 752–754, July 2010.

[14] C. Wu, R. Huang, Q. Huang, J. Wang, and Y. Wang, "Design guideline for complementary heterostructure tunnel FETs with steep slope and improved output behavior," *IEEE Electron Device Letters*, vol. 37, pp. 20–23, Jan. 2016.

[15] S. Datta, R. Bijesh, H. Liu, D. Mohata, and V. Narayanan, "Tunnel transistors for low power logic," in *IEEE Compound Semiconductor Integrated Circuit Symposium (CSICS)*, pp. 1–4, IEEE, Oct. 2013.

[16] R. Li, Y. Lu, S. D. Chae, G. Zhou, Q. Liu, C. Chen, M. S. Rahman, T. Vasen, Q. Zhang, P. Fay, T. Kosel, M.Wistey, H. Xing, S. Koswatta, and A. Seabaugh, "InAs/AlGaSb heterojunction tunnel field-effect transistor with tunnelling in-line with the gate field," *Phys. Stat. Solidi (c)*, vol. 9, no. 2, pp. 389–392, 2012.

[17] V. Saripalli, S. Datta, V. Narayanan, and J. P. Kulkarni, "Variation-tolerant ultra low-power heterojunction tunnel FET SRAM design," in *IEEE/ACM International Symposium on Nanoscale Architectures (NANOARCH)*, pp. 45–52, IEEE, June 2011.

[18] A. W. Dey, J. Svensson, B. M. Borg, M. Ek, and L.-E. Wernersson, "Single InAs/GaSb nanowire low-power CMOS inverter," *Nano Letters*, vol. 12, pp. 5593–5597, Oct. 2012.

[19] L. Knoll, Q.-T. Zhao, A. Nichau, S. Trellenkamp, S. Richter, A. Schfer, D. Esseni, L. Selmi, K. K. Bourdelle, and S. Mantl, "Inverters with strained Si nanowire complementary tunnel field-effect transistors," *IEEE Electron Device Letters*, vol. 34, pp. 813–815, June 2013.

[20] S. Wang, A. Pan, C. O. Chui, and P. Gupta, "PROCEED: A pareto optimization-based circuit-level evaluator for emerging devices," *IEEE Transactions on Very Large Scale Integration (VLSI) Systems*, vol. 24, pp. 192–205, Jan. 2016.

[21] E. Baravelli, E. Gnani, A. Gnudi, S. Reggiani, and G. Baccarani, "TFET inverters with n-/p-devices on the same technology platform for low-voltage/low-power applications," *IEEE Trans. on Electron Devices*, vol. 61, pp. 473–478, Feb. 2014.

[22] Q.-T. Zhao, S. Richter, C. Schulte-Braucks, L. Knoll, S. Blaeser, G. Luong, S. Trellenkamp, A. Schafer, A. Tiedemann, J. Hartmann, and K. Bourdelle, "Strained Si and SiGe nanowire tunnel FETs for logic and analog applications," *IEEE Journal of the Electron Devices Society*, vol. 3, pp. 103–114, May 2015.

[23] R. Asra, M. Shrivastava, K. V. R. M. Murali, R. K. Pandey, H. Gossner, and V. R. Rao, "A tunnel FET for VDD scaling below 0.6 V with a CMOS-comparable performance," *IEEE Trans. on Electron Devices*, vol. 58, pp. 1855–1863, July 2011.

[24] Y. Khatami and K. Banerjee, "Steep subthreshold slope n- and p-type tunnel-FET devices for low-power and energy-efficient digital circuits," *IEEE Trans. on Electron Devices*, vol. 56, pp. 2752–2761, Nov. 2009.

[25] D. H. Morris, U. E. Avci, R. Rios, and I. A. Young, "Design of low voltage tunneling-FET logic circuits considering asymmetric conduction characteristics," *IEEE Journal on Emerging and Selected Topics in Circuits and Systems*, vol. 4, pp. 380–388, Dec. 2014.

[26] A. Sharma, A. A. Goud, and K. Roy, "Sub-10 nm FinFETs and tunnel-FETs: From devices to systems," in *Design,Automation and Test in Europe Conference and Exhibition (DATE)*, pp. 1443–1448, IEEE, Mar. 2015.

[27] U. E. Avci, D. H. Morris, and I. A. Young, "Tunnel field-effect transistors: Prospects and challenges," *IEEE Journal of the Electron Devices Society*, vol. 3, pp. 88–95, May 2015.

[28] M. Cotter, H. Liu, S. Datta, and V. Narayanan, "Evaluation of tunnel FET-based flip-flop designs for low power, high performance applications," in *Proc. International Symposium on Quality Electronic Design*, pp. 430–437, IEEE, Mar. 2013.

[29] J. Singh, K. Ramakrishnan, S. Mookerjea, S. Datta, N. Vijaykrishnan, and D. Pradhan, "A novel Si-tunnel FET based SRAM design for ultra low-power 0.3 V VDD applications," in *Design Automation Conference (ASP-DAC)*, pp. 181–186, IEEE, Jan. 2010.

[30] V. Saripalli, D. K. Mohata, S. Mookerjea, S. Datta, and V. Narayanan, "Low power loadless 4T SRAM cell based on degenerately doped source (DDS) $In_{0.53}Ga_{0.47}As$ tunnel FETs," in *DRC*, pp. 101–102, IEEE, June 2010.

[31] S. Datta, R. Bijesh, H. Liu, D. Mohata, and V.Narayanan, "Tunnel transistors for energy efficient computing," in *IEEE International Reliability Physics Symposium (IRPS)*, pp. 6A.3.1–6A.3.7, IEEE, 2013.

[32] Y.-N. Chen, M.-L. Fan, V. P.-H. Hu, P. Su, and C.-T. Chuang, "Evaluation of stability, performance of ultra-low voltage MOSFET, TFET, and mixed TFET-MOSFET SRAM cell with write-assist circuits," *IEEE Journal on Emerging and Selected Topics in Circuits and Systems*, vol. 4, pp. 389–399, Dec. 2014.

[33] L. Chang, D. M. Fried, J. Hergenrother, J. W. Sleight, R. H. Dennard, R. K.Montoye, L. Sekaric, S. J. McNab, A. W. Topol, C. D. Adams, K. W. Guarini, and W. Haensch, "Stable SRAM cell design for the 32 nm node and beyond," in *Digest of Technical Papers Symposium on VLSI Technology*, pp. 128–129, IEEE, June 2005.

[34] K. Noda, K. Matsui, K. Takeda, and N. Nakamura, "A loadless CMOS four-transistor SRAM cell in a $0.18-\mu m$ logic technology," *IEEE Transactions on Electron Devices*, vol. 48, pp. 2851–2855, Dec. 2001.

[35] S. G. Zang, X. Y. Liu, X. Lin, L. Liu, W. L. D. W. Zhang, P. F. Wang, and W. Hansch, "Applications of tunneling FET in memory devices," in *International Conf. on Solid-State and Integrated Circuit Technology (ICSICT)*, IEEE, Nov. 2010.

[36] A. Biswas and A. M. Ionescu, "1T capacitor-less DRAM cell based on asymmetric tunnel FET design," *IEEE Journal of the Electron Devices Society*, vol. 3, May 2015.

[37] M. F. andA. Heigl, G. Wachutka, G.Knoblinger, and D. Schmitt-Landsiedel, "Complementary multi-gate tunnelling FETs: Fabrication, optimisation and application aspects," *International Journal of Nanotechnology*, vol. 6, pp. 628–639, 2009.

[38] P. Guo, Y. Yang, G. Samudra, C. Heng, and Y. Yeo, "Temperature independent current biasing employing TFET," *Electronics Letter*, vol. 46, pp. 786–787, May 2010.

[39] B. Senale-Rodriguez, Y. Lu, P. Fay, D. Jena, A. Seabaugh, H. G. Xing, L. Barboni, and F. Silveira, "Perspectives of TFETs for low power analog ICs," in *IEEE Subthreshold Microelectronics Conference (SubVT)*, pp. 1–3, IEEE, 2012.

[40] A. Mallik and A. Chattopadhyay, "Tunnel field-effect transistors for analog/mixed-signal system-on-chip applications," *IEEE Trans. on Electron Devices*, vol. 59, pp. 888–894, Apr. 2012.

[41] A. R. Trivedi, S. Carlo, and S. Mukhopadhyay, "Exploring tunnel-FET for ultra low power analog applications: a case study on operational transconductance amplifier," in *Proceedings of the 50th Annual Design Automation Conference*, p. 109, IEEE, June 2013.

[42] V. Vijayvargiya and S. K. Vishvakarma, "Effect of drain doping profile on double-gate tunnel field-effect transistor and its influence on device RF performance," *IEEE Trans. on Nanotechnology*, vol. 13, pp. 974–981, Sept. 2014.

[43] K. Ghosh and U. Singisetti, "RF performance and avalanche breakdown analysis of InN tunnel FETs," *IEEE Trans. on Electron Devices*, vol. 61, pp. 3405–3410, Oct. 2014.

[44] G. Kaushal, K. Subramanyam, S. N. Rao, G.Vidya, R. Ramya, S. Shaik, H. Jeong, S. O. Jung, and R. Vaddi, "Design and performance benchmarking of steep-slope tunnel transistors for low voltage digital and analog circuits enabling self-powered SOCs," in *International SoC Design Conference (ISOCC)*, pp. 32–33, IEEE, Nov. 2014.

[45] B. Sedighi, X. S. Hu, H. Liu, J. J. Nahas, and M. Niemier, "Analog circuit design using tunnel-FETs," *IEEE Trans. on Circuits and Systems-I*, vol. 62, pp. 39–48, Jan. 2015.

[46] P. G. D. Agopian, M. D. V. Martino, S. D. dos Santos, F. S. Neves, J. A. Martino, R. Rooyackers, A. Vandooren, E. Simoen, A.-Y. Thean, and C. Claeys, "Influence of the source composition on the analog performance parameters of vertical nanowire-TFETs," *IEEE Trans. on Electron Devices*, vol. 61, pp. 16–22, Jan. 2015.

[47] C. Bordallo, V. B. Sivieri, J. A. Martino, P. G. D. Agopian, R. Rooyackers, A. Vandooren, E. Simoen, A. Thean, and C. Claeys, "Impact of the diameter of vertical nanowire-tunnel FETs with Si and SiGe source composition on analog parameters," in *Joint International EUROSOI Workshop and International Conference on Ultimate Integration on Silicon (EUROSOI-ULIS)*, pp. 253–256, IEEE, Jan. 2015.

[48] L. Barboni, M. Siniscalchi, and B. Sensale-Rodriguez, "TFET-based circuit design using the transconductance generation efficiency Gm/Id method," *IEEE Journal of the Electron Devices Society*, vol. 3, pp. 208–216, May 2015.

[49] B. Rawat and R. Paily, "Analysis of graphene tunnel field-effect transistors for analog/RF applications," *IEEE Trans. on Electron Devices*, vol. 62, pp. 2663–2669, Aug. 2015.

[50] M. S. Kim, X. Li, H. Liu, J. Sampson, S. Datta, and V. Narayanan, "Exploration of low-power high-SFDR current-steering D/A converter design using steep-slope heterojunction tunnel FETs," *IEEE Transactions on Very Large Scale Integration (VLSI) Systems*, pp. 1–11, Nov. 2015.

[51] V. Vijayvargiya, B. Reniwal, P. Singh, and S. Vishvakarma, "Analogue/RF performance attributes of underlap tunnel field effect transistor for low power applications," *Electronics Letters*, Feb. 2016.

[52] S. Kanungo, S. Chattopadhyay, P. S. Gupta, and H. R. and, "Comparative performance analysis of the dielectrically modulated full-gate and short-gate tunnel FET-based biosensors," *IEEE Trans. on Electron Devices*, vol. 62, pp. 994–1001, Mar. 2015.

[53] P. S. Gupta, S. Chattopadhyay, P. Dasgupta, and H. Rahaman, "A novel photosensitive tunneling transistor for near-infrared sensing applications: Design, modeling, and simulation," *IEEE Trans. on Electron Devices*, vol. 62, pp. 1516–1523, May 2015.

[54] M. Born, K. K. Bhuwalka, M. Schindler, U. Abelein, M. Schmidt, T. Sulima, and I. Eisele, "Tunnel FET: A CMOS device for high temperature applications," in *25th International Conference on Microelectronics*, pp. 124–127, IEEE, 2006.

[55] D. B. Abdi and M. J. Kumar, "Dielectric modulated overlapping gate-on-drain tunnel-FET as a label-free biosensor," *Superlattices and Microstructures*, vol. 86, pp. 198–202, Oct. 2015.

[56] S. O. Koswatta, M. S. Lundstrom, and D. N. and, "Performance comparison between p-i-n tunneling transistors and conventional MOSFETs," *IEEE Trans. on Electron Devices*, vol. 56, pp. 456–465, Mar. 2009.

[57] M. Taylor, "Is dark silicon useful? Harnessing the four horsemen of the coming dark silicon apocalypse," in *Design Automation Conference (DAC)*, pp. 1131–1136, IEEE, June 2012.

[58] K. Swaminathan, E. Kultursay, V. Saripalli, V. Narayanan, M. Kandemir, and S. Datta, "Steep-slope devices: From dark to dim silicon," *IEEE Micro*, vol. 33, pp. 50–59, Sept. 2013.

[59] S. Saurabh and M. J. Kumar, "Estimation and compensation of process induced variations in nanoscale tunnel field effect transistors (TFETs) for improved reliability," *IEEE Trans. Devices and Materials Reliability*, vol. 10, pp. 390–395, Sept. 2010.

[60] S. Datta, H. Liu, and V. Narayanan, "Tunnel FET technology: A reliability perspective," *Microelectronics Reliability*, vol. 54, pp. 861–874, Mar. 2014.

[61] C. Pan and A. Naeemi, "Interconnect design and benchmarking for charge-based beyond-CMOS device proposals," *IEEE Electron Device Letters*, vol. 37, no. 4, pp. 508–511, 2016.

10

Future Perspectives

10.1 Introduction

In the previous chapters, we discussed how TFETs offer distinctive advantages over conventional MOSFETs due to their capability to achieve a subthreshold swing lower than 60 $mV/decade$ at room temperature. However, TFETs suffer from low ON-state current (I_{ON}) and researchers have extensively explored techniques to boost the I_{ON} in a TFET. In the previous chapters, we reviewed different TFET architectures that are being investigated to address the problem of low I_{ON} and to improve other electrical characteristics of TFETs. We have also discussed potential applications of TFETs in different kinds of circuits and highlighted the merits and demerits of employing TFETs in CMOS circuits. In this chapter, we examine the future prospects of TFETs and assess their capability of being deployed for wide-scale commercial applications.

In this chapter, we start by reviewing the current state of research on TFETs as manifested by the best reported subthreshold swing and the highest I_{ON} for TFETs. We discuss the competitive advantages of TFETs over conventional MOSFETs. We also examine the advantages and disadvantages of other emerging devices such as impact ionization MOS (IMOS) and nanoelectromechanical switches (NEMS), which are competing with TFETs for future commercial applications. Finally, we outline our perspective on the road ahead for TFETs, highlighting the challenges and the possible solutions.

10.2 State-of-the-Art TFETs

During the last ten years, researchers have extensively explored novel TFET architectures employing different materials such as silicon, germanium, III-V semiconductors, carbon nanotubes, graphene, and nanowires. With the advancement in the research on TFETs and the exploration of newer material systems to realize TFETs, the best reported electrical characteristics of

TFETs are continuously improving. Since the first experimental demonstration of a CNT TFET with a subthreshold swing of 40 $mV/decade$, there have been several experimental demonstrations of subthreshold swing smaller than 60 $mV/decade$ in TFETs [1–4]. The primary focus area for research on TFETs has been to simultaneously achieve an average subthreshold swing smaller than 60 $mV/decade$ and ON-state current comparable to a MOSFET.

To make a fair assessment of the progress of research on TFETs and the possibility of TFETs being employed for future applications, it is important that various TFETs that have been reported are compared quantitatively with the best-in-class MOSFETs. However, the reported electrical characteristics of different TFETs are often not directly comparable. For example, the electrical characteristics of TFETs are reported at different supply voltages, employing different gate work functions, and having different threshold voltages, channel lengths, etc. To ensure a fair comparison of the reported TFETs, the comparison of the electrical characteristics must be done after appropriate normalization [2, 5]. An excellent comparison of the subthreshold swing and the I_{ON} of the reported TFETs, after normalization, has been presented in [5].

Currently, most of the experimental results have been reported for TFETs based on silicon, germanium, and III-V semiconductors since the fabrication setups for these materials are already well developed. Although, there are several reports of silicon- and germanium-based TFETs exhibiting subthreshold swing lower than 60 $mV/decade$, the I_{ON} of the TFETs based on silicon and germanium are low, even when strain is applied to boost the I_{ON} of the TFETs. As a result, it is generally believed that silicon-based TFETs may not be able to deliver the ON-state current that is required for high-performance transistors. Therefore, integration of novel materials on a silicon platform might become necessary to realize high-performing TFETs [3].

TFETs based on III-V materials exhibit higher I_{ON}, especially the TFETs that utilize III-V heterojunctions and possess small effective bandgaps. The highest drain current at a subthreshold swing lower than 60 $mV/decade$ has been reported for an InGaAs-based heterojunction TFET and is in the range of a few nA/micron [2, 6]. The lowest subthreshold swing of 21 $mV/decade$ for three decades of drain current has been reported for a III-V/Si heterojunction nanowire TFET [2, 7]. For the III-V TFETs, obtaining a good-quality dielectric–semiconductor interface and abrupt source–channel junction continue to be the biggest challenge.

For carbon-based materials such as CNT and graphene, there are numerous simulation-based studies that predict that a subthreshold swing lower than 60 $mV/decade$ and a high I_{ON} can be achieved, simultaneously. However, lower subthreshold swings have been demonstrated for CNT TFETs only at low current levels. Therefore, it is essential that more experiments are carried out on carbon-based TFETs, especially in demonstrating a high ON-state current.

Recently, a TFET based on a two-dimensional material has been experimentally demonstrated that achieves a minimum subthreshold swing of 3.9 $mV/decade$ and an average subthreshold swing of 31 $mV/decade$ for four decades of drain current at room temperature [4]. It utilizes a highly doped germanium as the source and atomically thin molybdenum disulfide (MoS_2) as the channel. The TFET has a vertical strain-free heterojunction between Ge and MoS_2 with a staggered band alignment, which results in a low tunneling barrier height and a high tunneling area [4]. Additionally, the Ge source allows obtaining a high doping concentration with sharp doping profile that is not easy to obtain in a 2-D material. Currently, at a drain voltage $V_{DS} = 0.1\ V$, this is the best reported average subthreshold swing for an experimental TFET. It demonstrates the huge potential of TFETs in low-voltage energy-efficient circuits.

10.3 Comparison with MOSFET

In Chapter 9, we discussed the applications of TFETs in various types of circuits. The competitive advantages TFETs over the best-in-class MOSFETs in realizing circuits have been assessed in several studies [8–13]. As already pointed out, TFETs suffer from a low I_{ON} and this inhibits TFETs from being employed in high-performance circuits. In spite of this limitation, TFETs have advantages over MOSFETs in low-power applications because of low subthreshold swing. It is shown that for the same performance level, the device that is able to meet the required constraint of I_{ON}/I_{OFF} ratio at lower supply voltage is more energy efficient [14]. At low to medium performance levels, TFETs are capable of meeting the required I_{ON}/I_{OFF} ratio at lower supply voltages due to their lower subthreshold swing compared to a MOSFET. Therefore, a TFET can be more energy efficient than a MOSFET at low to medium performance levels.

At lower supply voltages, due to the lower subthreshold swing in a TFET, a TFET circuit can operate at a higher speed compared to a similar MOSFET circuit [8, 11, 12]. For example, it has been demonstrated that an inverter is 4× faster when implemented using TFETs instead of MOSFETs at $V_{DD} = 0.4\ V$, with the process-induced variations taken into account [12]. It should be noted that the supply voltage range over which a TFET-based circuit outperforms a MOSFET-based circuit depends on several factors, such as switching activity, topology, and the interconnect capacitances of the circuit. For example, when the switching activity in the circuit is low, TFET outperforms MOSFET when the supply voltage is less than 0.55 V [12]. However, when the switching activity is high, TFET outperforms MOSFET when the supply voltage is less than 0.33 V [12].

It should be noted that most of the studies that have predicted higher energy efficiency of TFETs compared to MOSFETs are either based on first-order analysis or numerical simulations. However, due to the peculiar electrical characteristics of TFETs such as exponential onset of current in the output characteristics, unidirectional current flow, and delayed saturation in the output characteristics, achieving the same functionality in a circuit realized using TFETs instead of MOSFETs can require modifications in the circuit. As a result, the number of transistors and the topology of the circuits realized using TFETs can be quite different from that realized using MOSFETs. Additionally, when TFETs are employed in actual circuits, then the impact of wire capacitances, signal integrity issues due to coupling between the signals, and increased impact of process-induced variations at smaller device dimensions will play a crucial role in determining the energy efficiency of the circuits. Therefore, it will be interesting to find out if the higher energy efficiency of TFETs predicted using simulations can actually be realized in integrated circuits.

10.4 Other Low-Subthreshold Swing Devices

As pointed out in previous sections, the most interesting characteristic of a TFET is its low subthreshold swing, which allows realizing energy-efficient circuits. Therefore, several other devices that exhibit subthreshold swing lower than the conventional MOSFETs are also being actively investigated. It is important to appreciate that TFETs not only compete with the conventional MOSFETs for future wide-scale commercial applications, but also compete with other low subthreshold swing devices. Therefore, it is worth evaluating the advantages and disadvantages of other emerging low subthreshold swing devices. In this section, we will briefly describe the operating principle of some of the low subthreshold swing devices and evaluate their merits and demerits.

10.4.1 Impact-Ionization MOS (IMOS)

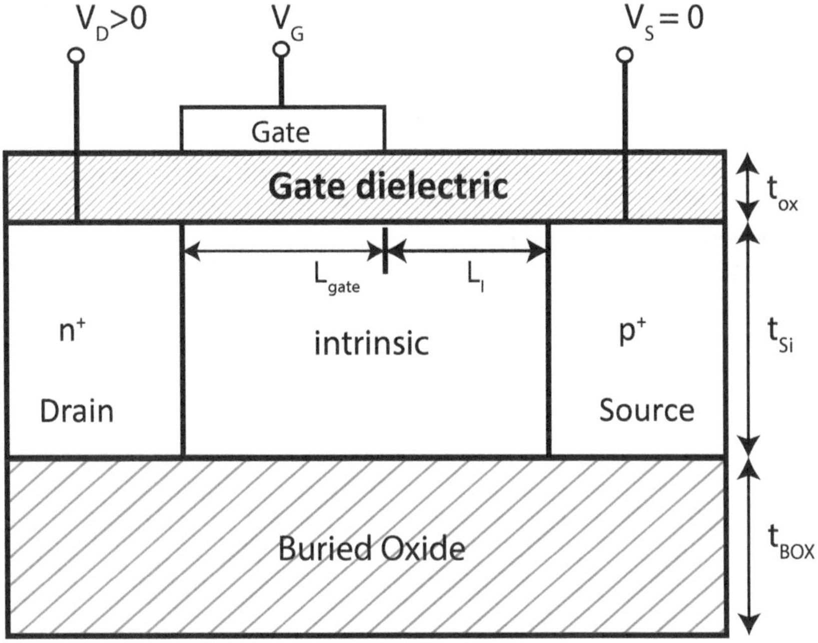

FIGURE 10.1
Schematic diagram of an IMOS [15].

An IMOS is a low subthreshold swing device that works on the principle of modulation of avalanche breakdown voltage by the gate of the transistor [15–19]. Figure 10.1 shows a schematic diagram of an IMOS implemented in SOI. The structure of an IMOS is similar to a gated p-i-n structure of a TFET. However, the gate covers only a part of the intrinsic channel region. In the normal mode of operation, the source is grounded and a positive bias is applied to the drain. When the gate voltage is increased, a channel is formed under the gate and the intrinsic region L_I outside the gate remains intrinsic. As a result, the drain-to-source voltage drop occurs mainly across the intrinsic region L_I of the TFET and the lateral electric field in the intrinsic region L_I increases. The increase of gate voltage also results in the increase of the transverse electric field and eventually the IMOS breaks down when the ionization integral approaches unity. The turn-on characteristics of the IMOS due to avalanche breakdown are very steep. Since the impact-ionization coefficients strongly depend on the electric field and there is feedback inherent in the avalanche multiplication process, a subthreshold swing much lower than 60 $mV/decade$ is achieved in an IMOS [15]. However, IMOS suffers from high supply voltage requirements, reliability issues due to hot-carrier-induced degradation, non-rail-to-rail output voltage, and high active power dissipation [15, 16]. Recently, a bipolar-based IMOS was proposed in [17]. It employs the current gain of the parasitic bipolar junction transistor (BJT) in a MOSFET to reduce the supply voltage by 60%, obtain a small subthreshold swing, and improve the reliability of the device [17].

10.4.2 Nanoelectromechanical Switch (NEMS)

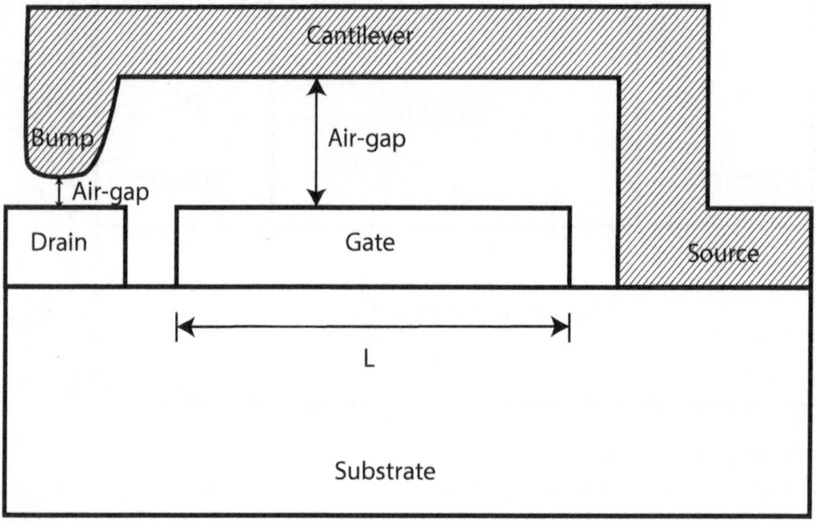

FIGURE 10.2
Schematic diagram of a cantilever-type of NEMS in the OFF-state [20].

Nanoelectromechanical switches operate by employing electrostatic force to move a structure such that a conductive path is established between opposite electrodes [3, 21–26]. There are different types of NEMS reported in the literature. Figure 10.2 shows a cantilever-type NEMS with three terminals: gate, source and drain. The drain contact is fixed and the source contact having a cantilever can be pulled in by the electrostatic force created due to the gate voltage. The switch is turned on when the gate voltage creates sufficiently high electrostatic force to deform the cantilever such that the source and the drain are in contact. The switch is turned off by decreasing the electrostatic force created by the gate and allowing the deformed cantilever to regain its original shape. To turn off the switch, the elastic restoring force of the deformed cantilever must be sufficient to overpower the adhesive force at the source–drain contact.

The biggest advantage of NEMS is that it provides a zero leakage current and zero subthreshold swing. Therefore, NEMS circuits have improved energy efficiency compared to CMOS circuits [3]. However, NEMS has a higher switching delay because it is dominated by the mechanical delay that is determined by the movement of electrodes rather than the electrical delay that is determined by the charging and discharging of capacitors. The stability of contact resistance is a key requirement for NEMS for which more research needs to be carried out [3]. Additionally, the operation of NEMS at smaller supply voltages ($V_{DD} < 1\ V$) is a challenge. At smaller supply voltages, NEMS are prone to stiction issues due to decreased restoring force [21]. The scalability and the reliability of NEMS at smaller device dimensions are other important challenges for future applications.

10.4.3 Negative Capacitance FETs (NCFETs)

By replacing the conventional insulator in the gate stack of a MOSFET with a ferroelectric insulator of appropriate thickness, an effective negative gate capacitance can be obtained in a MOSFET [27–34]. Figure 10.3 shows a schematic representation of a negative capacitance FET (NCFET) with a ferroelectric material in the gate stack. Due to the negative capacitance at the gate of the MOSFET, a step-up voltage transformer effect can be observed. This amplifies the gate voltage and a subthreshold swing smaller than 60 $mV/decade$ is obtained [27, 28]. The biggest advantage of the NCFET is that it is a direct replacement for a conventional MOSFET since only the gate stack needs to be modified. The operating principle, ON-state current, and the scalability of the NCFET is similar to that of a MOSFET [3]. The biggest challenge for the NCFET is the identification of an appropriate ferroelectric material that yields a low subthreshold swing and minimum hysteresis at room temperature and the integration of that material into the conventional CMOS technology [3]. Additionally, the minimum subthreshold swing in an NCFET is determined by the stability constraints and realizing hysteresis-free operation [34].

There are other emerging research devices that are being investigated for replacing MOSFETs [3, 16, 35]. Interested readers can refer to [3] for an overview of the emerging research devices.

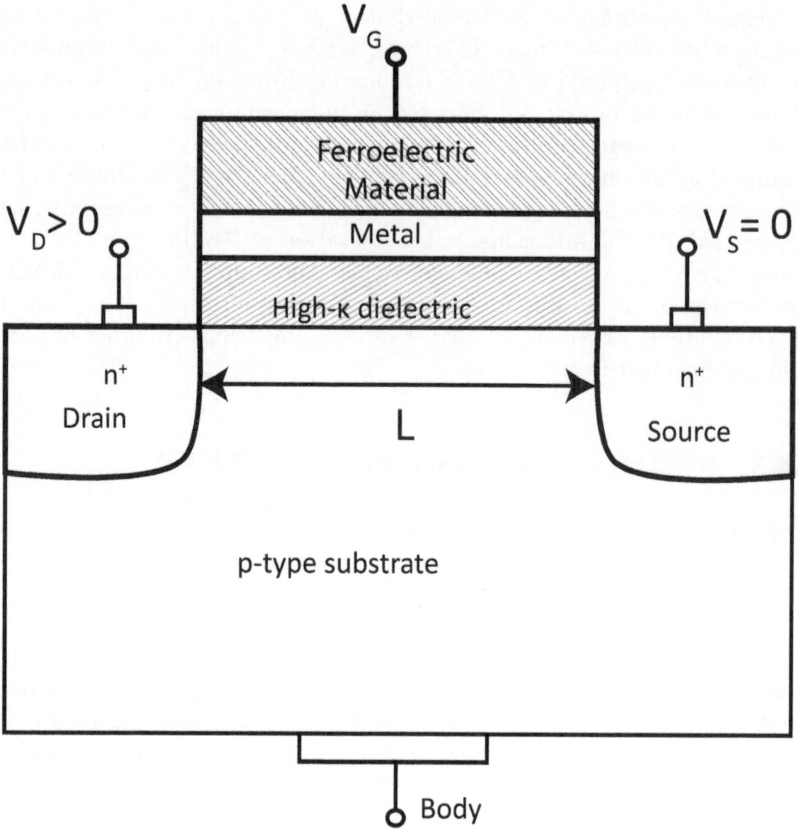

FIGURE 10.3
Schematic diagram of an NCFET.

10.5 Challenges and the Road Ahead

As mentioned above, the biggest challenge for a TFET is to deliver a high I_{ON} with a sustained subthreshold swing lower than 60 $mV/decade$ for at least four decades of current [3]. Based on simulation results, it appears that this condition can be satisfied by TFETs based on III-V semiconductor heterostructures, CNTs, graphene, or nanowire heterostructures. In the near future, with appropriate device and process optimizations, it seems quite possible that TFETs meeting the above conditions will be experimentally demonstrated. Recently, subthreshold swing lower than 60 $mV/decade$ for at least four decades of current has been satisfied by a TFET based on MoS_2 at $V_{DD} = 0.1\ V$ [4]. This demonstrates the capability of TFETs for use in future low-voltage energy-efficient circuits.

To exploit TFETs for wide-scale commercial application, it is becoming imminent that novel non-silicon materials will be required to be integrated into the conventional CMOS technology. This will be an enormous challenge, not only from the technological perspective, but also from the economic perspective. The TFETs based on non-silicon materials must exhibit substantial gains in terms of energy efficiency and performance to justify the high cost that is expected to be incurred in the integration of non-silicon materials in CMOS technology, possibly by displacing silicon. TFETs are being targeted for advanced technology nodes below 10 nm. Several studies have shown that the subthreshold swing and the I_{OFF} in a TFET degrade appreciably when the channel length is reduced below 10 nm. Therefore, it needs to be explored if the competitive advantage of TFETs over MOSFETs can be maintained at smaller device dimensions. Additionally, the impact of process-induced variations, random dopant fluctuations, reliability issues, etc., in a TFET at smaller device dimensions and at a supply voltage below 0.5 V need to be examined. These factors will play a crucial role in determining the prospects of TFETs in advanced process nodes.

In spite of several challenges ahead in the commercialization of TFETs, recently there has been tremendous progress in the research on TFETs. Among the emerging research devices, TFETs seems to be the most promising device that can replace MOSFET in the CMOS technology. In the ITRS roadmap for emerging research devices, TFETs figure prominently and it is expected that the research on TFETs will continue into the future.

10.6 Summary

In this chapter, we assessed the current state of research on TFETs. We discussed the characteristics of state-of-the-art TFETs and examined the competitive advantages of TFETs over MOSFETs. We also discussed other emerging low subthreshold swing devices. Finally, we presented our perspective on the future of TFETs for wide-scale applications.

Theoretically, TFETs have been shown to possess tremendous potential for use in energy-efficient circuits. It seems quite possible that in the near future, TFETs can be employed in the mainstream CMOS technology overcoming the technological barrier that TFETs currently face. This offers an exciting opportunity to researchers working on TFETs to explore, invent, and realize novel TFET architectures using different material systems and employ TFETs in newer circuits and innovative applications. We hope that this book on the fundamental aspect of TFETs ignites reader interest in pursuing further research on TFETs and ushering in a new era of "green electronics" and a better world.

REFERENCES

[1] J. Appenzeller, Y.-M. Lin, J. Knoch, and P. Avouris, "Band-to-band tunneling in carbon nanotube field-effect transistors," *Physical Review Letters*, vol. 3, pp. 196805–1–196805–4, Nov. 2004.

[2] A. C. Seabaugh and Q. Zhang, "Low-voltage tunnel transistors for beyond CMOS logic," *Proceedings of the IEEE*, pp. 2095–2110, Dec. 2010.

[3] International Technology Roadmap for Semiconductors (ITRS), "Emerging research devices, 2013 edition." http://www.itrs.net, 2013.

[4] D. Sarkar, X. Xie, W. Liu, W. Cao, J. Kang, Y. Gong, S. Kraemer, P. M. Ajayan, and K. Banerjee, "A subthermionic tunnel field-effect transistor with an atomically thin channel," *Nature*, vol. 526, pp. 91–95, Oct. 2015.

[5] H. Lu and A. Seabaugh, "Tunnel field-effect transistors: State-of-the-art," *IEEE Journal of the Electron Devices Society*, vol. 2, pp. 44–49, July 2014.

[6] G. Dewey, B. Chu-Kung, J. Boardman, J. M. Fastenau, J. Kavalieros, R. Kotlyar, W. K. Liu, D. Lubyshev, M. Metz, N. Mukherjee, P. Oakey, R. Pillarisetty, M. Radosavljevic, H. Then, and R. Chau, "Fabrication, characterization, and physics of III-V heterojunction tunneling field effect transistors (H-TFET) for steep sub-threshold swing," in *IEEE International Electron Devices Meeting (IEDM)*, pp. 33.6.1–33.6.4, IEEE, 2011.

[7] K. Tomioka, M. Yoshimura, and T. Fukui, "Steep-slope tunnel field-effect transistors using III-V nanowire/Si heterojunction," in *Symposium on VLSI Technology (VLSIT)*, pp. 47–48, IEEE, 2012.

[8] Y. Khatami and K. Banerjee, "Steep subthreshold slope n- and p-type tunnel-FET devices for low-power and energy-efficient digital circuits," *IEEE Trans. on Electron Devices*, vol. 56, pp. 2752–2761, Nov. 2009.

[9] U. E. Avci, S. Hasan, D. E. Nikonov, R. Rios, K. Kuhn, and I. A. Young, "Understanding the feasibility of scaled III-V tfet for logic by bridging atomistic simulations and experimental results," in *Symposium on VLSI technology*, pp. 183–184, IEEE, June 2012.

[10] S. Datta, R. Bijesh, H. Liu, D. Mohata, and V. Narayanan, "Tunnel transistors for low power logic," in *IEEE Compound Semiconductor Integrated Circuit Symposium (CSICS)*, pp. 1–4, IEEE, Oct. 2013.

[11] E. Baravelli, E. Gnani, A. Gnudi, S. Reggiani, and G. Baccarani, "TFET inverters with n-/p-devices on the same technology platform for low-voltage/low-power applications," *IEEE Trans. on Electron Devices*, vol. 61, pp. 473–478, Feb. 2014.

[12] D. H. Morris, U. E. Avci, R. Rios, and I. A. Young, "Design of low voltage tunneling-fet logic circuits considering asymmetric conduction characteristics," *IEEE Journal on Emerging and Selected Topics in Circuits and Systems*, vol. 4, pp. 380–388, Dec. 2014.

[13] Q. Huang, R. Huang, C. Wu, H. Zhu, C. Chen, J. Wang, L. Guo, R. Wang, L. Ye, and Y. Wang, "Comprehensive performance re-assessment of TFETs with a novel design by gate and source engineering from device/circuit perspective," in *IEEE International Electron Devices Meeting (IEDM)*, pp. 13.3.1–13.3.4, IEEE, Dec. 2014.

[14] A. M. Ionescu and H. Riel, "Tunnel field-effect transistors as energy-efficient electronic switches," *Nature*, vol. 479, pp. 329–337, Nov. 2011.

[15] K. Gopalakrishnan, P. B. Griffin, and J. D. Plummer, "Impact ionization MOS (I-MOS)-Part I: device and circuit simulations," *IEEE Trans. on Electron Devices*, vol. 52, pp. 69–76, Jan. 2005.

[16] A. Tura and J. C. S. Woo, "Performance comparison of silicon steep subthreshold FETs," *IEEE Trans. on Electron Devices*, vol. 57, pp. 1362–1368, June 2010.

[17] M. J. Kumar, M. Maheedhar, and P. P. Varma, "Bipolar I-MOS : An impact-ionization MOS with reduced operating voltage using the open-base BJT configuration," *IEEE Transactions on Electron Devices*, vol. 62, pp. 4345–4348, Dec. 2015.

[18] S. Ramaswamy and M. J. Kumar, "Junctionless impact ionization MOS: Proposal and investigation," *IEEE Transactions on Electron Devices*, vol. 61, pp. 4295–4298, Dec. 2014.

[19] N. Kannan and M. J. Kumar, "Dielectric-modulated impact-ionization MOS transistor as a label-free biosensor," *IEEE Electron Device Letters*, vol. 34, pp. 1575–1577, Dec. 2013.

[20] K. Akarvardar, D. Elata, R. Parsa, G. C. Wan, K. Yoo, J. Provine, P. Peumans, R. T. Howe, and H. S. P. Wong, "Design considerations for complementary nanoelectromechanical logic gates," in *Electron Devices Meeting, 2007. IEDM 2007. IEEE International*, pp. 299–302, IEEE, Dec. 2007.

[21] O. Y. Loh and H. D. Espinosa, "Nanoelectromechanical contact switches," *Nature nanotechnology* 7, pp. 283–295, May 2012.

[22] J. Sun, M. E. Schmidt, M. Muruganathan, H. M. Chong, and H. Mizuta, "Large-scale nanoelectromechanical switches based on directly deposited nanocrystalline graphene on insulating substrates," *Nanoscale*, vol. 8, no. 12, pp. 6659–6665, 2016.

[23] J. L. Muñoz-Gamarra, A. Uranga, and N. Barniol, "CMOS-NEMS copper switches monolithically integrated using a 65 nm CMOS technology," *Micromachines*, vol. 7, no. 2, p. 30, 2016.

[24] K. Kato, V. Stojanovic, and T. J. K. Liu, "Non-volatile nano-electromechanical memory for energy-efficient data searching," *IEEE Electron Device Letters*, vol. 37, pp. 31–34, Jan. 2016.

[25] Y. Zhang, S. Irving, L. Peng, X. Fu, D. Koppelman, W. Zhang, and J. Ardonne, "Design space exploration for device and architectural heterogeneity in chip-multiprocessors," *Microprocessors and Microsystems*, vol. 40, pp. 88–101, 2016.

[26] G. Riusa and F. Perez-Muranob, "Nanocantilever beam fabrication for CMOS technology integration," *Nanocantilever Beams: Modeling, Fabrication, and Applications*, p. 3, 2016.

[27] S. Salahuddin and S. Datta, "Use of negative capacitance to provide voltage amplification for low power nanoscale devices," *Nano letters 8*, pp. 405–410, Feb. 2008.

[28] A. I. Khan, D. Bhowmik, P. Yu, S. J. Kim, X. Pan, R. Ramesh, and S. Salahuddin, "Experimental evidence of ferroelectric negative capacitance in nanoscale heterostructures," *Applied Physics Letters*, vol. 99, no. 11, p. 113501, 2011.

[29] A. Rusu, A. Saeidi, and A. M. Ionescu, "Condition for the negative capacitance effect in metal–ferroelectric–insulator–semiconductor devices," *Nanotechnology*, vol. 27, no. 11, p. 115201, 2016.

[30] C. I. Lin, A. I. Khan, S. Salahuddin, and C. Hu, "Effects of the variation of ferroelectric properties on negative capacitance FET characteristics," *IEEE Transactions on Electron Devices*, vol. 63, pp. 2197–2199, May 2016.

[31] J. Jo and C. Shin, "Negative capacitance field effect transistor with hysteresis-free sub-60-mv/decade switching," *IEEE Electron Device Letters*, vol. 37, pp. 245–248, Mar. 2016.

[32] G. Catalan, D. Jiménez, and A. Gruverman, "Ferroelectrics: Negative capacitance detected," *Nature materials*, vol. 14, no. 2, pp. 137–139, 2015.

[33] A. O'Neill, "Negative capacitance using ferroelectrics for future steep-slope MOSFETS," *ECS Transactions*, vol. 69, no. 10, pp. 171–177, 2015.

[34] A. Jain, M. Alam, *et al.*, "Stability constraints define the minimum subthreshold swing of a negative capacitance field-effect transistor," *Electron Devices, IEEE Transactions on*, vol. 61, no. 7, pp. 2235–2242, 2014.

[35] E. Gnani, E. Baravelli, P. Maiorano, A. Gnudi, S. Reggiani, and G. Baccarani, "Steep-slope devices: Prospects and challenges," in *Journal of Nano Research*, vol. 39, pp. 3–16, Trans Tech Publ, 2016.

Appendix A

Simulation of Double Gate Tunnel Field-Effect Transistor

A.1 Device Structure

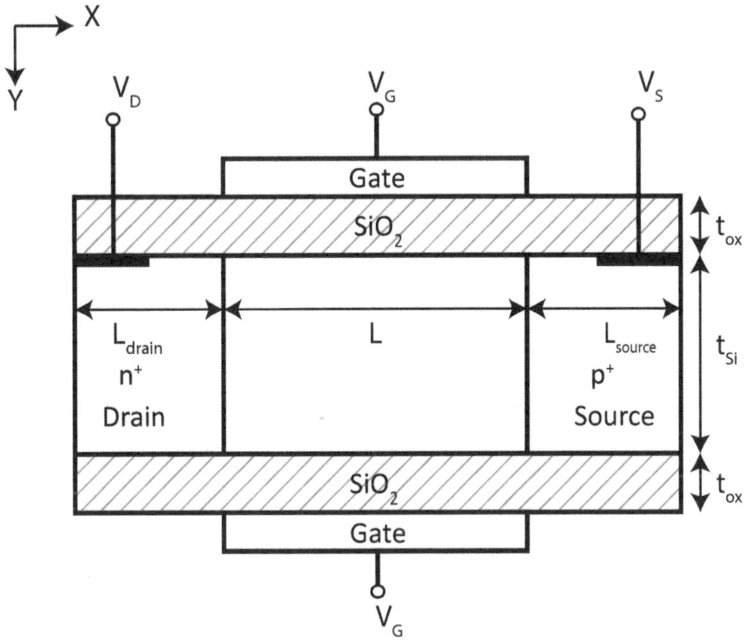

FIGURE A.1
Schematic cross sectional view of the DGTFET [1].

TABLE A.1
Device parameters of the DGTFET [1].

Device Parameter	**Value**
Source Doping (N_A)	1×10^{20} $atoms/cm^3$
Drain Doping (N_D)	5×10^{18} $atoms/cm^3$
Channel Doping (N_D)	1×10^{17} $atoms/cm^3$
Channel Length (L)	30 nm
Drain Length (L_{drain})	100 nm
Source Length (L_{source})	100 nm
Gate Oxide Thickness (t_{ox})	3 nm
Silicon Body Thickness (t_{Si})	10 nm
Gate Work Function (Φ_m)	4.5 eV

In this appendix, we provide a sample input file for simulation of a Double Gate Tunnel Field-Effect Transistor (DGTFET) using 2-D device simulator ATLAS [1, 2]. Figure A.1 shows the schematic cross-sectional view of the DGTFET. The device parameters for the DGTFET are shown in Table A.1.

The most important aspect in the simulation of a TFET is the choice of the tunneling model and its calibration. The simulation file in this appendix employs non-local tunneling model. The tunneling model has been calibrated using ATLAS (version 5.12.1.R). The tunneling model shows good agreement with the results presented in [1] and Eq. (2) in [3].

It should be noted that the example provided in this appendix is just an illustration of the technique to simulate a TFET. Depending on the tool being employed, the level of accuracy required and the physical phenomenon being modeled, the tunneling model may require re-calibration and other tool-specific parameter changes.

A.2 Sample Input File

```
# Input file for device simulation tool ATLAS
go atlas
#Defining meshes for the device: Meshes should be as fine as possible,
meeting the computer resource constraints.
mesh space.mult=1.0

x.mesh loc=0.000 spac=0.01
x.mesh loc=0.095 spac=0.001
x.mesh loc=0.100 spac=0.00005
x.mesh loc=0.105 spac=0.0005
x.mesh loc=0.115 spac=0.005
x.mesh loc=0.115 spac=0.0005
x.mesh loc=0.130 spac=0.00005
x.mesh loc=0.135 spac=0.001
x.mesh loc=0.180 spac=0.005
x.mesh loc=0.23 spac=0.01

y.mesh loc=-0.003 spac=0.0005
y.mesh loc=0.000 spac=0.0001
y.mesh loc=0.002 spac=0.0005
y.mesh loc=0.008 spac=0.0005
y.mesh loc=0.010 spac=0.0001
y.mesh loc=0.013 spac=0.0005

# defining another mesh for non-local BTBT modelling in ATLAS
qtx.mesh loc=0.05 spac=0.0005
qtx.mesh loc=0.13 spac=0.0002
qtx.mesh loc=0.20 spac=0.0005
qty.mesh loc=0.000 spac=0.0005
qty.mesh loc=0.005 spac=0.001
qty.mesh loc=0.010 spac=0.0005
```

Example continued on the next page ...

#Defining regions of the device
region num=1 y.max=0.000 material=Oxide
region num=2 y.min=0.000 y.max=0.010 material=Silicon
region num=3 y.min=0.010 material=Oxide

#Defining electrodes: channel length in this example is 30 nm
electrode name=gate x.min=0.1 x.max=0.13 top
electrode name=drain x.max=0.050 y.min=0.000 y.max=0.0
electrode name=source x.min=0.180 y.min=0.0 y.max=0.0
electrode name=gate2 x.min=0.1 x.max=0.13 bottom

#Defining the tunneling parameter for silicon: The value of me.tunnel in this example has been calibrated using [1, 2].
material material=Silicon me.tunnel=0.14 region=2

#Defining doping: The doping profile is assumed to be abrupt
doping uniform n.type conc=1e17 reg=2
doping uniform n.type conc=5e18 x.max=0.100 reg=2
doping uniform p.type conc=1e20 x.min=0.130 reg=2

#Defining contacts
contact name=gate workfunction=4.5
contact name=gate2 workfunction=4.5 common=gate

#Defining physical models: Band gap narrowing effect is included. The effect of concentration dependent mobility and SRH models are also include, though their impact in this example is found to be not appreciable.
models bbt.nonlocal bbt.forward qtunn.dir=1 bgn consrh conmob print

#included to improve the convergence
method newton
solve init

Example continued on the next page ...

```
#Applying bias for simulation
solve vsource=0.0
solve vgate=0.0

#in order to improve convergence, drain voltage is increased gradually
solve vdrain=0.03125
solve vdrain=0.0625
solve vdrain=0.125
solve vdrain=0.25
solve vdrain=0.5
solve vdrain=1.00

#Simulating the transfer characteristics
log outf=xfer_0.log master
solve vgate=0.0 vstep=0.1 name=gate vfinal=1.5

#plotting the transfer characteristics using the plotter of ATLAS
tonyplot xfer_0.log

#Storing relevant information of the device for further analysis
output val.band con.band charge e.lines
save outf=test.str
```

REFERENCES

[1] K. Boucart and A. M. Ionescu, "Double-gate tunnel FET with high-κ gate dielectric," *IEEE Trans. on Electron Devices*, vol. 54, pp. 1725–1733, July 2007.

[2] Silvaco International, Santa Clara, CA, *ATLAS User's Manual*, Mar. 2007.

[3] M. Schlosser, K. Bhuwalka, M. Sauter, T. Zilbauer, T. Sulima, and I. Eisele, "Fringing-induced drain current improvement in the tunnel field-effect transistor with high-κ gate dielectrics," *IEEE Trans. on Electron Devices*, vol. 56, pp. 100–108, Jan. 2009.

Appendix B

Simulation of SOI Tunnel Field-Effect Transistor

B.1 Device Structure

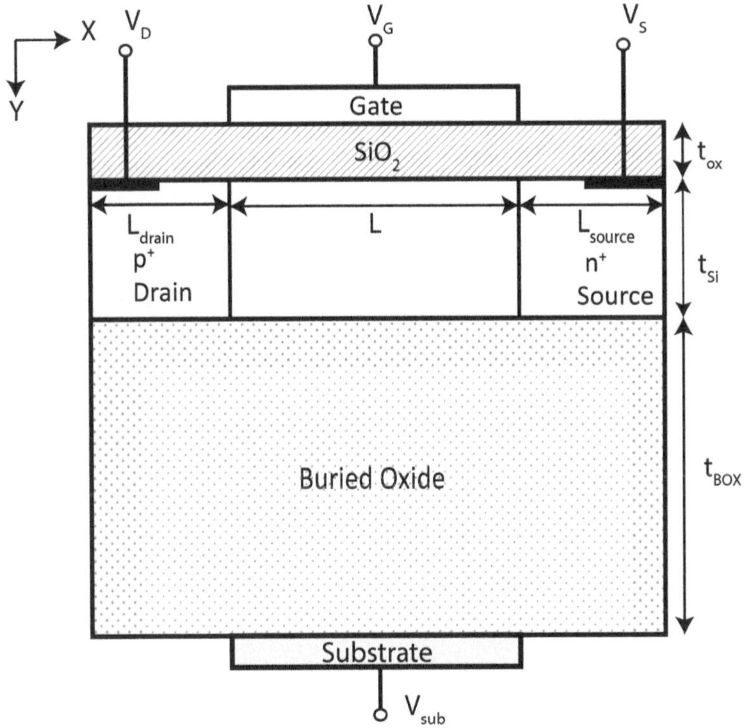

FIGURE B.1
Schematic cross sectional view of the SOI TFET.

TABLE B.1
Device parameters of the SOI TFET.

Device Parameter	**Value**
Source Doping (N_D)	1×10^{21} $atoms/cm^3$
Drain Doping (N_A)	1×10^{21} $atoms/cm^3$
Channel Doping (N_D)	1×10^{15} $atoms/cm^3$
Channel Length (L)	100 nm
Drain Length (L_{drain})	100 nm
Source Length (L_{source})	100 nm
Gate Oxide Thickness (t_{ox})	2 nm
Buried Oxide Thickness (t_{BOX})	190 nm
Silicon Body Thickness (t_{Si})	10 nm
Gate Work Function (Φ_m)	4.8 eV

In this appendix, we provide a sample input file for simulation of a Silicon on Insulator (SOI) TFET using 2-D device simulator ATLAS [1]. Figure B.1 shows the schematic cross-sectional view of the SOI TFET. The device parameters for the SOI TFET are shown in Table B.1.

The SOI TFET in this example is a p-type TFET. The simulation file in this appendix employs Kane's tunneling model. The parameters for this model is based on [2]. It should be noted that the example provided in this appendix is just an illustration of the technique to simulate a TFET. Depending on the tool being employed, the level of accuracy required and the physical phenomenon being modeled, the tunneling model may require re-calibration and other tool-specific parameter changes.

B.2 Sample Input File

```
# Input file for device simulation tool ATLAS
go atlas
#Defining meshes for the device: Meshes should be as fine as possible,
meeting the computer resource constraints.
mesh space.mult=1.0
x.mesh loc=0.000 spac=0.025
x.mesh loc=0.08 spac=0.0002
x.mesh loc=0.12 spac=0.0002
x.mesh loc=0.15 spac=0.001
x.mesh loc=0.18 spac=0.0002
x.mesh loc=0.22 spac=0.0002
x.mesh loc=0.3 spac=0.025
y.mesh loc=-0.002 spac=0.001
y.mesh loc=0.000 spac=0.0001
y.mesh loc=0.005 spac=0.001
y.mesh loc=0.01 spac=0.001
y.mesh loc=0.06 spac=0.025
y.mesh loc=0.20 spac=0.025

#Defining regions of the device
region num=1 y.max=0.000 material=oxide
region num=2 y.min=0.000 y.max=0.010 material=Silicon
region num=3 y.min=0.010 material=oxide
#Defining electrodes: channel length in this example is 200 nm
electrode name=gate x.min=0.1 x.max=0.2 top
electrode name=drain x.max=0.05 y.min=0.0 y.max=0.0
electrode name=source x.min=0.25 y.min=0.0 y.max=0.0
electrode substrate
```

Example continued on the next page ...

#**Defining doping:** The doping profile is assumed to be abrupt
doping uniform n.type conc=1e15 reg=2
doping uniform p.type conc=1e21 x.max=0.100 reg=2
doping uniform n.type conc=1e21 x.min=0.200 reg=2
#**Defining contacts**
contact name=gate workfunction=4.8

#**Defining physical models:** Band gap narrowing effect is included. The effect of concentration dependent mobility and SRH models are also include, though their impact in this example is found to be not appreciable.
#**Defining the tunneling parameter for silicon:** The parameter for Kane's model is based on [2]. This is a local tunneling model.
models bbt.kane bgn consrh conmob auger fldmob print a.btbt=4e19 b.btbt=41e6 bbt.gamma=2
#included to improve the convergence
method gummel newton
solve init

#**Applying bias for simulation**
solve vsource=0.0
solve vgate=0.0
#in order to improve convergence, drain voltage is increased gradually
solve vdrain=-0.03125
solve vdrain=-0.0625
solve vdrain=-0.125
solve vdrain=-0.25
solve vdrain=-0.5
solve vdrain=-1.00

#**Simulating the transfer characteristics**
log outf=xfer_soi_tfet.log master
solve vgate=0.0 vstep=-0.1 name=gate vfinal=-3
#plotting the transfer characteristics using the plotter of ATLAS tonyplot xfer_soi_tfet.log

#**Storing relevant information of the device for further analysis**
output val.band con.band u.bbt charge e.lines band.param
save outf=test.str

REFERENCES

[1] Silvaco International, Santa Clara, CA, *ATLAS User's Manual*, 2012.

[2] R. Vishnoi and M. J. Kumar, "2-D analytical model for the threshold voltage of a tunneling FET with localized charges," *IEEE Transactions on Electron Devices*, vol. 61, pp. 3054–3059, Sept. 2014.

Index